Industrial
Process Control Systems
2nd Edition

Industrial
Process Control Systems

2nd Edition

Dale R. Patrick
And
Stephen W. Fardo

THE FAIRMONT PRESS, INC.

CRC Press
Taylor & Francis Group
Boca Raton London New York

CRC Press is an imprint of the
Taylor & Francis Group, an **informa** business

Industrial process control systems / Dale R. Patrick and Stephen W. Fardo. -- 2nd ed.

First published 2009 by The Fairmont Press and CRC Press

Taylor & Francis Group
6000 Broken Sound Parkway NW, Suite 300
Boca Raton, FL 33487-2742

First issued in paperback 2018

© 2009 by Taylor & Francis Group, LLC
CRC Press is an imprint of Taylor & Francis Group, an Informa business

No claim to original U.S. Government works

ISBN-13: 978-1-138-11330-5 (pbk)
ISBN-13: 978-1-4398-1576-2 (hbk)

Visit the Taylor & Francis Web site at
http://www.taylorandfrancis.com

and the CRC Press Web site at
http://www.crcpress.com

Library of Congress Cataloging-in-Publication Data

Patrick, Dale R.
    Industrial process control systems / Dale R. Patrick and Stephen W. Fardo. --
2nd ed
        p. cm.
    Includes index.
    ISBN-10: 0-88173-591-4 (alk. paper)
    ISBN-10: 0-88173-592-2 (electronic)
    ISBN-13: 978-1-4398-1576-2 (Taylor & Francis distribution : alk. paper)
        1. Process control. 2. Computer integrated manufacturing systems. I.
Fardo, Stephen W. II. Title.

    TS156.8.P25 2009
    670.42'7--dc22

                                                                    2009012960

While every effort is made to provide dependable information, the publisher, authors, and editors cannot be held responsible for any errors or omissions.

Contents

Preface

Manufacturing deals with the transformation of materials into marketable products. Industry is primarily responsible for this function in our society. Depending on the product being manufactured, a number of unique processes are generally grouped into systems that accomplish specific manufacturing operations. Systems that respond to temperature, pressure, flow, level, and analytical procedures are commonly used in an industrial setting to manufacture a product.

For many years, manufacturing was largely accomplished by manual control operations. These techniques have given way to automated manufacturing operations that control a system with very little human intervention. The computer now serves as a tool to assist in this operation. Computer integrated manufacturing is a buzzword that describes this operation. The computers of a manufacturing facility are often tied together in a large network. These networks are used to schedule and oversee material ordering and routing, parts production, assembly, quality evaluation, testing, and shipment to the customer. The entire operation depends on the control of manufacturing processes. Technically trained personnel must be aware of these operations. *Industrial Process Control Systems (2nd Edition)* was developed for these people.

In this presentation, the systems concept will serve as the basis of our approach to industrial process control. An operating system is first discussed. The system is then divided into a number of essential blocks or parts. The function or role of each block is then discussed. Functional blocks are then divided into discrete components and studied in more detail. Through this approach, one should be able to see how the pieces of a process system fit together.

The explanations used in this approach are essentially nonmathematical and apply to typical industrial equipment that is designed to achieve specific manufacturing operations. When math is used, it is presented only to show a practical relationship in the explanation of a process or a control procedure.

The organization of each chapter includes objectives, key terms, an introduction, the text, system applications, and a summary. For schools that use this material, the text lends itself well to discussions, written assignments, and associated laboratory activities. For independent study and industrial training programs, this same approach provides a very productive learning procedure.

The 2nd edition of this book has been expanded to include the topic of automated processes and robotic systems (Chapter 9). Illustrations have been reviewed and revised and content carefully checked. The authors hope you will find this book informative and easy to understand.

We want to thank Brian W. Fardo, who teaches at Berea College, for his assistance in the revision of this book. He has an extensive background in technology systems.

Chapter 1

Process Control Systems Overview

OBJECTIVES

Upon completion of this chapter, you will be able to
1. Explain the systems concept used in industry today.
2. Sketch a block diagram of a basic system.
3. Describe the function that process control systems have in industry.
4. List the basic functions of five systems used in industry.
5. Identify the various media used in pressure systems.
6. Describe the changes that computer technology has brought to the industrial process field.

KEY TERMS

In an investigation of process control systems, one frequently encounters a number of new and somewhat unusual terms that are in common usage. These terms are not necessarily unique to process control systems. They are, however, quite important, and in many cases serve as the basis of a control principle or instrument operation. A few of these terms have been singled out for study before the chapter proceeds. Since this chapter is an overview of the entire field of process control, the definitions presented here are very general. In many cases, a more explicit definition will be presented in a later chapter. Reviewing these terms will make the reading of the text more meaningful at the outset.

Automation —The technique of making a system, process, or apparatus operate without human intervention.
Colorimetry —An analytical procedure that identifies different chemicals by examination and comparison of their color.

Computer —An electronic device that performs mathematical or logical calculations and data processing functions automatically and repetitively in accordance with a predetermined program of instructions.

Control —A function that alters the operation of a system or maintains a setpoint at some desired value.

Cylinder —The chamber in which a piston moves in the control of an operation.

Energy— An expression that shows the capacity of something to do work.

Energy source —The part of a system that is responsible for producing operational energy.

Head—Pressure resulting from gravitational forces on liquids.

Hydraulics —A branch of science that deals with the transmission of energy through the use of liquids in motion.

Instrument —A device that will measure, record, indicate, or control an operation,

Load—A system function that refers to the amount of work being accomplished by components or the output of a system.

Mass —spectrometer An analytical instrument that identifies gas, vapor, or solid material by identifying its atomic content.

Microprocessor —An integrated circuit network or chip that has a central processing unit, memory, control, instructions, and input/output interfacing built into its structure.

Oscilloscope —An instrument that has a cathode-ray tube that makes a visual display of voltages with respect to time.

Pneumatics —A branch of mechanics that deals with the transmission and control of energy through the use of air, vapor, or gas.

Pressure —A measure of the force applied to a specific unit area.

Process —An assemblage of phenomena performed in and by equipment in which some manufacturing operation is altered. Some examples are heating, cooling, filling, cutting, boring, forging, etching, and milling.

System —An organization of parts that are connected together to form a functioning machine or operational procedure.

Transmission path —The part of a system that is responsible for providing a path for the transfer of energy.

Temperature —The degree of hotness or coldness of a body as determined by its ability to transfer heat to its surroundings.

Ultrasonic —Vibrations of the air similar to sound but above the range of human hearing.

Valve—A mechanical device in which the flow of liquid, gas, air, or loose material in bulk can be started, stopped, throttled, or regulated in the operation of a system.

Variables —Process quantities that are measured, controlled, or used in some type of production decision making.

Work—The transformation or expenditure of energy when a physical or mechanical operation is being performed.

INTRODUCTION

Within a short span of time, industry has made a transition in manufacturing: from production techniques that were accomplished largely through manual operations to sophisticated automatic procedures that need little human effort. Through this transition there have been some drastic changes that have resulted in better product quality and more economical production.

The initial transition to automation was envisioned as a means of improving production through reduced labor costs. In many parts of industry, this particular goal has been realized, to some extent. In other phases of industry, the savings from a reduced payroll have been offset by increased equipment acquisition costs. But with this comes the need for better trained technical personnel.

Automated production has brought about a number of features that are of far more importance than the original advantage of a smaller payroll. These features include reduced waste, improved tolerances, better product consistency, production convenience, and improved product development techniques. All of these features have led to very decided changes in all industrial manufacturing.

Automation of equipment and the improved technology that has resulted from its acceptance has caused industrial process control to become the fastest-growing field in industry today. This field has had a tremendous effect on industrial technicians and engineers. The basic concepts of industrial process control have become an essential tool for those who are pursuing a career in nearly any part of industry. In this book, the industrial process system will serve as a vehicle in the investigation of automatic control procedures.

THE SYSTEMS CONCEPT

At one time, industrial process systems were limited to a number of simple motor control applications and the devices that were needed to achieve control. Magnetic contactors, electrical switch gears, gaseous tubes, rheostats, and potentiometers generally served as the nucleus of the control field during its infancy. Developments in solid-state electronics and microcomponent design have brought important technological changes to the field. Automated machinery, for example, may now employ small dedicated computers that are designed to control specific system functions. Microprocessors, artificial intelligence, programmable controllers, and electronic devices may all be combined to control a single process. A person working with this type of equipment must have an understanding of the complete system to be able to locate faulty components when a malfunction occurs.

In this book, the systems concept will serve as the basis for the understanding of industrial process control. In this approach, the system will be divided into a number of essential blocks that are the bases for its operation. The role played by each block will become meaningful in the operation of the overall system. After the location of each block has been established, component operation related to each block will then become relevant. Through this approach, a person should be able to see how the pieces of an industrial process system fit together in a definite order.

Basic System Functions

The word *system* is defined as "an organization of parts that are connected together to form a functioning machine or operational procedure." In this respect, there is a wide variety of different systems used in industry today. An electrical power system, for example, is needed to produce electrical energy and distribute it to a particular location. Hydraulic and pneumatic systems are used in industry to accomplish one or more operations, and to control other system functions. Temperature systems are used to control heat, and to cool certain areas during manufacturing operations. Industrial processes are also controlled by pressure, flow, level, analytical, data acquisition, and transmission systems.

Obviously, each industrial system has a number of characteristics that distinguish it from other systems. However, there is a common set of elements found in each system. These parts play the same basic role in all systems. The terms energy source, transmission path, control, load, and *indicator* describe the parts of a system. A block diagram of these parts is shown in Figure 1-1.

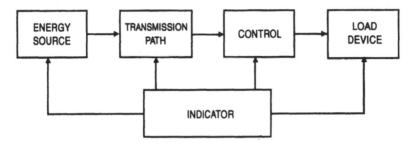

Figure 1-1. Block diagram of a basic system.

Each element of a basic system has a specific role to play in the overall operation of the system. This role becomes important when a detailed analysis of the system is to take place. Hundreds and even thousands of discrete components are sometimes needed to achieve a specific block function. Regardless of the complexity of the system, each block must achieve its basic function in order for the system to be operational. Being familiar with these functions and being able to locate them within a complete system is a big step in understanding the overall system operation.

The energy source of an industrial system is responsible for converting energy into something useful. Heat, light, sound, chemical, nuclear, and mechanical energy are primary sources of energy. A primary energy source usually goes through an energy transformation before it can be used in an operating system.

The transmission path of a system is somewhat simple when compared with other system functions. This part of the system provides a path for the transfer of energy. It starts with the energy source and continues through the system to the load device. In some cases, this path may be a single feed line, electrical conductor, light beam, or pipe connected between the source and the load. In other systems, there may be a supply line between the source and the load, and a return line from the load to the source. There may also be a number of alternate or auxiliary paths within a complete system. These paths may be series connected to a number of small load devices, or they may be parallel connected to the many independent devices.

The control section is by far the most complex part of the entire system. In its simplest form, control is achieved when a system is turned on or off. Control of this type can take place anywhere between the source and the load device. The term *full control* is commonly used to describe this operation. In addition to this type of control, a system may also employ some type of *partial control*. Partial control usually causes some type of an operational change in the system other than the on or off condition. Changes in electric current, hydraulic pressure, temperature, and air flow are some of the system alterations achieved by partial control.

The load of a system refers to a specific part or number of parts designed to produce some form of work. The term *work* is used when energy goes through a transformation or change. Heat, light, chemical action, sound, and mechanical motion are some of the common forms of work produced by a load device. Generally, a large portion of all energy produced by the source is consumed by the load device during its operation. The load is typically, the most obvious part of the entire system because of the work function.

The indicator of a system is primarily designed to display certain operating conditions at various points throughout the system. The indicator is an optional part in some systems, while in others it is essential for the operation. When an indicator is essential, system operation depends entirely on specific indicator readings. The term *operational indicator* is used to describe this application. *Test indicators* are also needed to determine different operating values. In this role, the indicator is temporarily attached to the system to make measurements. Test lights, pointer-deflection meters, oscilloscopes, chart recorders, digital display instruments, and pressure gauges are indicators used in this capacity. An indicator will normally add to the total load of the system.

PROCESS CONTROL SYSTEMS

Process control systems apply to the large section of industry that deals with those things that have direct influence on the manufacturing of a finished product. This involves the manipulation of a variety of process variables in order to achieve automatic control. The number of variables depends on the product that is being manufactured. Usually, more than one variable is altered before a manufacturing process is complete. Some representative variables that change with respect to time are *pressure, temperature, flow, level, conductivity, viscosity,* and *weight.* When one variable is altered, it often has a profound effect on another variable. Process control systems are used in branches of industry that deal with the manipulation of variables. These manipulated variables influence the quality of a product being manufactured.

PRESSURE SYSTEMS

Systems that respond to changes in pressure are one of the largest divisions in the process control field today. Pressure systems are designed to perform a particular work function that is the result of pressure being transferred through a network of pipes or tubes. Fluid, gas, steam, and air are commonly used in industry to control the flow of energy or power. Specifically, this process includes such things as hydraulics, pneumatics, steam or vapor systems, and liquid distribution.

Hydraulic Systems

Hydraulics is one major division of the pressure classification of industrial process control. This type of system is commonly found in automatic machinery control and in material-forming operations. The popularity of this type of system can be attributed to such things as operational simplicity, smoothness of operation, reliability, and adaptability. Figure 1-2 shows a double-acting hydraulic system used to control a punch-press ram.

The primary energy source of the hydraulic system in Figure 1-2 is an electric-motor-driven pump and a reservoir. The rotary mechanical energy of the motor is changed into fluid energy through this device. The pump itself does not pump hydraulic pressure. During its operation, fluid is set

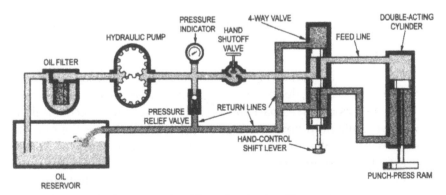

Figure 1-2. Double-acting hydraulic system.

into motion. Fluid entering the inlet port is released and forced through the outlet port. During each revolution of the pump rotor blade, a fixed amount of fluid is forced into the system. Fluid entering the system then encounters resistance to its flow. This resistance is what creates hydraulic pressure. Figure 1-3 shows a simplification of a vane hydraulic pump and its internal workings. The transmission path of a hydraulic system is typically solid pipe or some form of flexible tubing.

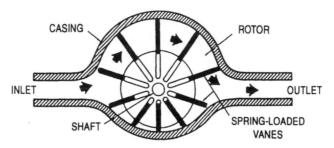

Figure 1-3. Dual-vane hydraulic pump.

The hydraulic system in Figure 1-2 has both full and partial control of system fluid. The hand shutoff valve permits full control of the system by stopping the fluid flowing through the transmission line. The pressure of the system can be partially altered by changing the operating speed of the pump. The four-way valve of the system also has an unusual feature as a control function. It can be positioned to have a form of partial control. The valve can restrict the amount of fluid reaching the cylinder. It may also alter the flow path of both high- and low-pressure fluid. Fluid flow can also be stopped completely by placing the shift lever of the valve in its off position. The pressure relief valve is a full control device that protects the system automatically. Running the pump with the hand valve closed would ordinarily cause the relief valve to open and return high-pressure fluid into the reservoir.

The load of a hydraulic system is that part of the system that does work. In Figure 1-2, the double-acting cylinder serves as the primary load device of the system. This part functionally changes the mechanical energy of hydraulic fluid flow into linear motion which moves the ram of a press. Pressure applied by the ram to its outside work alters the load to some extent. The composite load of this system includes transmission-line resistance, cylinder resistance, and the outside work load. Hydraulic motors are also used as a load in some systems to produce rotary motion.

The indicator of the hydraulic system in Figure 1-2 is an optional item, as in other systems. In this application, the indicator is used to show system pressure under normal load conditions. Monitoring fluid pressure and maintaining it at a constant level ensures consistent operating action of the system.

Pneumatic Systems

Pneumatic systems used in industry range from the powering of hand tools to the lifting and clamping of products during machining operations. The energy source of this type of system is a pump or a compressor and a storage tank to hold the air. The pump of a compressor maybe driven by an electric motor or by a portable internal combustion engine. Figure 1-4 shows a simple pneumatic system driving a double-acting cylinder ram.

Pneumatic systems are designed to use the air from the room where the compressor is located. Through the action of the pump, outside air is forced into a tank. Compressed air is stored or passed through the system. The storage tank of compressed air serves as the reservoir of the system.

After air has been compressed, it must be conditioned before it can be effectively used by the system. Conditioning initially calls for the removal of dirt and moisture. This is achieved by an air filter with a condensation trap and drain. In some pneumatic systems, a fine mist of oil is then added to the compressed air. This provides lubrication for all parts throughout the system. Oil-free pneumatic systems omit the oil mist and operate without lubrication. This type of system is generally found in clean-environment applications. Oil-free pneumatic systems represent a new division of the pneumatic field.

The air pressure of a pneumatic system must be adjusted to a specific level by an air regulator valve. Constant pressure must then be maintained

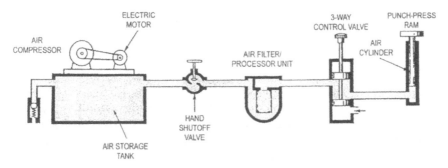

Figure 1-4. Pneumatic system simplification.

during system operation. Motor-driven air compressors are designed to operate only when the system pressure of the storage tank drops below a predetermined level.

The transmission path in a pneumatic system is slightly different than in its hydraulic counterpart. Solid pipes, tubing, and flexible hoses are used as feed lines from the compressor to different parts of the system. Return lines to the storage tank are not used in pneumatic systems. Air is simply dumped into the atmosphere. Pneumatic systems are somewhat simplified because of this feature.

The pneumatic system in Figure 1-4 has both full and partial control of system air. The hand shut-off valve and air-pressure relief valve provide full control of air circulating through the transmission path. Air flow can also be altered by the regulator and the three-way control valve. Pneumatic and hydraulic controls are similar in many respects.

The load of a pneumatic system, as in other systems, is designed to perform a work function. The basic load device in Figure 1-4 is the pneumatic cylinder. This part functionally changes the mechanical energy of air into linear motion that drives a punch-press ram. Pressure applied by the ram to outside work also influences the primary load. The composite load of the entire system includes transmission-line resistance, control resistance, and the outside work load. Pneumatic load devices can also be used to produce rotary motion and striking blows for hammering.

Static Pressure Systems

Static pressure systems are commonly used in industry to distribute liquids in a variety of processing applications. This type of system utilizes pressure that is developed as a result of the source being elevated above other system components or by liquids being stored in a large tank. *A static head* is developed and used to force liquid through a network of distribution lines. The height or depth and the density of the liquid develop a force at the bottom of a tank. This force is used to drive liquid from the tank into a network of pipes or tubes. Figure 1-5 shows a hydrostatic head, developed by water in a reservoir, that is used to generate electricity.

Static pressure systems represent an unusual part of process control. The pressure developed by this type of system is used in filling operations, for mixing and blending, and in liquid distribution. Generally, an obvious form of mechanical work is not achieved by the system load. The system load is essentially anything that causes a change in the demand of the system. In this case, it is a composite of the total opposition encoun-

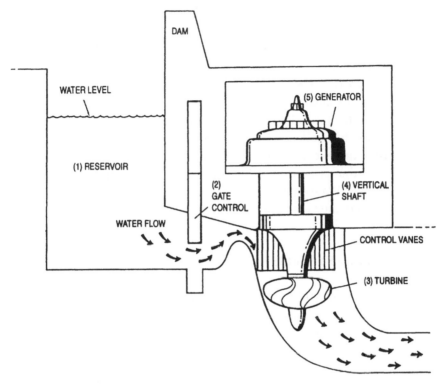

Figure 1-5. Hydroelectric power system. The force of water in reservoir (1) plunging through control gate (2) spins turbine (3) connected by vertical shaft (4) to generator (5), producing electricity.

tered by the fluid as it moves through the system. In effect, heat developed by the flow of the liquid is representative of the system work function. The components in this system are essentially the same as those used in a hydraulic or pneumatic system.

Steam Pressure Systems

Steam pressure is used in industry as a heat source, in refining operations, in food processing, and in chemical production processes. This type of system develops pressure-by transforming liquid into a vapor when heat is applied. The resulting pressure occurs when steam or vapor is forced to pass through the resistance of confined pipes or tubes.

In practice, fossil fuel or electricity serves as the primary source of heat for a steam pressure system. The applied heat causes liquid placed inside of a container to expand and eventually vaporize. Industrial boilers are used to perform this operation. *Fire-tube boilers* are constructed so that the heat source runs through pipes or tubes in the center of the liquid source. In operation, steam is taken from the boiler through a water jacket at the top of the unit.

Water-tube boilers are designed to heat water inside of tubes that are placed around the heat source. Heat from the source passing over the outside of the water tubes is transferred to the water inside. The resulting steam is ultimately collected at the top of the tubes, where it is distribute through pressure to the remainder of the system. Figure 1-6 shows a simplified boiler that is used to drive a steam turbine for the generating of electrical power. Water-tube boilers are usually quite large and are erected at the site where they are to be used.

TEMPERATURE SYSTEMS

Of all the industrial processes utilized by industry today, temperature is by far the most common. Over 50 percent of all measured variables of concern to industry involve some form of temperature measurement. These systems must generate heat, provide a path for its distribution, ini-

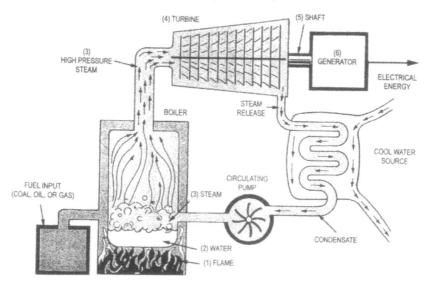

Figure 1-6. Simplified boiler/steam turbine system. Heat from burning fuel (1) changes water in boiler (2) into steam (3), which spins turbine (4) connected by shaft (5) to generator (6), producing electrical energy.

tiate some type of control, utilize this energy to do work, and ultimately record temperature through some precise measurement.

The energy source of a thermal system is primarily concerned with changing energy of one form into something of a different type. Electricity and chemical action are primarily used to perform this operation. When fuel is burned, a chemical change takes place. The carbon in the fuel unites with oxygen to produce carbon dioxide. Heat released by this action is the end result of the energy transformation process. Electricity is also used as an energy source, to produce temperatures of 3500°F to 5000°F (1926°C to 2760°C).

The control function of a thermal system is primarily concerned with the flow of heat between the energy source and the load device. Temperature control may be achieved manually by a human operator or automatically by a *system temperature controller*. Controllers of this type are used in industry to achieve automatic control. The temperature-sensing element is filled with a fluid that changes volume in response to variations in temperature. This change is used to physically alter the flow of heat to the system load.

The temperature of an operating system may employ a platinum *resistance temperature detec*tor (RTD) to sense temperature at different locations. The temperature can then be instantly observed on a liquid crystal display. This particular instrument is typically portable, with an accuracy of ±0.1 degrees Fahrenheit or Celsius for each digit of the display. Measuring applications include food and chemical processing, aerospace, and electronic component manufacturing.

Temperature monitoring can also be accomplished with a noncontact infrared thermometer. This portable instrument is described by one manufacturer as a "heatspy." Temperature is sensed by pointing the instrument viewing chamber in the direction of the surface to be evaluated and squeezing the trigger. The temperature appears on the display instantly. This type of instrument is used in general plant-maintenance operations, heat treating and processing, metal ore production, and building evaluation. Instruments of this type play a very important role in the overall operation of a thermal system.

LEVEL DETERMINATION SYSTEMS

The measurement of liquid or solid material levels in a tank or vessel is a process that applies to nearly all phases of industry, regardless of the product being manufactured. In practice, level systems are used to evaluate the amount of material in a tank, bin, or hopper that is available for accomplishing a particular process. The systems concept is quite unique when applied to the level process. In a strict sense, level evaluation is often considered as a control function that applies to a variety of different processes. When viewed as a system, however, it must employ a source, path, control, load, and indicator. In general, the function of the system is level determination, while the actual material being manipulated is incidental.

In a *liquid distribution system,* the supply source may be a large bulk storage tank placed at a convenient location in the industrial facility. In a sense, the tank serves as a reservoir that represents potential energy that can be released when the need arises. When the system is in operation, liquid flows through the distribution path to the remaining components. Control may be achieved manually by a human operator or automatically by a controller. The final component being filled by the bulk storage tank, plus the flow path of the system, serve as the system load. Indicators are usually attached to several parts of the system to evaluate its operation.

Level determination is similar to other process systems in that it involves an evaluation procedure before control can be implemented. Generally, evaluation is achieved by some type of measurement. A sensing element is primarily responsible for this part of the operation. It detects a change in level and generates a signal that is used to initiate the control function. Typically, these systems respond to mechanical, pneumatic, electrical, radiation, or ultrasonic information. The selection of a specific level determining process is based on level range, the material involved, operating pressure, accuracy, cost range, construction, and temperature.

Level determination is either measured directly or inferred from some indirect action. The *direct* method of evaluation uses an actual physical change in the material itself to obtain a measurement. An *inferential* method, by comparison, employs some outside variable, such as pressure, to indicate a specific change in level. Mechanical mechanisms, pressure gauges, and diaphragm elements are some of the common inferential instruments being used today. Figure 1-7 shows a simplification of a level determination system that is controlled by the weight of a container. When a desired level is achieved, the weight of the container applies a force to the pressure transducer. A corresponding output signal is developed and ultimately used to actuate the liquid control valve. When the operation is complete, another container is immediately positioned for filling.

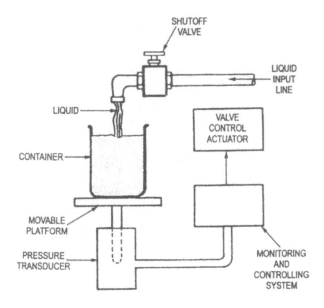

Figure 1-7. Weight level determination system.

In level determination, the detector or sensor is a vital element in the operation of the system. The response of the system is largely dependent on the sensor. Sensors respond to some form of floating mechanism, the capacitance effect, the electrical conductivity of material, radiation, ultrasonic signals, pressure, weight, and photoelectric energy. The *presence/absence detector* responds to the level of fluid in a pipe. The fluid level passing through a sensor flange is detected by a change in capacitance. If fluid stops flowing, the detector senses this condition. The output of this sensor could be used to trigger an alarm or to turn off a pump for dry-run protection. Applications of this type of sensor are in food, beverage, pharmaceutical, and cosmetic industries. The capacitance principle can also be used to detect the level of material in a tank or container.

FLOW PROCESS SYSTEMS

The flow of material through a system represents an important industrial process that has a wide range of applications. In their simplest form, *flow process systems* are used to monitor and control water, natural gas, solvents, chemicals, and an infinite variety of other materials. Operation is based on the actual material that is flowing, its density, the volume being processed, accuracy, and the type of control required. Measurement and control of flow are the major functions of a flow process.

The systems concept that has been used with other processes applies equally well to the flow principle. Flow process systems are unique because they are mainly used to initiate some type of control operation or to measure specific values. These principles are component functions more than composite system functions.

Flow equipment is classified according to the primary function that is serves, and is categorized as either flow *rate or totalizing flow*. Flow rate instrumentation is an evaluation that occurs at a specific instant. Totalizing flow is a measurement that is used to evaluate the total, amount of material that moves past a given point during a specific period. In practice, the equipment is classified according to the principle of operation utilized in its design. Such things as *head pressure, area changes, electromagnetic effects, flow velocity,* and *positive displacement* are included in the totalizing flow classification. Figure 1-8 shows three in-pipe methods of measuring flow.

The term *laminar* is used to describe flow that moves in a straight line. Instruments used to perform this measurement are called *laminar flowmeters*. This type of instrument is primarily designed for clean-gas applications. Some instruments of this type employ an area of small parallel passages in the structure of the meter. These passages cause a difference in pressure to appear between the input and output of the meter. Differential pressure measured across an instrument can be directly equated to flow. Laminar flowmeters are an important addition to flow instrumentation.

An obstruction bar or strut placed in the flow path of a meter produces a predictable number of swirls or vortices when liquid passes through a pipe or tube. This type of instrument is called a *vortex shedding flowmeter*. The number of vortices that appear on the downstream side of the strut in a given unit of time is directly proportional to the liquid flow rate.

The *variable area flowmeter* is one of the simplest methods of measuring flow rate. It essentially consists of a tapered tube that is larger in diameter at the top, and a metering float. Flow entering the tube at the bottom lifts the float, which indicates the rate on a calibrated scale. Variable area flowmeters come in a variety of styles and sizes. The instrument shown here uses a weighted ball as the float. The type of float used and the area and length of the flow path all influence the flow rate of this instrument.

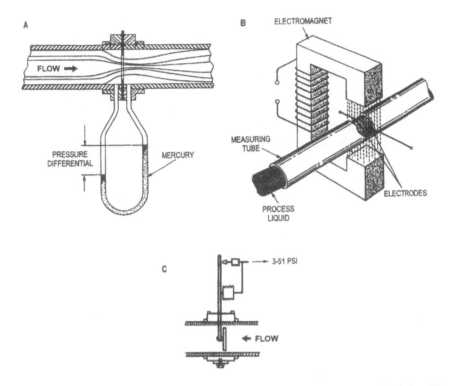

Figure 1-8. In-pipe flow measuring principles: (A) Head pressure principle; (B) Electromagnetic principle; (C) Target principle. (Courtesy *Foxboro Co.*)

ANALYTICAL PROCESS SYSTEMS

Analytical systems are used to evaluate manufacturing processes. This includes the testing of physical and chemical properties of a product at various steps during its development. Quality control, specification testing, and inspection are usually the goals of analytical process systems.

The number of analytical processes used by industry is almost endless when compared with other system applications. *Density, specific gravity, viscosity,* and *alkalinity* are only a few of these processes. The type of equipment used to evaluate these processes also makes a rather formidable list.

In practice, analytical processes are divided into four basic energy/matter interaction groups. These groups are classified according to an interaction that takes place between the substance being tested and an external source of energy. The external source of energy may be derived from an electric or magnetic field, thermal or mechanical energy, electromagnetic radiation, or chemical energy. When studying analytical process systems, these groups will serve as the basis of our classification.

Electric or *magnetic field instruments* represent a powerful method of analytical evaluation. These instruments are used to analyze a current, voltage, or magnetic flux that results when a test sample is exposed to an electric or magnetic field. Through this type of processing, a number of distinguishing chemical and physical features of the sample material can be evaluated. Mass spectrometers, moisture analyzers, gas detectors, and oxygen analyzers are some of the instruments included in this classification.

Thermal or mechanical energy can be used in a number of analytical instruments as an external source to produce an interaction with the substance under test. These instruments respond to such things as energy transmission, work done, or changes in the physical state of a sample. Thermal conductivity, dew point, density, specific gravity, viscosity, vibration, and sound velocity are some of the effects achieved through this type of analysis.

Instruments that respond to an interaction between electromagnetic radiation and matter represent the third major division of analytical process systems. The energy interaction measured by these instruments involves photons. Photons are emitted or absorbed whenever changes occur in the quantized states occupied by the electrons of atoms and molecules. This analysis technique depends on the frequency of the excitation energy source. Energy of the highest frequency, or shortest wavelength, tends to produce high energy levels that are suitable for penetration into matter. *Gamma rays, X rays, visible light, ultraviolet light, infrared radiation,* and *microwaves* are used as excitation energy sources in this type of instrumentation.

The *electromagnetic interaction principle* is basically the same for all instruments in this classification. These instruments are frequently grouped according to the type of energy source being employed during operation. X rays, ultraviolet energy, radiation, infrared light, and visible light are used in different instruments. These instruments are used to analyze the elemental and molecular composition of gases, liquids, and solids.

The final energy interaction division used with process analysis system instrumentation is based on chemical energy reactions. Systems of this type respond to such things as *reactant or sample consumption, measurement of the reaction product, thermal energy liberation,* and *solution equilib-*

rium. The relationship of one chemical element to another and the thermal and physical behavior of certain materials permit a positive method of identification and analysis.

Applications of the chemical energy interaction principle include automatic in-line process evaluation and laboratory analysis equipment. In practice, instruments of this classification are normally tailored to the samples being tested and the information that is needed. A large variety of specialized processing equipment appears in this classification. Colorimetry, combustion analysis, reaction product analysis, and pH measurement are some functions of this group of instruments.

MICROCOMPUTER PROCESSING SYSTEMS

Computer technology has brought significant changes to the industrial process instrumentation field. Calculations can be performed quickly, data can be stored and retrieved for analysis, and deductions can be made that can be used to control a process automatically. All of this has caused industry to become more dependent on computerized equipment. Industry has also placed a higher premium on its operational time.

Minicomputers were developed to provide technology for industrial applications not requiring the capacity of a full-scale computer. These units were less costly than a main computer and permitted operations to be performed at the actual test site. The new technology of large-scale integration ultimately brought about the development of the microcomputer. The entire central processor unit of a computer can now be built on a single integrated-circuit chip called a *microprocessor* (MPU). A microprocessor coupled with memory, input/output interface chips, and a power source forms a microcomputer system.

Personal computers have virtually revolutionized industrial process control. Faster development times, smaller equipment size, increased reliability, easy serviceability, and lower product cost are only a few major advantages. These systems are often described as *dedicated* units because they are designed to perform specific tasks in controlling processes or the operation of instruments. They are also called *smart* instruments because data are collected and analyzed, and decisions are automatically made to influence the final outcome. The inexpensive cost of microprocessors has caused them to appear in all kinds of industrial process equipment.

A microprocessor chip performs the arithmetic logic and control functions of a computer. This chip is designed to receive digital data in the form of 1s and 0s. It may store these data for future processing, perform arithmetic and logic operations in accordance with previously stored instructions, and deliver the results to an output device. In a sense, the microprocessor is actually a computer built on a single integrated circuit (IC) chip. The *arithmetic logic unit, accumulators, data registers, address registers, program control, instruction decoder, sequence controller,* and *memory* are included in a chip measuring 1 cm^2. The potential usefulness of this chip has not yet been fully realized in industry.

There are certain functions that are basic to almost all microprocessor systems. Included in these operations are *timing, fetch and execution, read memory, write memory, input/output, transfer,* and *interrupts.* An understanding of these basic functions is important in the operation of a microcomputer system.

Programming is a series of instructions that are developed for a microcomputer that will permit it to perform a prescribed operation or function. These instructions may be placed in read-only memory or may be supplied through keyboard information. Instructional sets are built into the read-only memory or a microcomputer system and cannot be changed. The instructions are normally called *addressing modes.* Included in this are the *inherent mode, immediate mode, relative addressing, indexed addressing, direct addressing,* and *extended addressing.* A program utilizes these instructions to manipulate data for control processing operations.

Microcomputer processing systems represent one of the most significant developments and innovations that has taken place in industrial process control. A gas chromatograph/mass spectrometer may be connected to a computer. The computer performs a number of data acquisition functions by being interfaced with the system. These data develop an output signal that is used to analyze the sample. Data may be stored in memory, displayed on the cathode-ray tube, applied to a system network, or placed on a graphic recorder for further analysis. The computer makes all this possible.

SUMMARY

Automation has caused process control to be the fastest growing field in industry today. This will have a tremendous effect on the careers of industrial technicians and engineers who must work with this equipment.

The systems concept is a method used to explain the operation of industrial equipment and instrumentation. In this approach, the system is divided into a number of blocks that are the basis of its operation. The role played by each block, and its function in the overall system, then becomes more meaningful.

A system is an organization of parts that are connected together to form a complete operating unit. The terms energy source, transmission path, control, load, and indicator are the basic parts of a system.

Process control systems are used to automatically control variables that have some influence on the quality of a product being manufactured.

Pressure systems perform a particular work function that is the result of pressure being transferred through a network of pipes or tubes. Fluid, gas, steam, and air are used to control the flow of energy. Hydraulics, pneumatics, and steam, vapor, and liquid distribution are all examples of pressure systems.

Temperature systems are designed to generate heat, provide a path for its distribution, initiate some type of control, utilize it to do work, and measure values. Electricity and chemical energy are typical primary energy sources for temperature systems.

Level systems are used to evaluate and control the amount of material in a tank, bin, or vessel that is available for accomplishing a particular process. This type of system involves an evaluation procedure before some type of control can be implemented. Measurements are achieved directly or inferred by some indirect action.

Flow process systems dare used to monitor and control such things as water, natural gas, solvents, chemicals, and an infinite variety of other materials. System operation is based on the actual material involved, its density, the volume being processed, accuracy, and the type of control required.

Analytical systems are used to evaluate manufacturing processes and to test the physical and chemical properties of a product at various steps during its development. These systems are divided into four energy/matter interaction groups: electric or magnetic field energy, thermal or mechanical energy, electromagnetic radiation, and chemical energy.

Computer technology has produced a number of significant changes in the industrial process field. Calculations can be made quickly, data can be stored or retrieved, and deductions can be made from data to control a process automatically. Microcomputer systems are now being designed to perform specific tasks in the processing field. Microprocessor chips are responsible for most of the innovations in this field. A single MPU chip uses an arithmetic logic unit, accumulators, data registers, address registers, program control, instruction decoding, sequence controllers, and memory. Microprocessors are essentially computers built on a single chip.

Chapter 2

Process Control

OBJECTIVES

Upon completion of this chapter, you will be able to
1. Describe the primary function of process control.
2. Explain the basic concept of control.
3. Define open-loop and closed-loop system operation.
4. Describe the operation of a PID controller.
5. Give some examples of process time lag, dead time, resistance, and capacity.
6. Define commonly used controller terms, such as steady-state error, transient response, stability, modes of control, and sensitivity.
7. Describe OSHA and profile its importance.

KEY TERMS

Anticipation function—The ability of a circuit to determine or visualize a future action.

Automation—The technique of making an apparatus, a process, or a system operate without human intervention.

Capacity—The ability of a system or device to receive, store, hold, or accommodate some material or substance in its structure.

Closed-loop system—A control operation in which some of the output is fed back and compared with the input to generate an error signal.

Controlled variable—The process value that is being manipulated by a system,

Controller—An instrument that evaluates the controlled variable error of a process and initiates a corrective action.

Dead time—The elapsed time between the instant a deviation occurs and when the corrective action first occurs.

Derivative control—A mode of control that provides an output that is related to the time rate of change that occurs between a process variable and its setpoint.

Differential amplifier—An amplifier designed so that its output is proportional to the difference between signals applied to its two inputs.

Dynamic variable—A process variable that changes from moment to moment due to input from unspecified or unknown sources.

Error signal—The algebraic difference between the setpoint value and the process variable value applied to the input of a controller.

Final control element—The part of a control system that has direct influence on the process and brings it to the setpoint condition.

Inertia—A condition that refers to the ability of a process to continue to remain in a certain state after a change occurs.

Integral control—A mode of control which develops an output that increases at a rate that is proportional to the difference between the setpoint and the value of the process variable being measured.

International System of Units (SI)—A system that sets measurement standards.

Mode of control—A method or type of control.

Occupational Safety and Health Act (OSHA)—Legislation that insures safe and healthful working conditions for workers.

Offset—A characteristic of proportional control that produces a residual error in the operating point of the controlled variable when a change occurs.

On-off control—A controller mode of operation that alters its output by switching it on and off for periods of time.

Open-loop system—A control operation whose input is set to achieve a desired output without monitoring the status or condition of the output.

Process—The dynamic variables of a system used in manufacturing and production operations.

Process time lag—The time it takes a system to correct itself and seek a condition of balance after a process variable has changed.

Process variable—Any process parameter such as temperature, flow, liquid level, or pressure that changes its value during the operation of a system.

Proportional band (PB)—A band or range of values that a process variable operates within when a proportional controller is required to produce a full-scale change in output. If a 10 percent change in error causes a 1OQ percent change in controller output, the PB is 10.

Proportional control—A controller mode of operation in which the output is a linearly related function to the difference between as setpoint value and the process variable being controlled by a system.

Proportional plus derivative (PD) control—A mode of control which provides an output that is a linear combination of proportional and derivative control modes in a single instrument.

Proportional plus integral (P1) control—A mode of control which has proportional and integral operations combined in a single instrument.

Proportional plus integral plus derivative (PID) control—A mode of control that has proportional, integral, and derivative operations combined in a single instrument.

Rate action—Another name for the derivative mode of control.

Reset action—Another name for the integral mode of control.

Resistance—A condition that opposes energy transfer when a process change occurs.

Sensor—A transducer that measures some physical quantity and changes it into a different quantity.

Setpoint—A desired value setting of a process variable.

Steady-state error—The difference between controller output and input after the input has been applied for a prolonged period of time.

Summing amplifier—An operational amplifier circuit that adds two or more of the inputs applied to its summing input point.

Transient response—The behavior of a system or device prior to reaching a steady-state condition in response to a sudden change in an input quantity.

Variable—A process condition, such as pressure, temperature, flow, or level, which is susceptible to change and which can be measured, altered, and controlled.

INTRODUCTION

Process control is a unique part of industry that deals with the control of variables that influence materials and equipment during the development of a product. It may range from a relatively simple operation, such as filling bottles, to maintaining a proper level in an analytical procedure that determines the content of a complex chemical solution. The end result of the operation, in most cases, is a procedure that initiate's some type of *control function*. Controlling manufacturing processes is the basis of industrial automation today.

In Chapter 1, we described control as one of the primary functions of an operating system. In this regard, control refers to those things that maintain a desired system output by altering the flow of energy from the source to the load device. Control ranges from simple on-off operations to a number of partial or gradual changes that are achieved with sophisticated equipment. In manufacturing, control may be initiated by human energy, mechanical energy, electrical energy, a computer, or any combination of these. The end result may be to control the liquid level in a tank or the uniform flow of a chemical through a distribution system.

Processes

Manufacturing is an organized plan for developing or making a product from raw materials. A *process* is an activity or operation performed on raw materials or workpieces that converts them into a finished product. *Processes* are manufacturing operations performed on a product that bring it one step closer to becoming a finished product. Manufacturing processes depend on the type of material being manipulated, the equipment performing the operation, the quality of the finished product, and the quantity being produced. Some common manufacturing processes are heating, cooling, distilling, baking, soaking, and milling. These processes are grouped into areas that respond to changes in temperature, pressure, material flow, level changes, and analytical operations. Processes are continuously changed or altered during the manufacturing of a product. *Process control* is a term used to describe this function of industry.

THE CONTROL CONCEPT

Control, a common term used in industry, has many different meanings. In its simplest form, control may refer to something that causes a process to be turned on or off. In a testing laboratory, control may apply to an evaluation procedure used to see if specifications are being met. In computer systems, control usually refers to some type of data manipulation function. In addition, there are such things as *inventory control, machinery control, numerical control, programmable control,* and *quality control.*

In spite of the diverse meanings associated with the term control, there is one basic idea that tends to apply to all situations: Control has the primary influence on the final outcome of a process or operation. As an example, control must be performed before the final outcome is decided.

We are in control of something only when the final outcome is modified or changed in some way.

There are a number of significant factors that have direct influence on the effectiveness with which control is achieved. *Control time* or *response time* is a prime example of effectiveness. In practice, response time is a variable factor that depends on the application. In some situations, control may take place at periodic intervals, and then only for short durations. Control may also take place over long spans of time by initiating changes of a short duration. In general, the storage capacity of the system determines the response time of the control function.

Another important control factor is the effectiveness with which a system achieves control. This factor is based on the ability of a system to measure a value and compare it with a desired operating value. The precision of the measuring device determines the effectiveness of this type of control. Control effectiveness is primarily determined by the needs and application of the system.

There are a number of basic factors that have direct influence on the control of an operating process system. In a *manual control system*, these factors are normally performed by a human operator. *Automatic* systems achieve the same basic functions but through the manipulation of self-regulating controls. As a rule, automatic control operations are much more complex and difficult to achieve than those of a manual control system.

The basic functions of system control include *measurement, comparison, computation*, and *correction*. Measurement is essentially an estimate or appraisal of the process being controlled by the system. Comparison is an examination of the likeness of measured values and desired values. Computation is a calculated judgment that indicates how much the measured value and desired operating value differ. Correction is the adjustments which are made in order to alter operating values to a desired level. These functions must all be achieved by an automatic system during the normal course of its operation.

CONTROL SYSTEMS

Control of industrial processes is a broad field of concern that cuts across virtually every area of science and technology. The basic operating principles utilized by the control function are so specialized and unique that a whole new field of study has been created. Such terms as *process control automation, computer integrated manufacturing, computer numerical control, process instrumentation,* and *flexible manufacturing systems* are commonly used to describe this particular function in industry. Control is an essential function of manufacturing.

With the wide diversity of control used by industry today, there is need for further classification of the term into more workable divisions. Control is first classified as being either manual or automatic. This division generally refers to the amount of human effort needed to achieve a common function. Manual control is voluntarily initiated within the system with very little human effort.

The terms *open-loop* and *forward-feed* are frequently used to describe manual control systems. Valve adjustments and switching functions are examples of manual control operations. In general, this type of control is achieved by some degree of physical effort on the part of a human operator.

Automatic control, by comparison, applies to those things that are achieved, during normal operation, without human intervention. This type of control is used where continuous attention to system operation would be demanded for a long period without interruptions. Automatic control does not, however, necessarily duplicate the type of control achieved by a human operator. Equipment that employs automatic control is limited to only those things that can be forecast by the input data. Terms such as *closed-loop control* and *feedback* are commonly used to describe automatic control functions.

Open-loop Control

Open-loop control is relatively easy to achieve because it does not employ any automatic equipment to compare the actual output with the desired output. In manufacturing, open-loop operations are achieved by adjustment of the system to some predetermined setting by a human operator. The system then responds to this setting without any modification. Any changes made in operation are based entirely on some outside human judgment to correct the desired output. The open-loop system in Figure 2-1 is composed of a process energy source, a transmission path, a controller, and an *actuator* or final control element. The process energy source represents input variables such as time, temperature, speed, pressure, flow, displacement, acceleration, and force. The transmission path is responsible for transferring input energy to the remainder of the system. The controller provides intelligence for the system and governs the action of the actuator. The *manual setpoint adjustment* attached to the controller is used to alter the operating range of the controller. The actuator implements the response of the controller to the final controller element. The final control element can be a motor, pneumatic cylinder, solenoid, or hydraulic valve. The final control element is responsible for altering the process energy passing through the system. The output is considered to be the controlled process. Examples of controlled processes are water temperature, the pH of a chemical solution, the viscosity of crude oil, the temperature of molten aluminum in a furnace, or the path of a cutting tool on a milling machine.

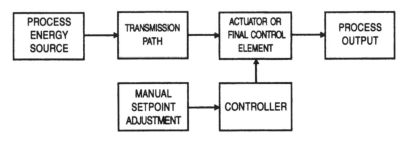

Figure 2-1 Open-loop control system.

The steam heating unit in Figure 2-2 is an example of an open-loop control system. Steam applied to the coils of the heat exchanger unit is used to alter the temperature of the tank water to a desired value. To start the heating process, the steam control valve is turned on manually by the operator. The temperature of the water flowing from the output line is then measured by touching the heat exchanger unit. This determines the effectiveness of the control process. If the water is too hot, the steam valve is turned off. If the water is not hot enough, the steam valve is permitted to remain on for a longer period or opened wider. The measurement function of the water heating unit is achieved by the operator's sense of feel. A drop in water temperature would be compared by memory with a desired output value.

A more sophisticated system of this type might employ a temperature measuring instrument in the output water line, as in Figure 2-3. This addition to the system would cause it to have improved accuracy. Any detected change in water output temperature could now be compared with a desired temperature value. The operator mentally compares the indicated value with a desired value and computes or decides if the steam input

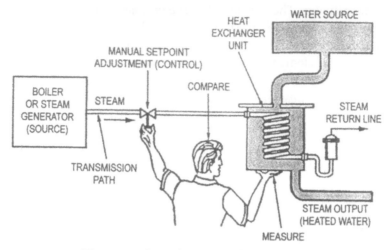

Figure 2-2. Open-loop steam heating system.

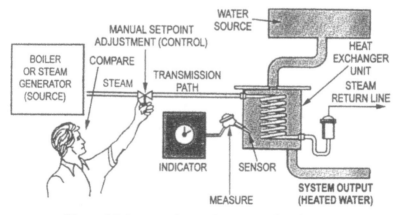

Figure 2-3. Improved open-loop steam heating system.

should be altered. A drop in temperature could be corrected by opening the steam valve to increase system temperature. A period of time is required before any corrective action takes place. If the amount of correction is too great or not enough, the process will need to be repeated. In manual systems of this type, control involves repeated steps of measurement, comparison, computing, and correction. This means that manual control usually demands continuous supervision by the operator, which is a decided disadvantage of this type of system.

The primary advantage of open-loop or manual control is simplicity of operation and low-cost installation. The intended accuracy of the process being controlled determines its suitability for manual control. In a strict sense, the operator of a manual control system forms the feedback path from output to input that closes the loop of the control system.

Closed-loop Control

Closed-loop refers to a type of system that is self-regulating. In this type of system, the actual output is measured and compared with a predetermined output setting. A feedback signal generated by the output sensing component is used to regulate the control element so that the output conforms to the desired value. The term *feedback* refers to the direction in which the measured output signal is returned to the control element. In a sense, the output of this type of system serves as the input signal source for the feedback control element. Closed-loop control is so named because of the return path created by the feedback loop from the output to the input.

A block diagram of a simple closed-loop or automatic feedback system is shown in Figure 2-4. The basic component layout of this system is similar to that of the open-loop system. The closed-loop system has a process energy source, actuator/final control element, process output, sensor, feedback circuit, controller, and setpoint adjustment. The feedback circuit connected to the process output is a distinguishing feature of this system. The origin of the feedback signal is a sensor attached to the process output. The feedback circuit returns a signal to the controller. The controller is fed by a *summing circuit* that compares the setpoint input and feedback signals. The input or process energy source is responsible for establishing the setpoint value of the system. The setpoint value is adjusted by the operator according to the needs of the system. The actual value of the process output is determined by the sensor. If the actual output is the same as the setpoint value, the controller indicates system balance, and the actuator or final control remains unchanged. If the actual output deviates from the setpoint value, signals applied to the controller are no longer in balance, and this causes a correction signal to be developed by the controller. The correction signal is directed to alter the actuator or final control element. This type of response causes the system to be self-correcting.

An example of a simple closed-loop or automatic feedback system is shown in Figure 2-5. The functional elements of this system are positioned to show their relationship in the closed-loop control path. These parts could apply to any one of a wide variety of automatic process systems. A

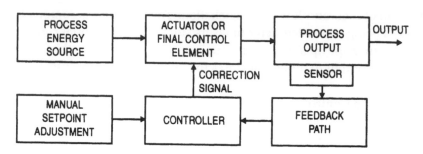

Figure 2-4. Automatic feedback system.

comparison of Figure 2-6 with the open-loop system in Figure 2-5 shows that the two systems perform the same basic control operations.

Closed-loop Variables

In closed-loop control, a number of variables are used to describe the operation of a system. Four of these variables are used as criteria to evaluate system performance: *transient response, steady-state error, stability,* and *sensitivity.*

Transient response is a condition that is used to describe how the output of a closed-loop system responds. This condition occurs before the

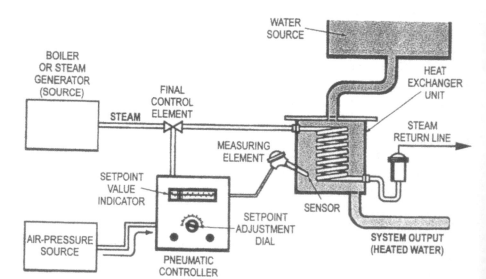

Figure 2-5. A Automatic steam heating system.

system reaches a steady state of operation. The output of a closed-loop system will follow any one of the three paths represented by the curves in Figure 2-6. Curve 1 shows an *underdamped* response. The output of this type of system oscillates a few cycles above and below the setpoint value before reaching the new steady-state condition. *Overdamping* is indicated by Curve 2. This shows the period required for the output to reach a steady-state condition to be quite long. Curve 3 is an example of a *critically damped* response. This condition occurs when the output reaches its steady-state condition without oscillations and in the shortest possible time. The control function of a system is designed, selected, or adjusted for a particular type of transient response. Some manufacturing processes are unaffected by oscillation of the output while others cannot tolerate it. Likewise, some processes are not affected by it taking a long period to reach a steady-state condition while others must achieve control almost immediately. Transient response is an important condition of operation that must be taken into account when selecting a particular system for control of a manufacturing process.

Steady-state error is a closed-loop variable that refers to a value attained by the output after a change in input has occurred. As can be seen in Figure 2-6, a slight difference exists between the actual output value and the desired output of a closed-loop system after the transient response occurs. This value difference is the steady-state error. It shows that the

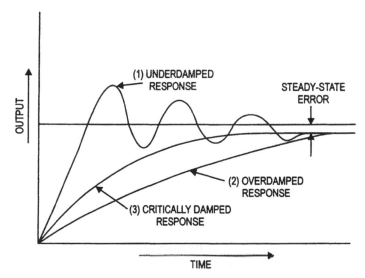

Figure 2-6. Transient-response curves.

output value of a closed-loop system is offset by a certain amount, an important consideration in the operation of a controller.

Stability is a closed-loop condition that deals with the ability of a system to attain steady-state control of a process variable after it has responded to a change. Instability causes the output value to change above and below the desired output. The underdamped response in Figure 2-6 shows an example of poor stability, permitting a system to produce uncontrolled oscillations. If this condition continues for a prolonged time, it can be damaging to some system components.

Sensitivity is a closed-loop variable that refers to the ratio of the percentage change in output to the percentage change in system input. The system input may be normal or affected by some unwanted disturbance. This variable identifies the effectiveness of the system in correcting for value changes in the input.

PROCESS CONTROL PRINCIPLES

The basic principles of process control are broad and far-reaching ideas that can apply to all types of industrial manufacturing systems. These ideas are not unique to process control. They are important and often serve as the basis of a number of specific operations or applications. A brief explanation of some common terms associated with process control is presented here as a general review.

Process—Activities performed on raw materials or workpieces to convert them into a finished product are called *processes*. A process could also be described as an operation utilized to achieve an industrial manufacturing function, such as pressure, temperature, flow, liquid level, mechanical motion, numbers, weight, specific gravity, viscosity, and numerous analytical values.

Controlled Variable—*Controlled variables* are the basic process values being manipulated by a system. These values may vary with respect to time, as a function of other system variables, or both.

Controllers—A *controller is* a hardware piece of equipment that employs pneumatic, electronic, and/or mechanical energy to perform a system control operation. These units are designed to maintain a process variable at a predetermined value by comparing its existing value to that of a desired system value.

Setpoint—A *setpoint* is a prescribed or desired value to which the controlled variable of a process system is manually adjusted. For example, if the temperature of the controller in Figure 2-7 is adjusted to a value of 95°C, it is indicated on the horizontal setpoint indicator. The controller output meter indicates the temperature value of the system output.

Sensor—A *sensor* is a piece of equipment that is used to measure system variables. Sensors are normally transducers that change energy of one form into something different. Sensors serve as the signal source in automatic control systems.

Control Element—A *control element* is a part of the process control system that exerts direct influence on the controlled variable to bring it to the setpoint position. This element accepts output from the controller and performs some type of operation on the process. The term *final control element* is used interchangeably with control element.

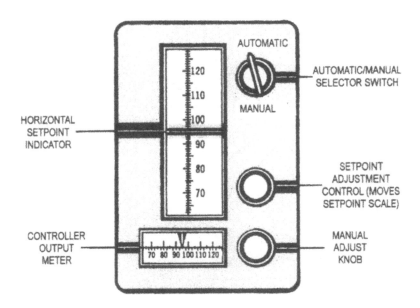

Balanced Condition—Ideally, it would be desirable for a process system to develop output energy that would be equal to the input energy. When this occurs, the process reaches a steady-state condition, or is said to be in balance. Any change in either the input or the output energy will upset a condition of balance. Since most process variables are in constant change, the condition of balance is in a continuous state of change.

Self-regulation—Some systems have a self-regulation capability that is designed to produce continuous balance. In Figure 2-8, the outflow of liquid tends to equal the inflow applied to the control valve. If the valve is opened wider, the liquid level of the tank will increase. This causes a corresponding increase in pressure at the bottom of the tank. Through a wide range of limits, liquid flow in this system will self-regulate. The primary limiting factor of the system is the depth and the size of the storage tank. Self-regulation systems are designed to adjust themselves toward a balanced condition.

Figure 2-9 shows a non-self-regulating type of control system for comparison. In this case, outflow is limited to a fixed value because of

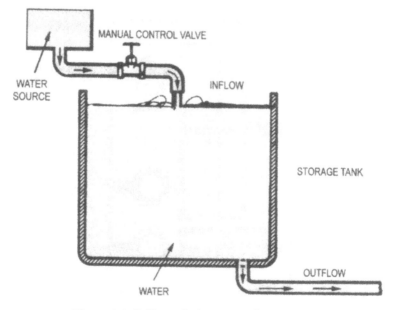

Figure 2-8. Self-regulating water flow system.

the restricted flow permitted by the constant-speed positive displacement pump. Unless the inflow and outflow are equal, the tank will eventually either become empty or overflow. Control applied to this type of system is often described as *negative self-regulation*. An additional control process must be employed by this system in order to maintain tank level. It is interesting to note that a negative self-regulation system always has a tendency to move in the direction of imbalance.

Process Time Lag

Most of the processes used in manufacturing operations perform well when variables are held within certain limits. When a process variable is subjected to some type of change, it takes a certain amount of time for the process to correct itself. *Process time lag* refers to the time it takes a system to correct itself and seek a condition of balance after a variable has changed. *Inertia, capacitance resistance,* and *dead time* are typical causes of process time lag.

Inertia is a common time-lag condition that refers to the ability of a process to continue in a certain state even after a change has occurred. The

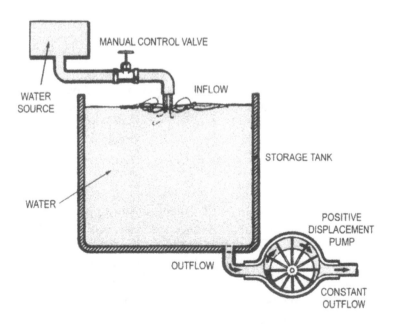

Figure 2-9. Non-self-regulating water flow system.

heat exchanger unit in Figure 2-4 will, for example, continue to retain heat even after the steam source has been removed. An object in motion has a tendency to resist the change to slow it down. In general, system inertia must be overcome before a process control operation can become effective.

Capacitance resistance has two components. *Capacity* is a time-lag condition that refers to the ability to store energy or a certain quantity of material. The walls of a hot-water storage tank, for example, tend to retain some of the heat energy initially applied to the water. Any type of energy-storing capacity retards the ability of the system to make a change. *Resistance* refers to those things that tend to oppose a transfer of energy or material when a process change occurs. The water inside a tank has different layers of resistance that must be overcome before the tank temperature reaches a prescribed value. When resistance and capacity are combined, the resulting condition is called a *time lag*.

Dead time is a property which refers to the time required for a process to change from one value to another. If the temperature of the incoming water of a heat exchanger unit drops to a lower value, a certain amount of time will elapse before the tank can sense that a change is needed. Dead time is not a retarding effect, but refers specifically to the actual time in which no change occurs. Dead time is dependent on the speed with which a change is transported through the system. Capacity, resistance, and process inertia all influence dead-time delay.

Setpoint Regulator Systems

Any process system that maintains its developed output at a consistent level in spite of load changes is classified as a *setpoint regulator system.* In practice, the setpoint of the controlled variable of this type of system is seldom changed. Heating systems, pressure regulator units, and voltage regulator circuits are examples of this type of control.

Setpoint Follow-up Systems

A *setpoint follow-up system is* a feedback control process in which the setpoint is in a constant state of change. The basic function of this type of system is to keep the controlled variable as close as possible to any value changes in the setpoint. In systems of this type, the setpoint value is called the *reference variable.* Industrial applications employing this type of control are found in graphic recording equipment and valve positioning equipment. Figure 2-10 shows a block diagram of a self-balancing setpoint follow-up system for a chart recording instrument.

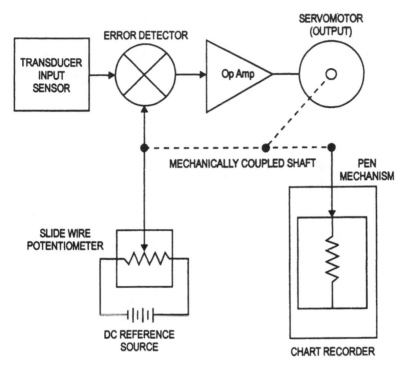

Figure 2-10. Self-balancing setpoint follow-up system.

CONTROLLER FUNDAMENTALS

The controller of an industrial process control system is an instrument that is primarily responsible for automation. A controller employs pneumatics, electronics, digital changes, and/or mechanical energy to automatically perform control operations. This instrument is designed to maintain an industrial process at some predetermined value. It compares the actual process value to that of a desired process value (the setpoint) and alters the final control element in response to this difference. A block diagram of a controller is shown in Figure 2-11. The controller maintains the process output at some predetermined value by accepting two signals. One is the setpoint input. This is manually adjusted to some desired process input value by the operator. The other signal is an indication of the actual process value. This is developed outside of the controller by a sensor attached to the process output. The resulting signal forms a feedback return path to the controller. The output of the controller drives an actuator

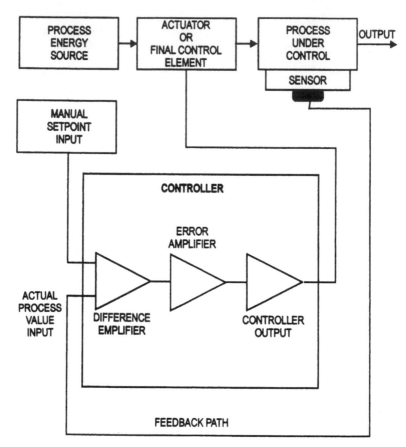

Figure 2-11. Functional block diagram of a controller.

or the final control element. The composite assembly permits the process to be controlled automatically through a self-correcting operation.

Modes of Control

The operational response of a controller is often described as its *mode of control*. Several different types of control are available. In some cases, only a single mode of control is needed to accomplish an operation. This is described as a *pure control operation*. *On-off, proportional, integral,* and *derivative* are examples of pure control. More sophisticated control is achieved by combining two or more pure modes of operation. This is described as a *composite mode. Proportional plus integral, proportional plus derivative,* and *proportional plus integral plus derivative* are examples of composite control modes. The specific mode of operation utilized by an instrument is determined by the control procedure of the application.

On-off Operation

An *on-off* or *two-state* controller is the simplest of all process control operations. The actuator or final control element driven by the output of the controller is automatically switched on or off. It does not have any intermediate level of operation. Control of this type is popular and inexpensive to accomplish.

The home heating system in Figure 2-12 is a common example of on-off control. The thermostat of this system serves as the controller. Should the interior temperature of the building drop below the setpoint value of the thermostat, the furnace will turn on. In a gas-furnace heating system, a solenoid gas valve serves as the final control element. It is energized by the thermostat and turns on the source of heat. Heat from this assembly causes the interior temperature of the home to rise. When the temperature rises above the setpoint value, the thermostat tells the furnace to stop producing heat. The solenoid gas valve closes and turns off the heat source. The heat source continues to be off until the temperature again drops below the setpoint of the thermostat. When this occurs, the thermostat again turns on the gas valve. This on-off cycling action continues according to the demands of the system as long as the system is operational. This response is typical of an on-off controller.

The operation of an on-off controller is determined by the position adjustment of its setpoint value. As a rule, the output continually cycles or oscillates above or below the setpoint value. Figure 2-13 shows the cycling response of an on-off temperature controller. Note the location of the setpoint and how the temperature of the system rises and falls above this value. The final control element of this system is the gas solenoid valve. The source of gas is turned on and off with respect to time. When gas is applied, it produces heat. Turning off the gas stops the process. In systems where continuous on-off cycling occurs, there can be some damage to the final control element after prolonged operational periods. On-off control is generally used in systems where precise control is not necessary. These systems must have a large capacity that changes slowly in order to be effective.

A graph showing the position of the gas valve with respect to system temperature in an ideal home heating system is illustrated in Figure 2-14. Assume that the setpoint of the thermostat is adjusted to 70°F. If the temperature is less than 70°F by a tiny amount, the gas valve will be fully on, or 100 percent open. This causes an immediate increase in temperature.

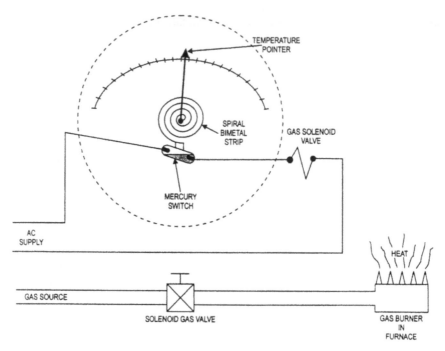

Figure 2-12. Thermostat-controlled gas heating system.

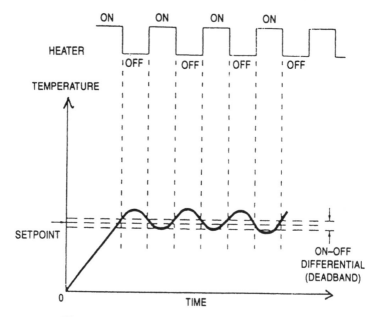

Figure 2-13. On-off temperature control action.

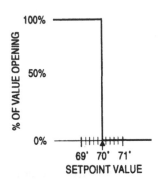

Figure 2-14. Valve position versus measured temperature of an ideal heating system.

When the temperature rises above 70°F by a tiny amount, it causes the gas valve to be fully off, or closed. This means that the gas valve of our ideal heating system continually cycles on and off, which can cause the gas valve to be damaged after a prolonged operational period.

A practical on-off controller has a *deadband,* or a *differential gap* in its operation. Deadband refers to the smallest possible temperature change which can drive the valve from open to closed in the operation of an on-off system. A graph illustrating the differential gap of an on-off controller is shown in Figure 2-15. Assume that the setpoint of the thermostat is adjusted to a value of 70°F. This is indicated at point A on the graph. When the system temperature drops to 69°F, it causes the gas valve to be fully on, or 100 percent open. This is indicated at point B on the graph. The temperature then rises to a value of 71°F. This is indicated at point C on the graph. Because 71°F is 1°F above the 70°F setpoint value, the gas valve is turned off, or fully closed, as shown at point D. The deadband or differential gap of this system is –1°F to +1°F, or 2°F.

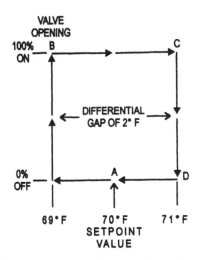

Figure 2-15. On-off controller with a differential gap.

Proportional Control

In on-off control, the final control element was either on or off. If the control element were a valve, it would have been fully open or closed. There is no intermediate adjustment of the valve. In proportional control, the final control element can be adjusted to any value between fully open and fully closed. Its value is determined by a ratio of the setpoint input and the actual process value of the system. In a valve-controlled system, operation is arranged so that the valve is normally adjusted to some percentage of its operating range.

Proportional control is defined mathematically as

$$V_{out} = K_p V_e$$
where

$$V_{out} = \text{controller output}$$
$$K_p = \text{proportional controller gain}$$
$$V_e = \text{error signal, or } V_{sp} - V_{pv}$$

V_e is the error signal that is determined by the difference in setpoint voltage (V_{sp}) and process variable voltage (V_{pv}). V_{pv} represents the actual value of the process variable being controlled. V_{out} of the expression is accurate for all steady-state conditions of operation.

Refer to the water temperature control system in Figure 2-16. In this system, cool water enters the tank at a constant temperature. It exits the tank at a higher but rather constant temperature. Steam applied to the heating coil increases the temperature of the water to some predetermined value. This value is determined by the setpoint adjustment. The controller senses the temperature of the water in the tank and compares the actual temperature of the water (V_{pv}) with the predetermined setpoint value (V_{sp}). The motor-driven steam valve is actuated by the proportional controller output (V_{out}). Valve opening is determined by the difference in temperature between actual and setpoint values (V_e).

The graph in Figure 2-17 shows a relationship between valve opening percentage and water temperature. The valve can be 100 percent open or fully closed. The water temperature range is 150°F to 200°F. The difference between temperature values for the valve to be fully open or closed is called the *proportional band of control*. The proportional band of this system is 200°F – 150°F, or 50°F.

The proportional band of a controller is generally expressed as a percentage of the full-scale span of operation. As a rule, the setpoint adjustment range of the controller usually identifies the operational span of the proportional band. The operational span of the setpoint is usually greater than the value of the proportional band. For our water-heating controller, the proportional band was 50°F. The setpoint adjustment span could be from 250°F to 100°F, or 150°F. Within the 200°F to 150°F band, the valve response is proportional to the temperature change. Outside of this band, the valve ceases to respond because it has reached its upper or lower limit. The proportional band range divided by the setpoint adjustment span is an expression of the percentage of full-scale operation. In this case, the proportional band is 33.3 percent, because 50 divided by 150 equals 0.333 or 33.3. Most proportional controllers have an adjustable proportional band that varies from a few percent up to several hundred percent.

For our water heating system to be operational, the steam valve must be opened to some percentage of its span. To start the system, assume that

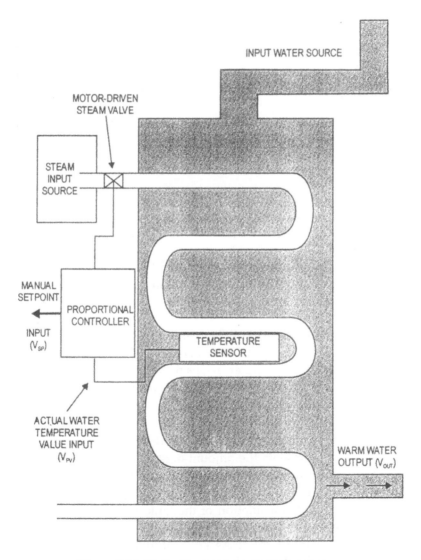

Figure 2-16. Water temperature control system.

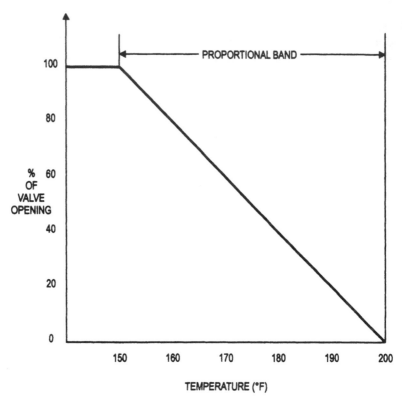

Figure 2-17. Graph showing the relationship of valve opening percentage to water temperature.

the setpoint is adjusted to 180°F. This generally causes the steam valve to be fully open. A full charge of steam is applied to the tank heating coil. As the temperature of the water begins to rise, the steam valve begins to close. The percentage of valve opening is proportional to the difference between the setpoint value and the actual temperature of the tank water. As the temperature of the water moves closer to the setpoint value, the percentage of valve opening is reduced. At 180°F, the valve will be 40 percent open. This represents the amount of steam needed to maintain the water temperature at 180°F. In a functioning system, this valve setting is somewhat unpredictable. Such things as ambient temperature, heat consumption of the water, insulation of the tank, temperature of the water source, and the temperature of the steam source are some of the conditions that must be taken into account. As a rule, proportional control works well in systems where the process changes are quite small and slow.

A disadvantage of proportional control is that it does not respond well to long-term or steady-state changes in the process being controlled. A change or disturbance will not let the process return exactly to its pre-disturbance value. This means that the process will have a difference in its new value. This is called an *offset*. It represents a new process value that is slightly less than the setpoint value. Offsets may or may not be acceptable for some industrial systems. The offset problem can be reduced by combining other modes of control with proportional controllers.

Integral Control

The output of a controller is used to actuate the final control element to eliminate any difference between the setpoint and actual values of an operating system. This difference is commonly called the *error signal*. A controller is designed to eliminate system error. In proportional control, the output was adjusted proportionally to correct the system error. As a rule, this type of control produces an offset problem in the output. An *integral controller* has an output whose rate of change is proportional to the system error signal. As long as there is an error, the output will continue to change to correct it. When the error is zero, the integral controller maintains the output at this value until a new error occurs. This means that an integral controller has inertia: It has a tendency to hold the output which was necessary to eliminate the error signal applied to its input.

Integral control is continuous, and the output changes at a rate proportional to the magnitude and duration of the error signal. When there is a large error signal, the output changes rapidly to correct the error. As the error gets smaller, the output changes more slowly. This action is done to minimize the possibility of overcorrection. As long as there is an error, the output will continue to change. Mathematically, this is expressed as

$$\Delta V_{out} / \Delta t = K_i \bullet V_e$$

or

$$\Delta V_{out} = K_i \bullet V_e \bullet \Delta t$$

where

ΔV_{out} = controller output voltage
Δt = time rate of change
K_i = integration constant
V_e = error voltage

The response of an integral controller is shown by the graph in Figure 2-18. A large error occurs between points a and b. This causes the output to increase very rapidly. Decreasing values of error between b-c and c-d cause the output to increase more slowly. The output continues to rise for these decreasing error values. When the error goes to zero at point d, the output does not change. It holds the output at the value that dropped the error signal to zero. When the error goes negative between points e-f, it causes the output to drop. Between points f-g, the error signal returns to zero. This causes the output to hold the value that it had before dropping to zero. At point g, the error signal increases, causing the output to ramp from its previous value.

Proportional Plus Integral Control

Pure integral control is rarely used by itself in process control instrumentation. In general, it responds poorly to *transients*. This means that the output responds slowly to rapid changes in the error signal. It does, however, respond well to long-term or steady-state errors, and is therefore primarily used in conjunction with proportional control. By combining proportional and integral principles in a single unit, the advantages of each operation can be fully realized. Proportional control reacts quickly but cannot reduce the error signal to zero. This is the offset problem of proportional control. Integral control responds slowly over a period of time. It does not respond to rapid changes but can reduce an error signal to zero. A combination of these two principles in a single unit is called *proportional plus integral (PI) control*.

The control action of

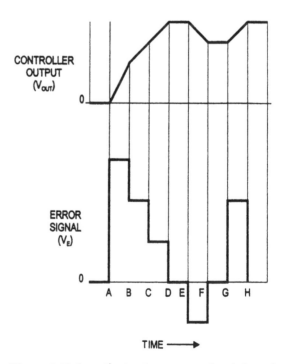

Figure 2-18. Input/output response of an integral controller.

a proportional plus integral controller is defined mathematically by the expression

$$V_{out} = K_p \bullet V_e + K_i \int V_e \bullet \Delta t$$

This is achieved by adding the proportional control formula to the integral control expression.

The response of a PT controller shows that when an offset is present the final control element is in a constant state of change. It changes at a rate that is proportional to the error signal until a correction is attained.

The output of a PT controller has two problems that must be addressed. First, the settling time of a PT unit is much greater than that of a proportional-only controller. Second, when a negative error signal is applied to an integrator, it moves the proportional band above the setpoint value. This means that the output will not start to change until the system overshoots the setpoint value. To reduce these two problems, a PT controller has its control actions independently adjustable.

The control functions can be combined in one of two ways. The integral and proportional circuits can be arranged in series or parallel. Figure 2-19 shows the blocks of a PI controller achieved by operational amplifiers. Note that the error amplifier has the setpoint (V_{sp}) and actual process variable voltage (V_{pv}) signals applied to it. The output of the integral and proportional functions are added together in a summing amplifier. The primary difference in the two circuits is the speed of the response. In the series arrangement, the integral circuit receives a larger amplified error signal because of its connection to the proportional amplifier. This causes the output to change more quickly than the parallel circuit. The series arrangement tends to force the error to zero more rapidly.

PI control is used on processes that have large load changes or frequent setpoint adjustments. Proportional control alone is not capable of altering this type of process by reducing the offset to an acceptable level. When the reset function of integral control is combined with the proportional mode, it automatically eliminates the offset problem. This permits the process to be controlled without prolonged oscillations, no permanent offset, and quick recovery after disturbances.

Derivative Control

Many controllers have an inertia or *hysteresis* problem. In the water temperature control system in Figure 2-16, it takes some time for steam

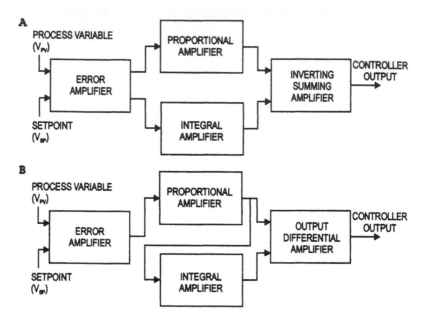

Figure 2-19. Proportional integral control arrangements. (A) Parallel arrangement; (B) Series arrangement.

to increase the temperature of the water. This means that there is a delay between the application of steam and the water temperature rising to a new value. The significance of this is that an error will not cause an immediate deviation from the setpoint value. When an error is detected by the system, it responds just as slowly to the corrective action. To overcome this sluggish characteristic, some exaggerated corrective action must be taken. If a controller produces a large corrective signal in response to a minute error, the system will be brought into control more quickly. This occurs even if the system has a large amount of inertia. However, if the corrective action remains large, the controller will overcompensate for the error. This could cause the unit to break into oscillation. A more desirable corrective action is one that is initially large but drops off with time. This is a characteristic of the *derivative controller.*

A basic element of the derivative controller is a *differentiator circuit.* A differentiator works in proportion to the rate of change of its input. Electronically, this is determined by the product of circuit resistance and capacitance. Mathematically, the output of a derivative controller is expressed as

$$V_{out} = K_p \bullet \Delta V_e / \Delta t$$

where

K_d = derivative gain
ΔV_e = change in error voltage
Δt = change in time rate

In this expression, $\Delta V_e / \Delta t$ is the derivative of the rate change of the error signal with respect to time. This means that if V_e is zero or a direct current value, the output of the differentiator will be zero. If the input occurs in steps, a rapid change in input will drive the differentiator into saturation. If the error voltage has a linear rise time or ramps, the output will maintain a constant value that equals the slope of the input. This is usually expressed in volts/second. In an electronic differentiator, this is determined by the RC components of *an operational amplifier* (opAmp) circuit. Figure 2-20 shows the circuitry of a basic opAmp differentiator.

Figure 2-21 shows the output response of an error signal applied to *a* derivative controller. A step in the error signal at point a produces an abrupt change in the output. This is due to the change in error and time rate or $\Delta V_e / \Delta V$ values of the controller. The output or V_{out} rises to its peak value. At error points a-b, b-c, and e-f, the error value is constant but not zero. The derivative of the error signal is zero, so that the output of the controller is also zero at these times. A gradual increase or ramp in the er-

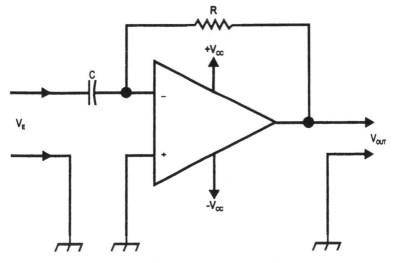

Figure 2-20. OpAmp differentiator.

ror signal produces a constant output. This is shown by the time between points c and d. An increase in the ramp of the error signal increases the value of the output. This is shown by the time between points d and e. A decreasing slope in the error signal causes the output to go negative. This is shown by the time between points f and g.

It is important to remember that the derivative controller only responds to changes caused by the applied error signal. If there is a steady-state error, the output will take no corrective action. If the rate change is fast, the output tends to oscillate or hunt its level. Because of this, pure derivative control is not used alone in controller operation. It is, however, used in conjunction with other control actions. A common application is in *proportional plus derivative* and *proportional plus integral plus derivative control*.

Proportional Plus Derivative Control

Proportional and derivative modes of control are often combined to reduce the tendency for oscillations and to allow for a higher gain setting. This combination is called *proportional plus derivative (PD) control*. The proportional mode of this controller produces a change in output that is a percentage of the error signal. Derivative control provides a change in

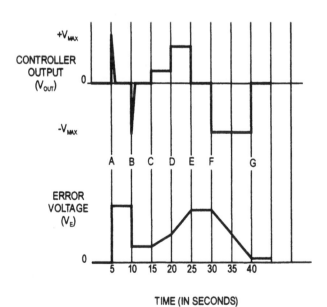

TIME (IN SECONDS)

Figure 2-21. V_e/V_{out} response of a derivative controller.

output that is due to a rate of change in the error signal. This response differentiates error-signal changes. It tends to maintain a level as long as change continues at a given rate. To some extent, PD controllers can anticipate condition changes. This anticipating action makes the controller very useful in controlling processes that have sudden changes in load. Electronic PD controllers are, however, susceptible to noise and transients. This may cause the output to saturate or go to its highest level. PD controllers are frequently used in motor servo systems and in systems that have small but quick process parameter changes.

The mathematical relationship of a proportional plus derivative controller is expressed as

$$V_{out} = K_p \bullet V_e + K_d \bullet \Delta V_e / \Delta t$$

where

K_p = proportional gain
K_d = derivative gain
V_e = error signal voltage
ΔV_e = change in error signal voltage
Δt = change in time rate

Note that the output voltage of this controller is the sum of the proportional and differential expressions discussed earlier. The combined expression can be easily achieved electronically with operational amplifiers.

Figure 2-22 shows a graphic response of a PD controller. An increase in the error signal at point a causes an abrupt rise in the output. This is largely due to derivative action. After this initial jump, the output continues to rise in a ramp between points a and b. Proportional action causes this response. At point b, the error signal becomes level. The output drops abruptly, due to the derivative function. Between points b and c, the output remains constant. The proportional unit causes this condition. At point c, the error signal drops in a decreasing ramp. Initially, this causes an abrupt change due to the derivative action, but the proportional unit takes over during points c and d, Between points d and e, the error signal is zero. In this sequence of events, the error signal dictates output time changes. The direction and action of change is dependent on specific proportional and derivative characteristics.

Proportional Plus Integral Plus Derivative Control

When proportional, integral, and derivative control operations are

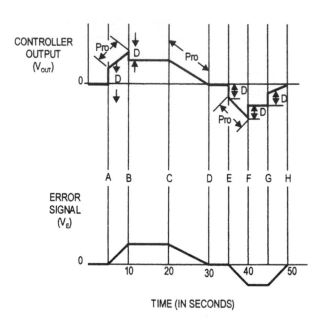

CONTROLLER
OUTPUT
(V_{out})

ERROR
SIGNAL
(V_e)

TIME (IN SECONDS)

Figure 2-22. Graphic response of a PD controller.

combined in a single instrument, it is considered to be a *proportional plus integral plus derivative (PID) controller.* This type of instrument combines the desirable characteristics of all three controlling elements in a single instrument. Each controlling unit receives the same error signal, and the outputs are then added through a summing amplifier. This type of control can be achieved electronically by operational amplifiers. Figure 2-23 shows a block diagram of the elements of a PID controller.

In a PID controller, each element has a unique responsibility in developing the output. The proportional unit provides a quick response for all system disturbances. The integral unit has an output whose rate of change is proportional to the length of time that the error signal is present. The derivative unit is primarily responsible for rate changes in the error signal. Combining these elements permits precision control of some rather difficult industrial processes.

Mathematically, the operation of a PID controller is considered to be the proportional unit plus the integral unit plus the derivative unit. Functionally, this is accomplished by adding the individual operations of each element together to produce a single output. The respective operations of each unit have already been defined. These are now combined into a single expression as

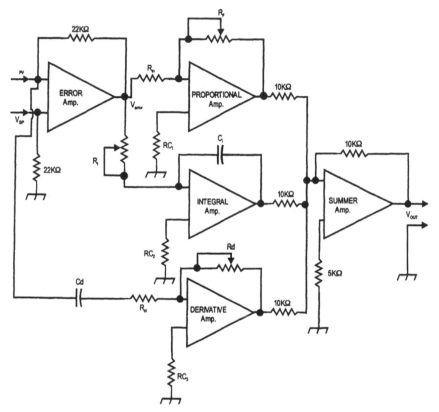

Figure 2-23. Block diagram of a PID controller.

$$V_{out} = K_p \bullet V_e + K_i \int V_e \bullet \Delta t + K_d \bullet \Delta V_e / \Delta t$$

Note that the terms of this expression have all been defined earlier.

Figure 2-24 shows the graphic response of a PID controller to an error signal. The output of each element is shown independently, and these are combined to form the controller's output. The horizontal part of this display is expressed in one-second intervals. Assume that the amplifiers of each block have a unity gain, and the RC time constant of the integral and derivative units are 1 second and 0.2 seconds, respectively. All three responses are added algebraically to obtain the combined output.

In the first second of operation, the applied error signal is zero. This causes all of the outputs to be zero in the same time frame. At the one-second point, the error signal has a value change. This causes a corresponding change in the output of each unit. The controller's output is

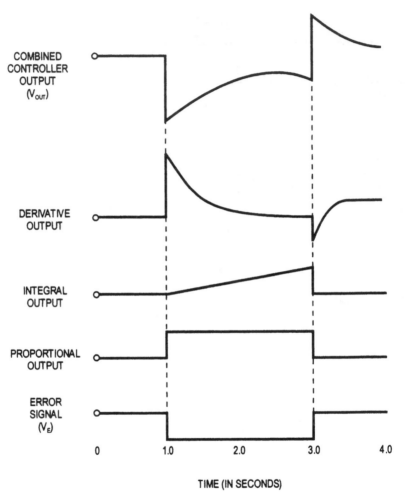

Figure 2-24. Graphic representation of a ID controller.

the inverted sum of each output unit. The error signal continues for three seconds. Note how the respective outputs change during this interval. The controller's output is a function of the individual outputs in a single operation. At the three-second point, the error signal returns to zero. This causes the output of each unit to change accordingly. These are combined to make the controller output. This shows that the collective output of a PID controller is fairly constant over a period of time.

The PID controller is a unique instrument that is widely used to control a number of difficult processes with a great deal of precision. This type of controller is generally more expensive than other units, and it is more difficult to prepare for operation. In some instruments, each mode of operation can be selected for independent use by programming in the desired operation. Each mode of control must be individually adjusted or *tuned* to make it functional. PT and PD control can also be accomplished by instrument programming. PID controllers are generally not used for all controller applications today. Controller selection is determined by such things as the amount of precision control needed, the difficulty of the process being controlled, the initial set-up and tuning procedure, the characteristics of the process being controlled, and the initial cost of the controller.

CODES, STANDARDS, AND SYMBOLS

As in other technical disciplines, process control has many codes, standards, and symbols that are used to describe its characteristics. Some of these are the result of historical usage, some are used for convenience, and others are the result of legislation. With the extensive growth that has taken place in this field, it has been extremely difficult to establish standards and values that find universal acceptance. As a result, there are many terms and values that are widely used in one part of process control but may not be used at all in other parts of the field. Personnel working with industrial processes must be familiar with a wide range of information in order to communicate effectively with others in the field. In this section, we will briefly present some of the units, codes, and standards that are used in process control.

Published Standards

There are several well-known organizations that develop and publish standards that are important to industry. Among these are the *American National Standards Institute* (ANSI), the *Institute of Electrical and Electronic Engineers* (IEEE), the *Instrument Society of America* (ISA), the *National Fire Protection Association* (NFPA), *Underwriters' Laboratories* (UL), the *Society of Manufacturing Engineers* (SME), and the *Society of Automotive Engineers* (SAE). The standards published by these organizations are for the use of manufacturing industries, consumers, and, in some cases, the general public. These standards are subject to periodical review, and industrial personnel must keep up to date with changes and revisions in order for industry to be in compliance with the regulations that control its operation.

Measurement Standards

In one way or another, all industries are involved in measurement and its associated instrumentation. The types of measurements performed may vary from the tolerances of manufactured products to the exact calibration of process control instruments. Regardless of how routine or how complicated the measurement may be, all industries should be aware of measurement standards for their own particular applications.

Safety and Health Standards

The *Occupational Safety and Health Act* of 1970 (OSHA) has received much attention from industry. Since these safety and health standards affect both industrial operation and the products manufactured by industries, there is a need for industrial personnel to be familiar with them. Several changes in the latest edition of the *National Electric Code*, for example, are a direct result of the requirements of OSHA.

The purpose of OSHA is to insure that, so far as possible, every working person in the nation has safe and healthful working conditions, and to preserve our human resources. The immediate effect of OSHA was to bring together scattered safety programs and to declare and enforce rules and regulations that would assure worker and workplace safety nationwide. The effect was immediate and profound, and, as a result of OSHA, today's workplace is a vastly safer place for all employees. The importance of safety and the health of industrial personnel cannot be stressed too much. We must rely on science and technology to provide methods to cope with safety and health problems in industry.

Many large industries employ safety engineers who are responsible for safety procedures and practices. These individuals try to detect unsafe practice and hazardous working conditions. They also recommend changes that could remove safety hazards. While some safety engineers work with industrial safety, others are responsible for product design to ensure that manufactured items meet required safety standards.

Metric System

Most nations today use the *metric* system of measurement. In the United States, the *National Bureau of Standards* began a study in August 1968 to determine the feasibility and cost of converting industry and everyday activity to the metric system. This conversion is now taking place.

The units of the metric system are decimal measures based on the physical quantities of length, mass, time, electrical current, temperature, and luminous intensity. Although this system is very simple, several countries have been slow to adopt it. The United States has been one of these reluctant countries due to the complexity and cost of a complete changeover of measurement systems. However, most U.S. industries have adopted the metric system because of its international acceptance.

The metric system was standardized in June 1966, when the *International Organization for Standardization* approved a system called *Le Système International d'Unites*. The abbreviation for this system is internationally recognized as SI. SI supplanted the older *meter-kilogram-second* (mks) and *centimeter-gram-second* (cgs) systems that recognized only three base units. These two systems are closely associated with the metric system, which recognizes six basic units: meter, kilogram, second, ampere, kelvin, and candela. The *International System of Derived Units is* shown in Table 2-1.

The coordination necessary to develop a standard system of units is very complex. The *International Advisory Committee on Electricity*, for example, makes recommendations to the *International Committee on Weights and Measures*. Final authority is held by the General Conference on *Weights and Measures*, which meets every few years. The laboratory associated with the International System is the *International Bureau of Weights and Measures*, located near Paris, France, but several laboratories in different countries cooperate in the process of standardizing units of measurement. One such laboratory is the *National Bureau of Standards* in the United States.

Conversion of SI Units

Sometimes it becomes necessary to make conversions of SI units so that very large or very small numbers may be avoided. For this reason, decimal multiples and submultiples of the basic units have been developed by using standard prefixes. These standard prefixes are shown in Table 2-2. As an example, we may express 1000 volts (V) as 1 kilovolt (kV), or 0.001 amperes (A) as 1 milliampere (mA). A chart that shows the relationship of these prefixes appears in Figure 2-25.

A value used to measure a specific quantity is often less than 1. Examples of this are 0.02 volts, 0.875 amperes, and 0.85 watts. When this occurs, prefixes are often used to simplify the expression. Some prefixes for values less than 1 are shown in the top portion of Figure 2-25. Notice that a milli value is 0.001 of a volt and a micro ampere is 0.000001 of an ampere. These prefixes may be used with any expression of measurement. The value is divided by the fractional part of the unit. For example, if 0.0008 A is changed to microamperes, 0.0008 A is divided by the fractional part of the

Table 2-1. International system of derived units.

Measurement Quantity	SI Unit
Area	square meter
Volume	cubic meter
Frequency	hertz
Density	kilogram per cubic meter
Velocity	meter per second
Acceleration	meter per second per second
Force	newton
Pressure	pascal (newton per square meter)
Work (energy), quantity of heat	joule
Power (mechanical, electrical)	watt
Electrical charge	coulomb
Permeability	henry per meter
Permittivity	farad per meter
Voltage, potential difference, electromotive force	volt
Electric flux density, displacement	coulomb per square meter
Electric field strength	volt per meter
Resistance	ohm
Capacitance	farad
Inductance	henry
Magnetic flux	weber
Magnetic flux density (magnetic induction)	tesla
Magnetic field strength (magnetic intensity)	ampere per meter
Magnetomotive force	ampere
Magnetic permeability	henry per meter
Luminous flux	lumen
Luminance	candela per square meter
Illumination	lux

unit (0.000001), The value of 0.0008 A is equal to 800 pA, because 0.0008 divided by 0.000001 equals 800.

When changing a basic expression to a value with a prefix, the decimal point of the number may be moved according to the number of places represented by the prefix. To change the voltage unit of 0.2 V to millivolts using this method, the decimal point in 0.2 V is moved three places to the right, because the prefix milli in this case represents three decimal places: 0.2 V equals 200 mV. This method may be used for converting any SI expression to a value with a prefix smaller than 1.

Values with a prefix may easily be converted back to the basic unit. For example, milliamperes may be converted back to amperes, or micro-

Table 2-2. SI standard prefixes.

Prefix	Symbol	Factor by Which the Unit is Multiplied
exa	E	$1,000,000,000,000,000,000 = 10^{18}$
peta	P	$1,000,000,000,000,000 = 10^{15}$
tera	T	$1,000,000,000,000 = 10^{12}$
giga	G	$1,000,000,000 = 10^{9}$
mega	M	$1,000,000 = 10^{6}$
kilo	k	$1,000 = 10^{3}$
hecto	h	$100 = 10^{2}$
deka	da	$10 = 10^{1}$
deci	d	$0.1 = 10^{-1}$
centi	c	$0.01 = 10^{-2}$
milli	m	$0.001 = 10^{-3}$
micro	μ	$0.000001 = 10^{-6}$
nano	n	$0.000000001 = 10^{-9}$
pico	p	$0.000000000001 = 10^{-12}$
femto	f	$0.000000000000001 = 10^{-15}$
atto	a	$0.000000000000000001 = 10^{-18}$

volts may be converted back to volts. When a value with a prefix is converted back to a basic unit, the prefix value must be multiplied by the decimal equivalent of the prefix value. For example, 52 mV is equal to 0.052 V.
When 52 mV is multiplied by the decimal equivalent of the unit (0.001 for
the prefix milli), this equals 0.052 V, because 52 multiplied by 0.001 equals
0.052.

When changing a fractional prefix value back to its basic value, the
decimal point in the prefix value is moved to the left the same number of
places represented by the prefix. To change 996 mV to volts, move the decimal point three places to the left, because the prefix milli in this example
represents three decimal places: 996 mV equals 0.996 V. This method may
be used to change any value with a fractional prefix back to its fundamental expression.

Industrial processes often incorporate measurement values that are
very large. Units such as 50,000,000 watts, 85,000 ohms, or 138,000 volts
are examples of very large value expressions. Prefixes may be used to
make these large numbers more manageable. Prefixes used to express
large values are shown in the bottom portion of Figure 2-25. To change a
large quantity to a smaller value, divide the large quantity by the prefix
decimal value. For example, to change 50 million watts (W) to megawatts
(MW), divide 50 million W by 1 million to get 50 MW. To convert 138,000
V to kilovolts, divide 138,000 V by 1,000 to get 138 kV.

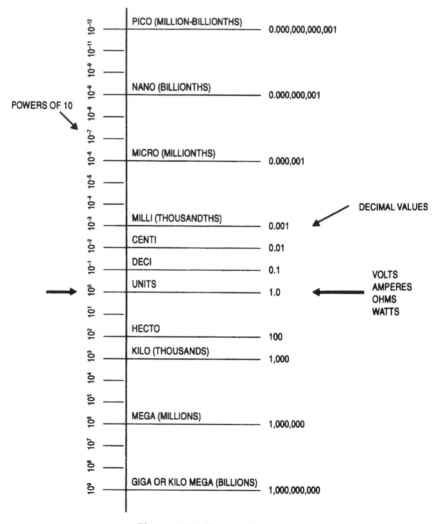

Figure 2-25. Conversion scale.

To change a large value to a prefix value, the decimal point of the large value is moved to the left according to the number of zeros of the prefix. Thus, 85,000 ohms equals 85 kilohms. To convert a prefix value back to a large number, the decimal point may be moved to the right by the same number of places in the prefix value.

The conversion scale shown in Figure 2-25 is useful when converting large and small units to values with prefixes. This scale may be used for powers of 10 or for decimal values. To use this conversion chart, follow these simple steps:

1. Find the position of the value as expressed in its original form.
2. Find the position of the value to which you are converting.
3. Write the original number as a whole number or in scientific notation.
4. Shift the decimal point the appropriate number of places in the direction of the term to which you are converting. Count the difference in decimal multiples from one value to the other.

The use of this step-by-step procedure is illustrated in the following examples:

A   Convert 100 picofarads to microfarads
1. Locate the prefix pico on the chart. Let this represent 100 pF.
2. Locate the prefix micro on the chart. This represents the converted value.
3. Note that the decimal point shifts six places to the left.
4. 100 pF converts to 0.0001 μF, or $100 \times 10^{-12} = 0.0001 \times 10^{-6}$.

B   Convert 20,000 ohms to kilohms
1. Locate units or 1.0 on the chart. Let this represent 20,000 ohms.
2. Locate the prefix kilo on the chart. This is the converted value.
3. Conversion occurs when the decimal point shifts three places to the left, or from units to a kilo value.
4. 20,000 ohms (units) converts to 20 KΩ, or $20,000\,\Omega = 20 \times 10^{3}\,\Omega$.

C   Convert 10 milliamperes to microamperes
1. Locate the prefix milli on the chart. Let this represent 10 milliamperes.
2. Locate the prefix micro on the chart. This is the converted value.
3. Note that the decimal point shifts three places to the right when changing a milli value to a micro value.
4. 10 mA converts to 10,000 μA.

English to Metric Conversions
     Since the United States uses both the metric and the English systems of measurement, it is imperative that a person working with physical quantities becomes familiar with some of the more common conversion

procedures. Conversions from English to metric and vice versa are calculations that one frequently encounters. The procedure consists of selecting
a known quantity and multiplying it by a conversion value to find a desired quantity. These conversion procedures deal with quantities of length,
area, mass volume, electrical measures, cubic measures, and temperature.
A partial listing of these conversions is shown in Figure 2-26 (A-G).

Industrial Symbols

The symbols used by industry today vary somewhat between different manufacturers, but a number of standard symbols are now being
recognized by industry. For example, it is easier to show symbols for a
battery connected to a lamp than to draw a pictorial diagram of the battery and lamp connected together, so symbols are used to represent various components and devices that are connected together in a *schematic
diagram.* Some standardization is needed in order to produce diagrams
that have the same meaning to all individuals involved. Troubleshooting
and maintenance are also simplified through the use of easily understood
schematic diagrams.

In the discussions of industrial process systems and circuit applications that follow, standard symbols will be used. The symbols shown in
Appendix A are commonly used in industry for electrical, electronic, hydraulic, pneumatic, and associated equipment control diagrams.

SUMMARY

Process control deals with the control of variables that influence materials and equipment during the development of a product.

Control is a system concept that takes on many different meanings,
but the basic idea of control is the primary influence that it has on the final
outcome of a process or operation. As such, control must be achieved before the final outcome is decided. Control effectiveness is primarily influenced by response time and accuracy of measurement, comparison, computation, and correction. Control of industrial processes can be achieved
by open-loop or closed-loop systems. Open-loop control involves a great
deal of physical effort by the operator. Closed-loop control employs a feedback path that samples the output to control the process automatically.

A process is a system variable being utilized by an industrial manufacturing operation. Controlled variables are the basic process values be-

Figure 2-26. Metric conversion: (A) Length; (B) Area; (C) Mass; (D) Volume; (E) Electrical; (F) Cubic measure; (G) Temperature.

A Length

Known Quantity	Multiply by	Quantity to Find
inches (in)	2.54	centimeters (cm)
feet (ft)	30	centimeters (cm)
yards (yd)	0.9	meters (m)
miles (mi)	1.6	kilometers (km)
millimeters (mm)	0.04	inches (in)
centimeters (cm)	0.4	inches (in)
meters (m)	3.3	feet (ft)
meters (m)	1.1	yards (yd)
kilometers (km)	0.6	miles (mi)
centimeters (cm)	10	millimeters (mm)
decimeters (dm)	10	centimeters (cm)
decimeters (dm)	100	millimeters (mm)
meters (m)	10	decimeters (dm)
meters (m)	1000	millimeters (mm)
dekameters (dam)	10	meters (m)
hectometers (hm)	10	dekameters (dam)
hectometers (hm)	100	meters (m)
kilometers (km)	10	hectometers (hm)
kilometers (km)	1000	meters (m)

B Area

Known Quantity	Multiply by	Quantity to Find
square inches (in^2)	6.5	square centimeters (cm^2)
square feet (ft^2)	0.09	square meters (m^2)
square yards (yd^2)	0.8	square meters (m^2)
square miles (mi^2)	2.6	square kilometers(km^2)
acres	0.4	hectares (ha)
square centimeters (cm^2)	0.16	square inches (in^2)
square meters (m^2)	1.2	square yards (yd^2)
square kilometers (km^2)	0.4	square miles (mi^2)
hectares (ha)	2.5	acres
square centimeters (cm^2)	100	square millimeters (mm^2)
square meters (m^2)	10,000	square centimeters (cm^2)
square meters (m^2)	1,000,000	square millimeters (mm^2)
ares (a)	100	square meters (m^2)
hectares (ha)	100	ares (a)
hectares (ha)	10,000	square meters (m^2)
square kilometers (km^2)	100	hectares (ha)
square kilometers (km^2)	1,000,000	square meters (m^2)

(Continued)

Figure 2-26. Metric conversion: (A) Length; (B) Area; (C) Mass; (D) Volume; (E) Electrical; (F) Cubic measure; (G) Temperature (*Continued*)

C Mass

Known Quantity	Multiply by	Quantity to Find
ounces (oz)	28	grams (g)
pounds (lb)	0.45	kilograms (kg)
tons	0.9	tonnes (t)
grams (g)	0.035	ounces (oz)
kilograms (kg)	2.2	pounds (lb)
tonnes (t)	100	kilograms (kg)
tonnes (t)	1.1	tons

Known Quantity	Multiply by	Quantity to Find
centigrams (cg)	10	milligrams (mg)
decigrams (dg)	10	centigrams (cg)
decigrams (dg)	100	milligrams (mg)
grams (g)	10	decigrams (dg)
grams (g)	1000	milligrams (mg)
dekagram (dag)	10	grams (g)
hectogram (hg)	10	dekagrams (dag)
hectogram (hg)	100	grams (g)
kilograms (kg)	10	hectograms (hg)
kilograms (kg)	1000	grams (g)
metric tons (t)	1000	kilograms (kg)

D Volume

Known Quantity	Multiply by	Quantity to Find
milliliters (ml)	0.03	fluid ounces (fl oz)
liters (L)	2.1	pints (pt)
liters (L)	1.06	quarts (qt)
liters (L)	0.26	gallons (gal)
gallons (gal)	3.8	liters (L)
quarts (qt)	0.95	liters (L)
pints	0.47	liters (L)
cups (c)	0.24	liters (L)
fluid ounces (fl oz)	30	milliliters (mL)
teaspoons (tsp)	5	milliliters (mL)
tablespoons (tbsp)	15	milliliters (mL)
liters (L)	1000	milliliters (mL)

(Continued)

ing manipulated by a system.

Controllers are hardware pieces that employ pneumatic, electronic, or mechanical energy to perform system control. A prescribed value to which a controller is adjusted is called the setpoint. A human operator

Figure 2-26. Metric conversion: (A) Length; (B) Area; (C) Mass; (D) Volume; (E) Electrical; (F) Cubic measure; (G) Temperature (*Continued*)

E Electrical

Known Quantity	Multiply by	Quantity to Find
BTU per minute	0.024	horsepower (hp)
BTU per minute	17.57	watts (W)
horsepower (hp)	33,000	foot-pounds per min (ft-lb/min)
horsepower (hp)	746	watts (W)
kilowatts (kW)	57	BTU per minute
kilowatts (kW)	1.34	horsepower (hp)
watts (W)	44.3	foot-pounds per min (ft-lb/min)

F Cubic Measure

Known Quantity	Multiply by	Quantity to Find
cubic meters (m³)	35	cubic feet (ft³)
cubic meters (m³)	1.3	cubic yards (yd³)
cubic yards (yd³)	0.76	cubic meters (m³)
cubic feet (ft³)	0.028	cubic meters (m³)
cubic centimeters (cm³)	1000	cubic millimeters (mm³)
cubic decimeters (dm³)	1000	cubic centimeters (cm³)
cubic decimeters (dm³)	1,000,000	cubic millimeters (mm³)

Known Quantity	Multiply by	Quantity to Find
cubic meters (m³)	1000	cubic decimeters (dm³)
cubic meters (m³)	1	steres
cubic feet (ft³)	1728	cubic inches (in³)
cubic feet (ft³)	28.32	liters (L)
cubic inches (in³)	16.39	cubic centimeters (cm³)
cubic meters (m³)	264	gallons (gal)
cubic yards (yd³)	27	cubic feet (ft³)
cubic yards (yd³)	202	gallons (gal)
gallons (gal)	231	cubic inches (in³)

(*Continued*)

makes this adjustment, which serves as one input to the controller. Another input is a measurement of the actual process value. These two values are then compared, and the developed output is the error signal. The error signal is then amplified or processed by the controller. The final output of the controller is used to alter an actuator or manipulate the final control element of the system.

Several different modes of operation are achieved by a controller. On-off control automatically switches the output above or below the

Figure 2-26. Metric conversion: (A) Length; (B) Area; (C) Mass; (D) Volume; (E) Electrical; (F) Cubic measure; (G) Temperature (*Concluded*)

G Temperature

Temperature conversions require specific formulas. These formulas follow:

1. To convert degrees Fahrenheit to degrees Celsius, use one of the following formulas:

$$°C = \frac{5}{9} (°F - 32°)$$

or

$$°C = \frac{°F - 32°}{1.8}$$

2. To change degrees Celsius to degrees Fahrenheit, use one of the following formulas

$$°F = \frac{9}{5} (°C) + 32°$$

or

$$°F = 1.8(°C) + 32°$$

3. Degrees Celsius may be converted to Kelvins by using this formula:
$$K = °C + 273.16°$$

4. Kelvins can be converted to degrees Celsius by simply changing the formula in Step 3 to this:
$$°C = K - 273.16°$$

5. To change degrees Fahrenheit to degrees Rankine, use this formula:
$$°R = °F + 459.7°$$

6. Through a change in the preceding, degrees Rankine can be converted to degrees Fahrenheit.
$$°F = °R - 459.7°$$

Through a combination of these formulas, temperature values gvien for any scale can be changest to values on one of the other three scales. Here is a comparison of the four scales:

Fahrenheit (°F)	212°	32°	−459.7°
Celsius (°C)	100°	0°	−273.2°
Kelvin (K)	373.2°	273.2°	0°
Rankine (°R)	671.7°	491.7°	0°
	Boiling point	Freezing point	Absolute zero

setpoint value. Proportional control responds to a ratio of the setpoint input and the actual process value of the system. It adjusts proportionally to different values to correct the problem. Integral control responds to large error signals. When the error signal gets smaller, the output changes more slowly. Derivative control responds to the rate of change of the applied error signal with respect to time.

Different combinations of control processes are often utilized. PT control is designed to provide automatic reset action which eliminates proportional offset of the output. PD control reduces the tendency for oscillations and allows for higher gain settings. PID control has a proportional response to signal error changes, with automatic reset to reduce output offset and an anticipating action that reduces errors caused by sudden load changes.

Process time lag is the time it takes a variable to correct itself after a change occurs. Inertia, capacity, resistance, and dead time are typical time lags.

The Occupational Safety and Health Act has added a new dimension of safety awareness to industry. Even though automated production through assembly line procedures is stressed by most industries, individual safety and health must be given top priority. OSHA provides for the safety of industrial workers through inspections which ensure that industries meet minimum safety standards.

The United States is now in the process of transforming industry from the U.S. or English system of measurement to the metric system. Most measurement is now based on the International System of Units. This system is based on the physical quantities of length, mass, time, electrical current, temperature, and luminous intensity. It is important to be able to convert U.S. units to SI units due to the changeover to the metric system.

Chapter 3

Pressure Systems

OBJECTIVES

Upon completion of this chapter, you will be able to
1. Define pressure and identify some systems that respond to this property.
2. Briefly describe the operation of a pressure system.
3. Define Pascal's law as it relates to pressure systems.
4. Describe the source, transmission path, control, load, and indicators of a pressure system.
5. Explain how to determine the amount of work accomplished by a system.
6. Define conditioning and indicate its importance to a pressure system.
7. Identify some differences between hydraulic and pneumatic systems.
8. Explain how fluid flow develops pressure in a hydraulic system.
9. Explain how pressure is developed in a pneumatic system.
10. Describe the purpose of an actuator.
11. Depict the source of linear and rotary motion within most pressure systems.
12. Discuss the reasons for the move to oil-free pneumatics.
13. Explain why oil-free pneumatics are not used in all circumstances.
14. Classify the three main groups of hydraulic fluids.
15. Discuss the reason why there has been a move away from petroleum-based hydraulic fluid.

KEY TERMS

Absolute pressure—Pressure that is above zero, or a perfect vacuum; indicated in pounds per square inch absolute (psia).
Actuator—A mechanism that physically drives, moves, or controls the operation of a system component.

Coalescing—An air filtering process in which liquid aerosols contact glass microfibers and form droplets that are drained away.

Compressor—A device that changes the volume of a gas or air and causes an increase in pressure.

Condenser—An instrument that plays a role in a steam pressure system by cooling spent steam in the return line.

Controller—A component that is used to alter a particular system process through some type of automatic operation.

Cylinder—A device that converts fluid power into linear mechanical force and motion; consists of a movable piston, connecting rod, and plunger operating in a cylindrical cavity.

Diaphragm—A thin, flexible partitioning element used to transmit pressure from one substance to another while keeping them from direct contact.

Emulsifiers—Chemicals that help stabilize and bring forth properties that would otherwise be useless.

Esters—A compound formed by the reaction of acid and alcohol with the elimination of water.

Filter—A device through which fluid is passed to separate matter held in suspension.

Friction—Resistance to relative motion when two or more bodies come into contact.

Gauge pressure—Pressure measured relative to ambient atmospheric pressure; indicated in pounds per square inch gauge (psig), or psi.

Head—An expression of pressure that is the result of gravitational forces on liquids; measured in terms of the depth below a free surface of the liquid, which is called the reference zero head.

Horsepower (hp)—A unit used to express mechanical power; one hp will lift 550 pounds one foot in one second or 33,000 pounds one foot in one minute.

HWCF—High water content fluid.

Hydraulics—A branch of science that is concerned with the behavior of liquids.

Hydrocarbons—An organic compound containing hydrogen and carbon.

Inertia—A condition that remains at rest or in uniform motion in the same straight line unless acted on by an external force,

Intake stroke—Air or oil that is brought through an inlet to power a piston or pump.

Manifold—A fluid conductor that provides multiple connection ports.

Nonpositive displacement pump—A pump that has no set amount of air or fluid that will be passed by or through the rotating member during its operation. \

Pascal (Pa)—The basic unit of pressure in the metric system.

Pascal's law—A law stating that pressure applied to a confined fluid is transmitted throughout the fluid and acts on each surface at a right angle.

Piston—A sliding piece that fits in a cylinder and transmits or receives motion by changes in fluid pressure.

Pneumatics—A branch of science that is concerned with the behavior of air or gas.

Positive displacement pump—A pump that moves a set amount of fluid or gas in a chamber between its inlet and outlet during one revolution of operation.

Pressure—The amount of force applied to a specific unit area; expressed in pounds per square inch or pascals.

Pressure system—A system that uses gas, fluid, steam, or air as a means of controlling the flow of energy or power.

Reciprocating motion—A mechanical operation based on back-and-forth straight-line motion.

Relief valve—A device that operates by pressure and bypasses pump delivery to the reservoir, limiting system pressure to a predetermined level.

Reservoir—A container that stores liquid in a fluid power system.

Resistance—A form of opposition that fluid encounters due to friction as it travels through different system components.

Servo—A mechanism that operates as if it were directly actuated by a controlling device; capable of supplying output power that is greater than that of the controlling device.

Solenoid—A coil of wire that, when energized with electricity, produces a magnetic field that resembles a bar magnet; this magnetic field causes a movable core to be drawn into the coil.

Spool—A term loosely applied to almost any cylindrically shaped part of a hydraulic component which moves to direct flow through the component.

Static head—An indication of force that is caused by the level or height of liquid in a container.

Torque—The turning effort or rotary thrust of a fluid motor; expressed in inch-pounds or foot-pounds.

Turbine—A rotary device that turns by the force of liquid, gas, or vapor moving against blades or vanes.

Vacuum—A space devoid of matter that represents negative gauge pressure.

Valve—A device that controls fluid flow, direction, pressure, and flow rate.

Valve actuator—A mechanical device that drives or positions a valve through its movement.

Vane—A sliding piece of material, as used in a rotary-vane pump, that moves with centrifugal force and conforms to the housing of the pump to capture fluid or air and move it through the system.

Vaporization—Converting fluid into gas with the application of heat or by spraying under pressure.

Viscosity—A measure of the internal friction that a fluid encounters when it flows.

Viscosity index—A change in viscosity with a change in temperature; the higher the index, the smaller the change in viscosity in relation to temperature.

Volute—A spiral or scroll-shaped form used in the housing of a pump or compressor.

Work—Exerting a force through a definite distance; measured in units of force multiplied by distance, or foot-pounds.

INTRODUCTION

Nearly all of the products manufactured by industry today are the end result of a process that involves *pressure*. Systems of this type are primarily designed to perform a particular work function that is the result of pressure being transferred through a confined network of pipes or tubes. Punch presses that move up and down in a reciprocating motion, rotating machinery, linear motion changes, container-filling operations, and liquid distributions are some representative applications of pressure systems. The value of this type of system to modern industry is almost beyond calculation.

The term *pressure system* is commonly used in industry to describe those things that employ some type of fluid, gas, steam, or air as a means of controlling the flow of energy or power. In practice, systems that respond to changes in pressure are placed in a common group. The basic principles of operation are similar in nearly all cases, regardless of the medium being processed. Specifically, this classification includes *hydraulics, pneumatics, steam or vapor systems,* and *liquid distribution.*

SYSTEM BASICS

A system that employs pressure in its operation is similar in many respects to all of the other industrial process systems. This primarily means that it must employ an energy source, a transmission path, control, a load device, and possibly one or more indicators in order to function. These parts are essential factors in the operation of all systems.

Energy Source

The *energy source* of a pressure system is unique in that it requires a degree of classification in order to distinguish it from other systems. Its primary function is still energy transformation, the basic purpose of all system sources. The unique feature of a pressure source, however, is the principles by which energy transformation occurs. Flowing liquids, for example, develop pressure as a result of fluid being forced through components that have some type of surface resistance. The source in this type of system is usually a pump that causes a continuous flow of fluid. Electricity is generally used as the primary energy source for this type of system. It is ultimately transformed into the mechanical energy of a flowing fluid through the action of a pump. This principle of operation is the basis of hydraulic pressure, which has widespread usage in industrial applications.

The energy source of a pneumatic system is another variation of the energy transformation principle of a pressure system. In this type of system, electricity also usually serves as the primary energy source. This form of energy is changed into mechanical energy through the rotary motion of a motor. The motor is ultimately used to drive a compressor, which increases pressure. Compressed air or gas is usually stored in a receiving tank and released when the need arises. Air pressure is used in a variety of mechanical applications and to achieve numerous control functions that apply to other systems.

Systems that distribute fluid for chemical processing may develop pressure as the result of storage tank position or elevation. The pressure of a *static head* at the bottom of a tank is always greater than that which occurs at other tank levels. The energy source, in this case, is simply potential energy that has been developed through the placement of liquids in an elevated tank. System pressure in this type of system is based on the weight of the liquid existing above a certain point. Water and chemicals are distributed through pipes and tubes as a result of head pressure developed by this type of source.

The term *head* is also used to describe pressure that is developed as a result of vaporization of liquids due to variations in temperature. Steam pressure, for example, is the result of boiler action, which causes a liquid to be changed into a vapor due to the application of heat. The increasing pressure forces the vapor through pipes to machines that use it to produce mechanical energy. The head pressure of this system is produced through the transformation of *thermal* energy into *mechanical* energy. Steam pressure may be used for industrial heating applications or for power to produce machine operations.

Transmission Path

The *transmission path* of a pressure system is by far the simplest part of the entire system. It essentially provides a flow path for a medium to follow. Hollow pipes, tubes, and flexible hoses are commonly used to achieve this operation. Hydraulic fluid, liquids, air, steam, and gases are used to transfer energy from the source to the load through this path. The primary function of the transmission path is simply to transfer either thermal or mechanical energy from the source to the remainder of the system.

Control

The *control* function of a pressure system is mainly responsible for altering the flow of energy along the transmission path. In its simplest form, control may be achieved by a two-position device that is used to turn the system on or off. More sophisticated control involves some type of variable changes in flow. Included in this are *regulation, proportioning control, check valves,* and *direction control elements.* Control maybe open-loop, which is altered by a human operator, or it may be closed-loop, which provides some form of automation. These control functions, as a general rule, apply equally well to both thermal and mechanical energy.

Loads

The *load* of a pressure system is responsible for some type of work function that occurs as a result of system operation. In some applications, this function may be obvious because it harnesses energy to perform some mechanical action. A punch press moving up and down, the rotary motion of a hydraulic motor, or the mechanical action of a valve actuator are obvious. The load of other systems may be hidden and rather difficult to define. In a fluid transmission system, the load is represented as a composite of all the opposition encountered by fluid flowing from the source to the point where it leaves the system. In this case, the mechanical energy of fluid flow is changed into heat energy which is ultimately lost to the system.

Indicators

Pressure system *indicators* represent one of the most important measurement applications in all of industry. Pressure measuring instruments outnumber all other system indicators.

In a functioning system, pressure must be monitored to insure proper operating efficiency. In addition to this, indicators are used to test system conditions during maintenance procedures. These instruments, as a general rule, respond in some way to the *elastic deformation principle* or as *electrical transducer elements*.

PRESSURE BASICS

Nearly everyone working in industry has, at one time or another, had an opportunity to see pressure do work. A classic example is commonly found in automobile service stations: hoists to lift cars for servicing. A system of this type uses both air and oil to develop the power needed to lift an automobile.

An automobile hoist operates on the principle that air added to the top of a long cylinder confined in an oil-filled tube can force the cylinder to move upward under pressure. The tube and cylinder are normally placed in the floor of the facility in such a way that the entire unit will retract when air pressure is removed. The cylinder will rise out of the floor when air is forced into the tube through the action of a control valve. The air source is developed by an electric-motor-driven compressor. Control of the hoist is achieved by manual manipulation of the air valve by the operator. Figure 3-1 shows how pressure is used to do work in an automobile hoist.

Pascal's Law

In 1653, the French scientist Blaise Pascal discovered a simple physical law: A pressure applied to a confined fluid is transmitted undimin-

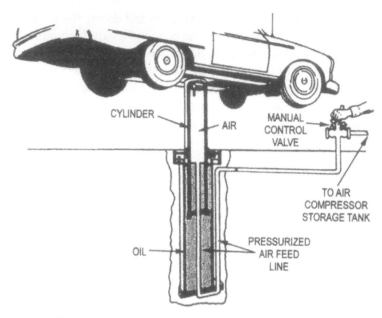

Figure 3-1. A fluid-power system automobile hoist.

ished throughout the fluid, and acts on all surfaces in a direction that is at right angles to those surfaces. This statement eventually became known as Pascal's law. The basic principles set forth in this law govern the operation of all industrial systems that use pressure as a process variable.

Figure 3-2 shows a graphic example of Pascal's law. The force applied to the fluid inside of the cylinder at piston A is instantly transferred to all parts of the cylinder. Piston B moves an amount that is equal to the movement originating at A if the pistons are the same size. The fluid in this system easily conforms to the inside shape of the cylinder. The same force acting on piston B is also applied to the inside walls of the cylinder. Thus, the cylinder walls must also be capable of withstanding this amount of pressure.

Force, Pressure, Work, and Power

The *force* applied to the piston in the system just described is defined as *any cause which tends to produce or modify motion*. To move a body or mass, an outside force must be applied to it. The amount of force needed to produce motion is based on the inertia of the body. Force is normally expressed in units of weight. Weight is defined as *the gravitational force exerted on a body (or mass) by the earth*. Since the weight of a body is a force

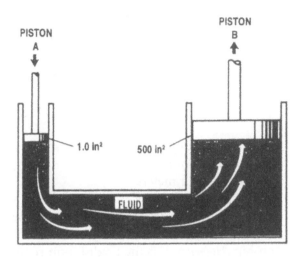

Figure 3-2. Illustration of Pascal's law.

(not a mass), units of force are expressed in both weight and force. The basic unit of force in the present English system is the *pound* (lb). In the metric system, the basic unit of force is the *newton* (N).

Pressure is a term used to describe the amount of force applied to a specific unit area, and is expressed in *pounds per square inch* (lb/in^2 or psi) in the English system, or newtons per square meter (N/m^2) in the metric system. The name *pascal* (Pa) has been assigned as the basic unit of pressure in the metric system ($1\ Pa = 1\ N/m^2$). This unit is relatively small, which often leads to expressions such as *kilopascals* (kPa) and *megapascals* (MPa) for larger values.

At sea level, the pressure of the atmosphere on the surface of the earth is 14.7 psi (101.36 kPa). In industrial applications, giant hydraulic presses are capable of squeezing metals with a pressure of 100 million psi (689.5 million kPa). Mathematically, pressure is expressed as

$$P = F/A$$

where
      P   = pressure in pounds per square inch or pascals
      F   = force in pounds or newtons
      A   = area in square inches or square meters

An interesting and important thing about force and pressure is that they only represent a measure of *effort*. A measure of what the system actually accomplishes is called work. In the fluid system just described, work is accomplished when the force applied to piston A causes it to move a certain distance. Work is commonly expressed in *foot-pounds* or *newton-meters (joules)*. The mathematical formula for this relation is

$$W = F \times D$$

where
      W = work in foot-pounds or newton-meters
      F = force in pounds or newtons
      D = distance in feet or meters

A more realistic concept of work must take into account the length of time that a force is acting. The term power is commonly used to express this relationship. *Horsepower* is the English unit commonly used in industry to express mechanical power. One horsepower can move 33,000 pounds a distance of 1 foot in 1 minute, or 550 pounds a distance of 1 foot in 1 second. Electrical motors are also rated in horsepower or fractional values of it.

Static Fluid Systems

A simple static fluid-power system is illustrated in Figure 3-3. In this system, a 100-lb force is applied to piston A, which has an area of 1.0 in^3. A pressure of 100 psi is developed by this action and is transferred through the fluid to piston B. The area of piston B is shown to be 100 in^3. Since the transfer of pressure through the fluid is equal on all parts of the cylinder, each square inch of piston B will receive 100 lbs of force. As a result, 100 psi times 100 in 2 equals 10,000 lbs of force applied to piston B.

The distance that piston B moves is directly proportional to the *ratio* of the areas of the two pistons. Moving the 1-in piston 4 inches into the cylinder forces 4 in^3 of fluid to be displaced. This displaced volume of fluid is based on the area of the piston times the distance it moves into the cylinder: Multiplying 1 in^2 by 4 inches gives 4 in^3 of fluid displacement. Spread over the 100 in^3 surface of piston B, this displacement causes the piston to move only 1/100 of the distance traveled by piston A. In this case, 1/100 of 4 inches equals only 0.04 in of motion. Piston B, therefore, has more force applied because of its size, but travels only a small distance. The amount of work done by pistons in a static fluid-power system shows an unusual relationship. The work done by piston A is 100 lbs times 4 in, or 400 in-lbs. At piston B, the amount of work achieved is also 400

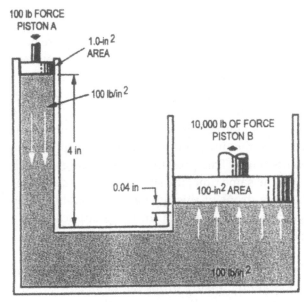

100 lb FORCE
PISTON A

1.0-in^2 AREA

100 lb/in^2

4 in

10,000 lb OF FORCE
PISTON B

0.04 in

100-in^2 AREA

100 lb/in^2

Figure 3-3. A static fluid-power system. When piston A moves 4 in, it causes piston B to move only 0.04 in.

in-lbs. This is determined by multiplying the applied piston force by the distance moved: 10,000 lbs times 0.04 in equals 400 in-lbs of work.

In actual practice, fluid systems do not show a 100 percent transfer of power from input to output. Fluid moving along cylindrical walls, for example, encounters a form of *surface friction*. The power loss resulting from this friction appears primarily as heat developed by the walls of the cylinder. In a static fluid system, the amount of power loss due to heat is so small that it would be considered negligible. In systems that move larger volumes of fluid over longer lines, losses of this type become an important consideration. As a general rule, excessive friction losses can be controlled by reducing the line length, keeping the number of bends in the transmission lines to a minimum, and preventing excessive fluid velocity by selecting proper-sized distribution lines. Designers generally take these things into account so that a high level of system operating efficiency can be achieved.

Fluid Flow Characteristics

As fluid flows through the components of a pressure system, it encounters a certain amount of opposition due to friction. In a fluid system, this opposition is called resistance. System pressure is developed as a result of fluid being forced against the surface areas of system components. Therefore, a direct relationship exists between system pressure and component resistance.

Figure 3-4 illustrates the friction/pressure relationship of a static fluid system. The pressure at point F is referenced at zero. A break in the system could cause this condition to occur. Other parts of the system show varying degrees of pressure change according to system resistance. Point B represents the highest-pressure point of the entire system because the full weight of the fluid appears at this point. Fluid flowing from point B to point F must change all of its potential energy into kinetic energy and heat energy. The moving fluid also causes a drop in pressure as it passes from points B to F. This pressure drop increases at each location point beyond B. At the same time, the source pressure decreases equally at each time. Pressure drops of this type are undesirable when power is being transferred through a system. In some applications, a drop in pressure is used to trigger the starting of a second operation in a sequential system.

Pressure drop is a much more significant characteristic in a *dynamic* or *continuous flowing* fluid system. Pressure drops can be caused by flow turbulence created by abrupt changes in direction, as in the corners of the system in Figure 3-5. System restrictions encountered by fluid flow are also a source of pressure drop. Control valves and tubing sizes are largely responsible for this type of drop. Small lines tend to increase the speed of fluid flow, which causes an increase in surface friction. Notice the pressure drops near each of the restricted areas in Figure 3-5.

The length of a system line is another important pressure-drop factor that must be considered: Fluid encounters more resistance when it travels

Figure 3-4. The friction-to-pressure relationship of a static fluid-power system.

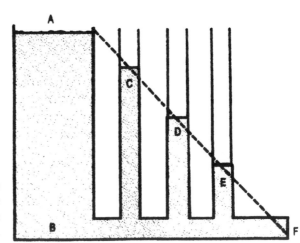

long distances. Proper system design usually minimizes this factor.

An interesting thing to note about pressure drop is what happens when the flow ceases. In this case, pressure drop stops and the pressure reaches a stable value throughout the system. A faulty pump or loss of electricity could cause this condition to occur. A break in the system line, by comparison, normally causes complete loss of pressure or a very pronounced change in pressure value. These two pressure conditions are very evident when compared with the characteristics of a normally operating system.

Compression of Fluids

Compression of system fluid represents a major difference between hydraulic and pneumatic systems. In general, all gases and liquids are compressible under certain conditions; yet oil, synthetic oil, or a combination used to transfer power throughout a hydraulic system is ordinarily not compressible except in extremely long transmission lines or under rather high pressures. A volume reduction of approximately 0.5 percent for every 1000 psi (5895 kPa) of pressure is typical. In most industrial ap-

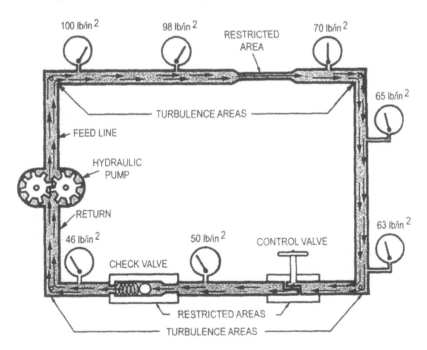

Figure 3-5. Illustration of pressure drops around a flowing fluid-power system.

plications, this amount of hydraulic fluid compression is not considered significant.

Fluid systems of the pneumatic type are primarily designed to respond to fundamental laws that apply to gaseous fluids. Industrial process applications of this type nearly always use air as the operating medium. Air must first be compressed by the system before it can be effectively used to transmit power. When air is reduced in volume by the compression process, its pressure increases. Compressed air is produced, stored, and then released into the system when it is needed. Pneumatic systems normally release air into the atmosphere after it has been used by the system.

Fluid compression represents a major difference between hydraulic and pneumatic systems. Hydraulic fluid is not compressed to any great extent under normal operating conditions. Pneumatic systems, by comparison, respond to compressed gases (primarily air). Because of this difference, there are some unique physical features in system components, but the basic function of each component is similar for both hydraulic and pneumatic systems.

INDUSTRIAL HYDRAULIC SYSTEMS

Most of the hydraulic systems used in industry are of the *circulating fluid* or *dynamic type*. This type of system employs a hydraulic pump as an energy source to move the fluid. The pump is driven by mechanical energy that has been produced electrically. Hydraulic fluid, under pressure, is made to circulate through the transmission lines, or pipes, of the system. System control is achieved by directional valves, metering devices, flow control valves, and regulators. Mechanical motion produced by the load is either rotary or linear. Cylinder actuators and hydraulic motors are typical components that achieve this function. A number of hydraulic indicators are also used to show different operating conditions and to test system components. Pressure, temperature, and flow are typical system values monitored by indicators.

Prepackaged hydraulic systems are often used in an industrial setting. The hydraulic pumping unit of this system is representative of those found in small- to medium-pressure applications. A unit of this type often houses an electric motor, a hydraulic pump, a pressure relief valve, gauges, and a reservoir. The primary energy source of this unit-electricity-must first be changed to mechanical energy through rotation of the motor. The motor, in turn, is used to drive the pump. Hydraulic fluid from the pump then circulates through the system and builds up pressure as it encounters system resistance.

Most of the hydraulic systems used in industrial applications today are equipped with a *filtering* device. Pressure-line filters are placed in the feed line, so that the hydraulic fluid is filtered before it has an opportunity to enter other system components. Any dirt particles coming from the pump or reservoir are removed from the fluid flow before entering the remainder of the system. As a general rule, filtering of this type reduces component breakdown a great deal.

Filtering devices can also be placed in the return line to remove dirt from the fluid after it passes through the entire system. One theory for placing the filter in this location states that fluid dirt is more prevalent in systems with moveable components. In complex hydraulic systems with numerous components, filters may be placed in both the feed and return lines to reduce contamination problems.

Control of the hydraulic system is usually achieved by a number of different valves. These include *check* valves, *reducing* valves, *directional* valves, *relief* valves, and *flow control* valves. These are manipulated manually by an operator, electrically by solenoid devices, or automatically by pilot pressure changes. In addition, valves can be tripped by an outside air source, temperature changes, or through the mechanical action of devices from different systems. Controllers are used to achieve automatic system control operations.

If the load of the hydraulic system was double-acting cylinders connected to the cylinder lines, high-pressure oil from the feed line would be applied to each cylinder, causing the piston to be thrust forward. Any oil on the front side of the piston would be forced into the return line by the stroke of the piston. Reversing the oil flow from a directional control valve would cause the action of the piston to be reversed. This mechanical motion would then be harnessed to do some type of work.

The primary function of a hydraulic system is the transfer of fluid from one location to another, causing some form of work to be accomplished. *Hydraulic fluids* are the vehicles used to accomplish this operation. These can be classified into three main groups: *petroleum-based, fire-resistant,* and *synthetic* fluids. Using the proper fluid in a hydraulic system permits optimum performance and a high level of efficiency for extended operational periods.

Petroleum-based Fluids

Typical *petroleum-based* hydraulic fluids contain a base oil and additives. In general, crude oil is used to produce the base oil of a hydraulic fluid. The type of crude always has some effect on the final product. Refining technology today is such that the character of the oil can be changed greatly by the degree and nature of the refining process.

The base oil of a petroleum-based hydraulic fluid makes very little difference in its overall performance, but additives supplied during the refining process have a significant effect on the response of the fluid used. Additives enable oil to protect from rust, enhance antiwear capabilities, prevent foaming, separate from water to prevent emulsions, inhibit oxidation, and lengthen its operating life. The additives in a hydraulic fluid range from less than 1 percent to 2 percent. In comparison, some multigrade motor oils contain up to 20 percent additives.

Petroleum-based fluids are selected to meet the needs of a variety of different applications. For example, in a system using piston pumps, the less active the additive is on metal surfaces, the more efficient the operation. Some fluids are selected because they inhibit rust and oxidation. Temperature, viscosity, compressibility, aeration, and corrosion prevention are some of the other factors to consider in the selection of a petroleum-based fluid for a specific application.

Fire-resistant Fluids

In the past several years, OSHA and ANSI have turned their attention toward the type of hydraulic fluids used in industry. In some applications, such as systems that operate in extremely high temperature surroundings, petroleum-based fluids have been totally ruled out because of the hazard of an explosion or fire. In general, it is simply not prudent to use petroleum-based fluids where fire resistance is necessary.

The major types of fire-resistant fluids are *phosphate esters, water glycols, invert emulsions,* and *high water-content fluids* (HWCFs). These fluids tend to be more expensive than their petroleum-based counterparts and have some performance deficiencies that must be tolerated. Thus, with the exception of HWCFs (which are similar to petroleum-based fluids), they are only used when fire resistance is necessary.

Phosphate esters, often called *snuffer fluids,* are a type of synthetic fluid. Considered to be a premium fire-resistant fluid, it is normally used where temperatures exceed the boiling point of water. These fluids are usually more expensive than other fire-resistant liquids.

Water-glycol fluids are less expensive than phosphate esters, provide good wear protection, and have fewer compatibility problems with seals and other components. Water content (generally 40 to 50 percent) must be maintained for proper performance. To prevent water loss, these fluids must not be used at a very high temperature.

Invert emulsions are the lowest-priced fire-resistant fluids. They have extremely fine water particles suspended in oil and other additives, giving fire resistance plus the viscosity and performance of oil. Their typical makeup is 60 percent oil and additives plus 40 percent water. Emulsifiers permit the addition of the same antiwear additives as in petroleum-based fluids, giving four to ten times the wear protection of soluble oils.

HWCFs are probably best in terms of fire protection. These thickened water-based fluids contain approximately 90 percent water and 10 percent additives. Because of these percentages, a great deal of water must be boiled away before a combustible mixture is reached; and much heat is absorbed, the steam blocks oxygen, and not much combustible residue remains when the water is gone. Originally, these fluids were not as good as invert emulsions in terms of protection. Today, some of the thickened fluids give better wear protection than straight petroleum-based fluids.

HWCFs are constantly improving. According to some officials in the industry, this fire-resistant fluid will be the replacement for petroleum-based fluids in the immediate future. This is largely because of improved performance levels and a price that is becoming less and less expensive.

Synthetic Fluids

Synthetic fluids or oils, generally consisting of esters or synthetic hydrocarbons, are considered to be premium hydraulic fluids. Characteristically, they have a high viscosity index, low temperature properties, low volatility, resistance to oxidation, and good lubrication properties. However, the high cost of these fluids generally dictates that they only be used in certain instances. These fluids tend to find a use where performance requirements are extremely stringent. Examples would be systems that generate high heat, have a relatively small reservoir, and require low maintenance and long fluid life, or in cases where startup temperatures are low. Synthetic fluids tend to find more widespread usage in new hydraulic system designs.

INDUSTRIAL PNEUMATIC SYSTEMS

The pneumatic system in Figure 3-6 is often used in small industrial manufacturing installations. The *air-compressor unit* serves as the source for the entire system. Typically, this unit employs electrical energy to produce rotating mechanical energy through the action of the motor. The *compressor* of this unit is driven by the motor. Compressed air is then stored in a *receiving tank*, from which it is eventually distributed to the system when needed. Receiving-tank pressures range from 100 to 150 psi (689.5 to 1034.25 kPa), as indicated on the tank pressure gauge. The feed line of the entire system is connected to the receiving tank for distribution throughout the plant.

Pneumatic systems must also employ some type of *air-processing* components in order to condition the air before it can be used. In systems with a small number of components, one conditioning unit is adequate; systems with a large number of components often connect conditioning units to each load device attached to the feed line. Conditioning the air involves filtering, pressure regulation, and lubrication. As a general rule, a conditioning unit is connected to the feed line between the compressor and the first control device.

Control of the pneumatic system of Figure 3-6 is achieved by a manually operated, four-way control valve. Air flow to the cylinder can be directed to cause forward motion or reverse motion, or it can be turned off by the control valve. One line to the cylinder will receive air and the other line will exhaust air into the atmosphere through the control valve. Exhausting air, instead of returning it to the system, is a unique characteristic of a pneumatic system.

Most pneumatic systems used in industrial process applications employ adjustable *air-flow valves* attached to each load device. These de ices are often placed in both lines, as indicated in Figure 3-6. Valves of this type are designed to regulate the actuating speed of a cylinder. By using a valve in each line, cylinder operation can be made to respond to a variety of action combinations. Without this control, the cylinder would receive maximum air pressure from the system. This would, of course, cause high-speed mechanical action of the cylinder from one position of the piston to the other. Adjustable air-flow valves add a great deal of versatility to the mechanical action of the load device.

The load device of the pneumatic system in Figure 3-6 is designed to produce linear mechanical motion. In this case, the cylinder is of the

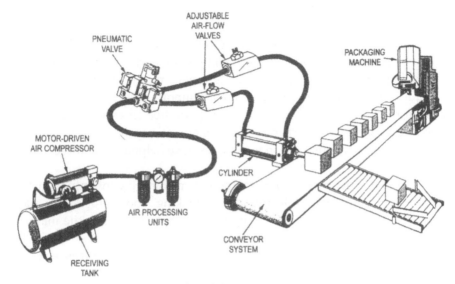

Figure 3-6. Industrial pneumatic system.

double-action type. It can be made to move in a forward or reverse direction, according to the air flow from the control valve. Pneumatic load devices are also designed to produce rotary motion. Air motors and rotary actuators are used to achieve this operation in industrial applications.

Most of the pneumatic systems used in industry employ several *indicators* placed at strategic locations throughout the system. As a general rule, system pressure is monitored by this type of indicator. The regulator and receiving tank of the pneumatic system both employ pressure gauges as indicators. Pressure drops along the system feed line can also be monitored by observing pressure readings at key locations. System maintenance and troubleshooting rely heavily on the use of indicators to locate faulty components.

OIL-FREE PNEUMATICS

Oil is one of the biggest sources of contamination in a pneumatic system. It can damage a finished product, destroy components, and pose a hazard to personnel. To overcome these problems, manufacturers have developed a wide range of components that can reliably operate without lubrication. Oil-free pneumatics represents a new trend in pressurized systems. This trend has brought about a number of changes in pneumatic technology.

A number of factors are responsible for the change to oil-free pneumatic systems. The rising cost of maintenance and system downtime are two important considerations. In addition, more stringent processing requirements for some products are an important issue. Stricter government regulation on the amount of oil permitted in exhaust air has also led to increased support for oil-free pneumatic systems. New materials and component construction techniques have helped make these systems a reality.

Some industries cannot tolerate even minute contamination from oil-based pneumatics. Electronic-chip making, food processing, paper making, health-related areas, temperature controls, and textile manufacturing are just a few that must avoid oil particulates at all costs. Oil-free pneumatics are responsible for a number of significant changes in component design and operation.

Components

The components of an oil-free pneumatic system are primarily the same as in a conventional lubricated pneumatic system. These include the basic source, path, control, load, and indicator. The primary difference in the two systems is in the construction of the components and the operational theory of the material.

The energy source of an oil-free pneumatic system is the compressor. Problems associated with an oil-free compressor are friction, heat, and higher noise levels. Friction has been reduced effectively through the use of specially designed self-lubricating materials. The temperature problem of an oil-free compressor limits its operation to applications of approximately 25 hp, although a number of higher-powered compressors with an extended life expectancy are presently being developed by some manufacturers. Compressor noise problems have been reduced by mounting the units in nonsensitive areas or by muffling the sound of the motor. In general, the problems with oil-free compressors have been confronted and reduced by new technology.

The control function of an oil-free pneumatic system is accomplished by valves and regulators. These system components have a number of problems with respect to material contamination and operational performance over long periods of time. As a rule, these problems have been corrected with the development of self-lubricating materials and coalescing filters.

The work function of an oil-free pneumatic system is accomplished with cylinders and motors. In general, these components have the same type of problems as those of the compressor and control valves. New developments in prelubricated materials and self-lubricating devices have reduced the problems associated with these components. The work function of an oil-free pneumatic system can now be accomplished with a great deal of efficiency for prolonged operational periods.

STATIC PRESSURE SYSTEMS

Static pressure systems are frequently used in industry to distribute liquids in a variety of different processing applications. These systems develop pressure as a result of the source being elevated to a position or head well above the remaining system components. A reservoir tower or standpipe is commonly used to hold fluids at a desired elevation. The height of the fluid and its density are used to develop static-head pressure. When fluid is released from the tower, it is forced to pass through system components due to head pressure.

A *water-tower system* is a common static-head pressure source used to develop pressure for water distribution. The tower serves as a reservoir to maintain water pressure at a constant level. Water is pumped up to a tower, where it is stored and released when the need arises. The static-head principle is also used to generate electricity, by releasing water that is stored in reservoirs or lakes developed behind dams. *Hydroelectric* power systems of this type utilize the developed head pressure to rotate turbines, which ultimately rotate large ac generators. Figure 3-7 shows a representative hydroelectric generating system.

The fluid of a static-head system usually has a limited attraction between its individual molecules. This attraction is such that when the fluid is poured into a container it is filled to a uniform level. By comparison, gas will completely fill the container and solids will retain their original form regardless of the shape of the receptacle. This means that fluids will conform to the inside shape of the container and develop an equal amount of pressure at the bottom. The height of the fluid and its density determine the head pressure at the bottom of the storage tank.

Static pressure systems that distribute fluids are primarily classified as *hydrostatic* systems. These systems are unique in their operation

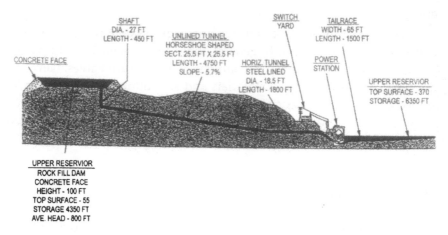

Figure 3-7. Pumped-storage hydroelectric power-generating system.

because pressure is not entirely confined to the network of pipes or hoses, as in pneumatic and hydraulic systems. Typically, hydrostatic operations include filling, mixing, or blending operations that only occur for limited periods. In this type of operation, pressure is needed to distribute liquid to a point of utilization. It generally does not do an obvious form of mechanical work, such as moving a piston or valve. The load of this type of system, however, is represented by the total opposition that the fluid encounters as it moves through the components. In practice, heat is often developed and represents the work function.

The components of a static pressure system are essentially the same as those used in either hydraulic or pneumatic systems. Control valves, controllers, and instruments are by far the most obvious parts being utilized. With the distribution of chemicals or liquids being the primary function, this type of system rarely employs cylinders or actuators. Pressure and flow measurement instrumentation is also of prime importance in a static pressure system in most process applications.

STEAM PRESSURE SYSTEMS

Steam pressure systems are widely used in industry for heating purposes, for a source of steam, or to perform a variety of work functions. As a rule, these systems develop pressure through the transformation of water into vapor when heat is applied. The expanding pressure of vapor forces it through pipes to machines that do work. A representative source, transmission path, control, load, and indicator are included in the physical makeup of this system.

The source of a steam pressure system is generally called a *boiler* because of the effect it creates when water is heated. A boiler usually consists of a metal container that is heated by some type of furnace unit. In practice, coal, natural gas, fuel oil, electricity, or, in some cases, an auxiliary source of steam is used as the primary heat source. The applied heat causes water inside the container to expand and eventually vaporize. Figure 3-8 shows a compact industrial steam generator/boiler unit that is heated by gas. Boilers such as this are essentially of the *fire-tube* type. The source of heat is applied to a tube or fixture placed in a container that is surrounded by water. Steam is extracted from the top of the water jacket.

Water-tube boilers are designed to heat water inside tubes that surround the heat source. Heat from the source that passes over the outside of the tubes is transferred to the inside water. Steam is collected at the top of the tubes, where it is distributed under pressure to the remainder of the system. Figure 3-9 shows a simplification of the water-tube boiler. These boilers are usually erected on the site and develop high levels of steam pressure for large systems. Water-tube boilers are frequently used to develop steam pressure for driving turbines in the generating of electrical power.

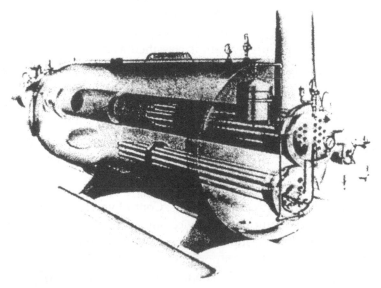

Figure 3-8. Gas-heated industrial steam generator/boiler.

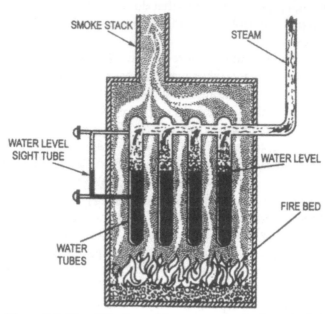

Figure 3-9. Simplification of the water-tube boiler principle.

Steam pressure systems are frequently used as heat sources in chemical-processing and food-production operations. These systems usually consist of a network of pipes and tubes similar to the hydraulic system. A steam pressure system is classified as a dynamic type of system. When it is used as a heating source, a liquid line is used to complete the transmission path. This type of flow is usually not forced by a circulating pump as it is in a hydraulic or pneumatic system. When some type of mechanical action, such as a machine operation, is being achieved, the system becomes much more dynamic in nature, as the spent steam is often recovered, condensed into water, and eventually reapplied to the boiler.

The components of a steam pressure system are essentially the same as those used in other pressure systems. Control valves, pressure gauges, controllers, and mechanical load devices are by far the most obvious parts being utilized. The application of the system and its intended work function primarily dictate the components needed to produce performance. The pressure of a steam system is often described as a *head,* which is similar to the hydrostatic pressure system.

PRESSURE-SYSTEM COMPONENTS

The components of a pressure system are quite unusual when compared with those of other industrial process systems. An interesting feature of a pressure system is the compatibility of its components. In this regard, pressure-actuated components can, in many instances, be used interchangeably. Because of this, discrete component characteristics that apply, in general, to all present system types will be discussed.

In some pressure applications there are component differences that should be noted. As an example, hydraulic components tend to be a bit larger and more rugged in construction than those used in pneumatic or steam systems. This difference is primarily attributed to the fact that air or steam is less dense than oil. It therefore takes larger volumes of air to transmit equal values of pressure and force. Hydraulic pressure, therefore, tends to be used more in applications that demand high pressure levels for heavy-duty machine operations. Except for the major differences in size, pressure, and ruggedness, all pressure components operate on the same basic principles and respond in a similar manner.

Pumps

The *pump* of a pressure-actuated system is often considered to be the heart of the entire system. It is designed to provide the system with an appropriate fluid flow that will develop pressure, just as the heart in our body does. The pump accepts mechanical power from the drive motor and converts it into an equivalent amount of fluid power.

Fundamentally, a pump is a device that accepts air or liquid, at an inlet port, forces it to move through a confined area, and expels it from an outlet port. Air and gas are compressed into smaller volumes through this process, which tends to increase pressure. Oil and liquids are forced to flow at a faster rate, which ultimately causes a pressure drop across system components. In practice, the specific system application of a pump determines the actual role that it must play.

Hydraulic pumps are primarily classified as *continuous operational* pumps. In this capacity, they must keep the fluid in a constant state of motion whenever the system is operational. A pneumatic pump, by comparison, is better described as an *air compressor*. In this role, the pump causes air to be squeezed into a smaller volume and forced into a receiving tank for storage. When the tank pressure builds up to a certain level, the compressor is turned off. Compressors are used for only short operational periods, when the demand for system air arises.

There are two general classifications of pumps used in fluid-system applications today, as shown in Figure 3-10. The first is called a *nonpositive displacement type*. This pump has no set amount of air or fluid that will be passed by the impeller blades during rotation. Flow is directly dependent on the speed of the impeller blades. The second classification is called a *positive displacement type*. This pump has a close clearance between the rotating member and the stationary components. As a result of this construction, a definite or positive amount of fluid will pass through the pump during each revolution.

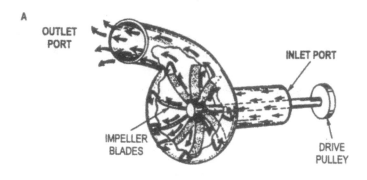

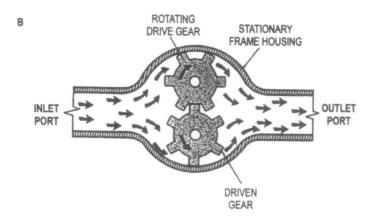

Figure 3-10. Pump classifications (A) Nonpositive displacement type; (B) Positive displacement type.

BASIC PUMP TYPES

The application of pumps in industry is so varied that it is difficult to list all the different types being used. A more realistic approach to this subject is to investigate the operation of a few basic pump types. An understanding of these applies, in a general way, to practically all industrial pump applications encountered. This section will consider four basic pumping methods commonly found in industrial pressure-system applications. The *reciprocating* motion of a piston is one method often used in many high-volume and high-pressure pumping applications. *Rotary* motion is another common method used to produce pumping action. *Gear* and *vane* pumps, which are available in a wide variety of designs and styles, operate on this principle and represent the most common of all pump types. The final type of pump uses *centrifugal force* to drive an impeller blade. This pump is more speed dependent than the other basic types. Each of these pumping methods is unique and has many industrial applications. The first three are of the positive displacement type, and the fourth is of the nonpositive displacement type.

Reciprocating Pumps

A common type of industrial pump used in both hydraulic and air-compressor applications operates on the *reciprocating* principle. This pump forces air or oil from a chamber by the reciprocating action of a moving piston (see Figure 3-11). During the intake stroke, air or oil is drawn into the chamber through an intake valve. A partial vacuum is created inside the chamber by the piston as it is being pulled to the bottom of its stroke, and the intake valve is pulled open, thus admitting air or oil into the chamber. The chamber is filled to capacity by the time the piston reaches the end of its stroke. Figure 3-11(A) shows this operational step.

When the piston reaches the bottom of its stroke, the action is reversed. In this case, the rotary motion of the motor causes the piston to change direction because of its eccentric connection to the drive disk. This action forces the discharge valve open and closes the intake valve. As a result, oil or air is forced out of the chamber. Figure 3-11(B) shows this operating condition with the piston near the top of its stroke.

For each revolution of the motor shaft, a reciprocating pump will make an intake and discharge stroke similar to the one just discussed. This would be classified as a *single-acting, single-stage pump*. Piston area and chamber volume are the key factors in determining the potential output

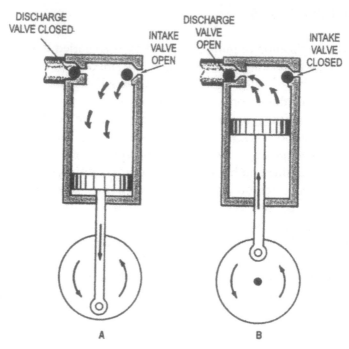

Figure 3-11. Operating principle of a reciprocating pump: (A) Intake stroke; (B) Pumping stroke.

of this pump. In some situations, two or more stages or cylinders may be driven by the same motor shaft.

Rotary-gear Pumps

The *rotary-gear* pump is a very popular method of changing rotary motion into fluid energy. Figure 3-12 shows the basic construction of an external type of rotary-gear pump. It contains two gears enclosed in a precision-machined housing. Rotary motion from the power source is applied to the drive gear. Rotation of this gear causes the second or driven gear to turn, with the teeth of the two gears meshing in the middle.

The basic operation of an *external-gear* rotary pump relies on unmeshed gears to carry fluid away from the inlet side of the pump. Fluid trapped in these teeth is transferred around the periphery of both gears to the discharge side. When the gears remesh on the discharge side, fluid is forced to pass through the discharge port. Very little fluid is permitted to return to the inlet side of the pump due to the close mesh structure of the gears.

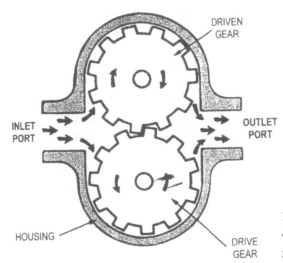

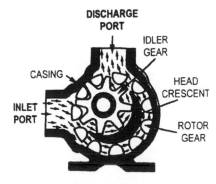

Figure 3-12. Basic construction of an external-gear type of rotary pump.

Another type of rotary-gear pump uses internal gears. In this type of pump, one gear rotates within another gear. (See Figure 3-13). The inner gear (*idler* gear) is designed so that it has less teeth than the driven outer gear (*rotor* gear). As the idler gear rotates within the rotor gear, the gear teeth unmesh at the inlet port and remesh at the discharge port. As the gears unmesh, fluid is drawn into the inlet port, filling the spaces between the gear teeth. The fluid moves smoothly around the *head crescent* and is expelled at the discharge port by the remeshing of the gear teeth. As a general rule, *internal-gear* rotary pumps can be operated equally well in either direction. The output capacity of this type of pump ranges from 0.5 to 1,100 gallons per minute (1.8925 dm^3 to 4.1635 m^3).

Figure 3-13. Operating principle of an internal-gear rotary pump.

Rotary-vane Pumps

Vane pumps also use rotary motion to circulate fluid or to compress it. This type of pump has a series of sliding vanes placed in slots around the inside structure of the rotor. As the rotor turns within its housing, *centrifugal force* or *spring action* forces the vanes of the rotor. These vanes conform to the internal shape of the housing and capture volumes of fluid as they pass by the inlet port. Rotation of the rotor/vane unit quickly moves fluid from the inlet port to the outlet port, which increases the flow or compresses the volume.

Figure 3-14 shows the basic construction of an *unbalanced, straight-vane* rotary pump. In this drawing, the rotor is offset toward the bottom of the housing. Through this type of construction, large volumes of oil or air can be made to move across the top with little or no return through the bottom. Offsetting the rotor in this way accounts for the term *unbalanced*, commonly associated with this pump.

Balanced-vane pumps, by comparison, are designed so that the rotor is in the center of its housing. Large volumes are made to appear at both the top and bottom of the housing. Separate inlet and outlet ports are developed by each side of the pump. Double pumping action of this type develops a smoother flow of air or oil than the unbalanced type of pump.

A typical unbalanced, straight-vane air compressor uses a rotary pump. Rotary-vane pumps, as a general rule, are designed for single-direction operation only. The drive motor and pump are housed in the same enclosure.

Centrifugal Pumps

A wide variety of *centrifugal* pumps are used to circulate fluids. This type of pump is classified as having a nonpositive displacement of fluids. Pumps of this type are primarily used for low-pressure, high-volume flow

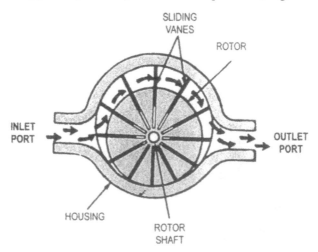

Figure 3-14. The basic construction of an unbalanced straight-vane rotary pump.

applications in industry. Displacement of the fluid passing through a centrifugal pump is of an indeterminate amount. The output of this pump is therefore dependent on the rotational speed of the driving source.

In Figure 3-15, two distinct types of centrifugal pumps are illustrated. The pump in Figure 3-15(A) is commonly known as a *volute-type* pump because of the spiral, or volute, shape of its housing. When fluid enters the inlet port of the volute pump, it is set into rotation by the revolving blades of the impeller. This generates a centrifugal force which causes the fluid to move outward toward the inner wall of the housing. The volute-shaped housing then causes the fluid to circulate in a spiral-like path toward the outlet port. In order to keep the flow continuous, the inlet port must replace all fluid expelled from the outlet port.

In the *axial-flow* type of centrifugal pump, shown in Figure 3-15(B), the propeller-type blades of the impeller maintain fluid flow in the direction of the axis of rotation of the drive shaft.

The structure of a centrifugal pump is such that there is always some

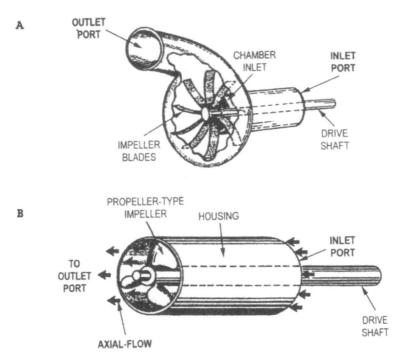

Figure 3-15. Two types of centrifugal pumps: (A) Volute type; (B) Axial-flow type.

clearance between the drive blades and the housing. This means that the amount of fluid displaced by the outlet port is not directly or positively related to the input. The volume of fluid delivered to the output is therefore dependent on the rotational speed of the pump and the resistance of the feed line connected to the outlet port. A build-up of resistance in the feed line may cause the fluid flow to slow down or even come to a complete stop. When this occurs, the operating efficiency of the pump drops to zero and no flow occurs. Any fluid in the pump simply rotates inside without being expelled. An increase in pump speed could be used to solve this problem. As a general rule, pumps of this type are only used for transferring large amounts of fluid at low pressure.

Conditioning Components

Pressure systems employ a number of devices to condition air or fluid before it is processed through other system components. These devices are normally placed in the system to prolong component life by reducing the flow of foreign particles. The number of conditioning components employed by a system is largely dependent on the system application. This ranges from a simple hydraulic system with a *line filter or strainer* to a rather sophisticated system that employs filters, strainers, and *heat exchangers. Pneumatic* systems, by comparison, condition air by filtering it to remove dirt and water, regulating the pressure to the proper level, and adding a mist of oil as a lubricant.

Hydraulic Conditioning

The number of components, type of control devices, and condition of the operating environment are the major factors to consider in determining the amount of conditioning needed in a hydraulic system. For systems with precision control valves that operate for long hours in a dirty environment, components such as micrometer filters and several strainers are considered a necessity.

Strainers are generally defined as *course element devices* placed in the pump inlet port or reservoir. These devices are usually of the *stainless-steel screen* type, with a wire mesh rating of 60 to 200 wires per in^2. Strainers are frequently placed in the reservoir filler opening, air breather, and pump inlet feed line.

Filters provide a finer grade of fluid conditioning. Typically, these devices are made of some porous medium, such as paper, felt, or fine wire mesh. Ratings range from 1 to 40 micrometers. A micrometer is one-millionth of a meter, or 0.00003937 inches. This rating refers to the particle size that is permitted to pass through the filter.

In-line and *T-type* filters are commonly used in many hydraulic systems. The T-type filter has a removable bowl or shell for element replacement. This type of filter usually employs a bypass relief valve that goes into operation when the element becomes contaminated and restricts flow. In-line filters must be removed from the line to clean or replace the filter element. Bypass relief valves are optional with this type of filter, depending on its application.

Some hydraulic systems employ heat-exchanger units to maintain the temperature of the fluid at a desired level. Machinery that is operating near a furnace, or that is used to control hot-metal rolling mills, often requires heat-exchanger units to cool the hydraulic fluid. Forced-air fan units, water-jacket coolers, and gaseous cooling devices are often used to achieve this function. As a general rule, applying heat is not a problem when a system becomes cold because the system produces heat during its normal operation. Fluid heating would be required in portable systems during cold starting conditions.

Pneumatic Conditioning

Pneumatic conditioning has some devices that are unique when compared with hydraulic conditioning. Filtering, for example, must remove moisture from the air as well as foreign particles. In-line and T-type filters frequently employ special chemical filter elements made of a desiccant. This substance is very dry and is designed to attract moisture. Elements of this type often require periodic recharging, which is a heating process to dry the element. Some T-type filters also employ a moisture trap at the bottom of the bowl. If inspection of the glass bowl shows an accumulation of moisture, it should be cleared by the drain valve.

Pressure regulation is a pneumatic conditioning process that is a necessary system requirement. After air passes through a T-type filter, it goes into a regulator valve. The movement of air passing through this valve can be changed by an adjustment screw. Through this adjustment it is possible to alter the system line pressure from the receiving tank to some desired operating level.

The pressure regulators in Figure 3-16 operate on a balance between atmospheric air pressure and system line pressure. Atmospheric pressure is applied to the top of the diaphragm, through a vent. System pressure is applied to the bottom of the diaphragm. Turning the adjusting screw adds mechanical pressure to the atmospheric pressure applied to the top of the diaphragm. When this pressure is greater than the system pressure, the diaphragm is forced down. This action opens the poppet valve, which admits more air from the receiving tank.

When the system line pressure becomes greater than the adjustment-screw/atmospheric-pressure setting, the diaphragm is forced upward. This action closes the poppet valve, thus maintaining the pressure at a set level. Figure 3-16(A) shows this condition of operation. Should the system line pressure drop, the process would be reversed, as shown in Figure 3-16(B). This causes the line pressure to be maintained at some predetermined value.

A regulator is simply a pressure-balancing device that maintains system line pressure at a preset level. In order for this device to function properly, pressure from the receiving tank must be greater than the pressure setting of the regulator valve. To increase the line pressure, the adjustment screw must increase the spring pressure on the diaphragm. Loosening the adjustment screw lowers the line pressure by reducing the diaphragm pressure. Regulators may

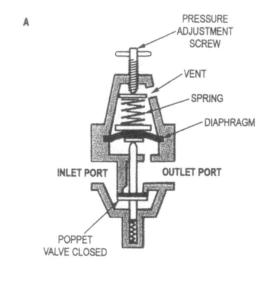

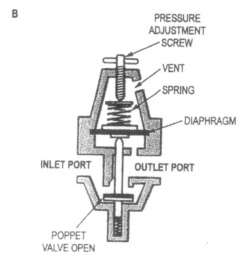

Figure 3-16. Operating principle of car pressure regulator valve: (A) System line pressure high; (B) System line pressure low.

appear at several places in a pneumatic system.

Lubricators are unusual conditioning devices found in some pneumatic systems. This type of device simply adds a small quantity of oil or mist to the air after it leaves the regulator. This conditioning process allows valves, cylinders, and pneumatic tools to last longer and operate more efficiently.

Figure 3-17 shows a typical lubricator used to supply a mist of oil to the transmission feed line of a pneumatic system. When air enters at the inlet port, it is directed into a narrowed area called the *venturi*. While passing through this area, the flow increases in velocity. Pressure developed in the narrow venturi area is lower than that of the larger area. As a result, oil is forced from the glass bowl into the oil tube and transported to the

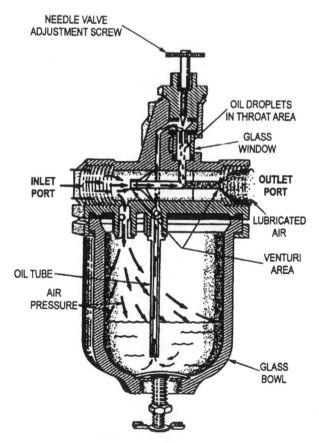

Figure 3-17. Operating principle of a pneumatic lubricator unit.

top of the unit. The needle valve is then adjusted to regulate the oil flow so that small droplets fall into the throat area. The air velocity at the bottom of the throat causes these droplets to break into a fine mist of oil and mix with the passing air. Ultimately, the lubricated air is forced to pass into the system through the outlet port.

In pneumatic systems, conditioning components such as the filter, regulator, and lubricator are often placed together in a combination unit. This combination is commonly called a *filter-regulator-lubricator* (FRL) conditioning unit. Since oil-free pneumatic systems do not necessitate this type of conditioning, FRL units are not frequently used in modern pneumatic systems.

Steam Conditioning

In small steam systems, spent steam is often exhausted directly into the atmosphere, and conditioning is not a very common practice. In general, this is not very practical because large quantities of heat also escape in the spent system. In addition, water is wasted, which causes a very serious economic problem when high-pressure steam is being developed. Steam pressure systems of a larger size therefore employ *condensers to* condition spent steam and improve system operation. Figure 3-18 shows a simplified illustration of a steam power system with a condenser in the return line.

Condensers are a unique part of a steam pressure system that are used to improve operating efficiency. In this regard, a condenser is often thought of as a heat exchanger. Cool water circulates around the exhaust line containing spent steam, which has already been cooled to some extent because of the work it has done. As the steam passes through the condenser pipes, it reaches the cool area and condenses into water. This action also creates a partial vacuum in the condenser side of the system, which improves circulation and creates additional thrust by the turbine. Some of the heat from the spent steam remains in the condensate water, which is ultimately returned to the boiler through a circulating pump. Through this condenser action, water is conserved, heat loss of the system is reduced, and circulation is improved.

Compressors

Hydraulic pumps and compressors are very similar in concept, and, in some cases, the design is exactly the same, yet there are a few differences. Some of the basic differences are as follows: A pump causes fluid

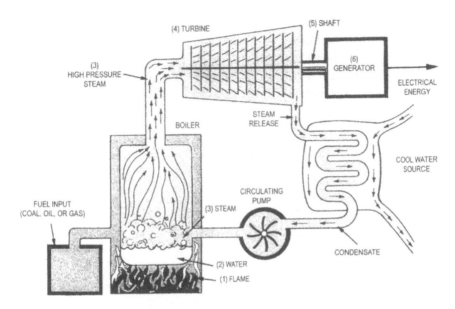

Figure 3-18. Simplified drawing illustrating the basic fossil fuel power system. Heat from burning fuel (1) changes water in boiler (2) into steam (3), which spins turbine (4) connected by shaft (5) to generator (6), producing electrical energy.

flow that develops pressure due to the movement, while a compressor causes air to be squeezed into smaller volumes and stored in a receiving tank. The air can be released immediately or used at a later period according to the demands of the system. Compressor operation is periodic and capable of long periods of maintenance-free operation when properly used. Hydraulic systems are powered by a single pump that runs continually when the system is operational. A majority of the pneumatic systems are powered by a compressor that develops pressure for the entire system. The compressor is a fairly simple device that develops pressure for small units or for complex systems that are tied together through a common manifold. Small compressors are typically of the positive-displacement type. Nonpositive-displacement compressors are usually larger, facility-type units.

Bellows Compressors

A *bellows compressor* consists of a flexible metal bellows assembly connected to the inlet and outlet ports with check valves. As shown in Figure 3-19, the bellows serves as a pumping chamber driven by a recip-

rocating piston that is energized by an electric motor.
This type of compressor
develops pressure up to 10
psig. It is used in pollution
control, pharmaceutical applications, and with other
delicate instruments where
precise measuring is used.
Lubrication is not needed
in the bellows assembly of
this compressor, which permits it to be used in oil-free
pneumatic systems.

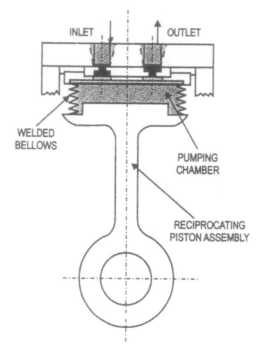

Figure 3-19.
Bellows compressor simplification.

Vane Compressors

A *vane compressor* is a
simple machine with few
moving parts. Similar to
their hydraulic counter-
parts, vane compressors
are inexpensive, have low
operating costs, and are
easy to adjust to the system. As shown in Figure 3-20, the vanes adjust to
the size of the internal chamber while turning on the rotor. This type of
compressor is compact and relatively free of vibration. The sliding vanes
are closely fitted in the rotor slots and wear very little during operation.
This type of compressor develops pressure to 150 psi.

Reciprocating Compressors

Reciprocating compressors consist of a piston moving within a cylinder
to trap and compress air. The operation of this compressor is similar to
that of an internal combustion engine. As shown in Figure 3-21, the piston
is attached to the crankshaft by a connecting rod, and moves up and down
in a chamber. The size of this compressor is generally rated in horsepower
values. Sizes range from less than 1 hp to 5,000 hp.

Diaphragm Compressors

Diaphragm compressors are a modification of the reciprocating com-

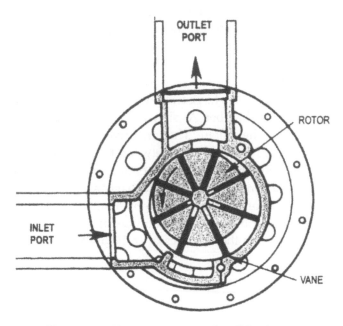

Figure 3-20. Vane compressor simplification.

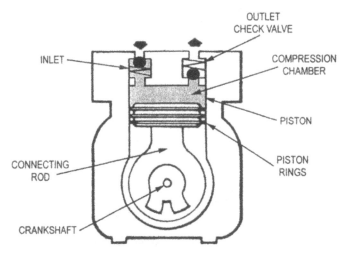

Figure 3-21. Reciprocating compressor simplification.

pressor. Compression is performed by the flexing of a metal or fabricated diaphragm. The reciprocating motion of a piston causes the flexing of the diaphragm. The space between the diaphragm and the piston may be filled with liquid.

Lobed-rotor Compressors

   Lobed-rotor compressors have two rotating elements that revolve in opposite directions in a chamber. As shown in Figure 3-22 the rotors do not actually touch when they rotate. They are driven by timing gears attached to each shaft. Air is forced to pass between the lobes of each rotor due to the turning motion. Because the rotors do not touch, air tends to leak between them at a constant rate. This leakage is called *slip*. For the best efficiency, these compressors should be operated at high speed. The air passage through the lobed-rotor assembly requires no lubrication during operation. This permits the compressor to be used in oil-free pneumatic systems.

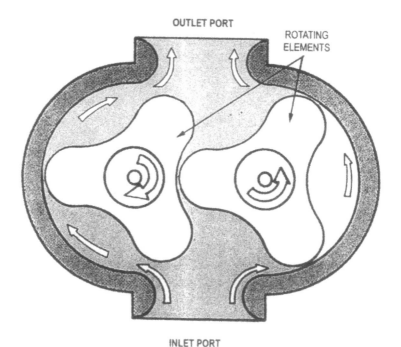

OUTLET PORT

ROTATING ELEMENTS

INLET PORT

Figure 3-22. Lobed-rotor compressor simplification.

Helical Compressors

The construction of a *helical compressor*, as shown in Figure 3-23, looks like two giant screws meshing together, and its operation is much like a hydraulic screw pump. These compressors may be either *oil flooded* or *dry*. The dry helical compressors need timing gears to maintain proper clearance between the rotating elements. These units are efficiently operated at continuous high speeds. The oil-flooded compressors do not require any timing gears, because the oil-laden screw surfaces can drive each other. However, oil separators are needed to remove the oil from the air as it leaves the compressor.

Single-screw Compressors

Operation of a *single-screw compressor* is very similar to that of a helical compressor. As shown in Figure 3-24, the drive shaft rotates the central screw, and air trapped between the screw teeth is compressed against the star-shaped rotors. This type of compressor tends to have low vibration, low noise levels, and low discharge pressures. Lubrication is required in the operation of this type of compressor. Single-screw compressors can produce a pressure of up to 125 psi in a lubricated pneumatic system. Two or more of these units can be ganged together to produce higher levels of compression.

Transmission Lines

The *transmission lines* of a pressure system may be either rigid metal tubing or a flexible thermoplastic hose. Rigid lines are used in applications that are usually free of vibration. As a general rule, rigid lines are more economical and provide less trouble than flexible lines.

Flexible transmission lines are made in a variety of types and sizes. The type of system and its application mainly determines which type of

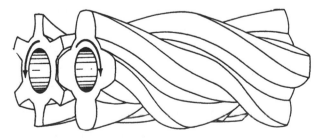

Figure 3-23. Helical compressor construction.

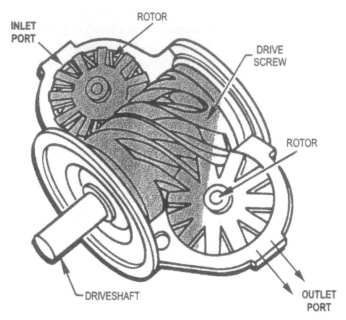

Figure 3-24. Single-screw compressor.

transmission line is to be used. The internal structure of a flexible hose varies a great deal between different manufacturers. Three basic elements of a flexible hose are considered when selecting it for a specific application: the tube or inner lining, the reinforcement material, and the outside cover material. These considerations determine pressure limits, temperature operating range, and outside exposure resistance. The transmission line of a pressure system is more susceptible to damage than any other component.

System Control

Control is an essential operation that applies to all pressure systems. Through this operation, changes in pressure can be made to do some type of useful mechanical work. Primarily, control is achieved by components that alter system pressure, direction, and volume of flow. Control may be of a partial nature, or it may completely stop a system operation, depending on its application.

Control devices are often placed at a number of different places within the basic system structure. The actual location of a specific device is mainly determined by the control function it must achieve. Pressure regulators, for example, may be attached to the output of the source or in the feed line attached to the load device. Direction control is typically found near the load device. Flow control maybe found in any part of the system, depending on specific needs.

Pressure Control

Pressure control refers to those functions that alter the pressure level of an operating system. Such operations include relieving, reducing, by-passing, sequencing, and counterbalance. As a general rule, control devices of this type are named according to the function they achieve.

Figure 3-25 shows the internal workings of an *orifice-disc relief valve*. When the line pressure attached to the right side of the valve exceeds the pressure settings of the adjustment screw, a flexible diaphragm is forced up from the bottom. This action forces the connector rod to move, which in turn causes the orifice disc to shift to the position shown in Figure 3-25(B). The flow path is completed through the holes in the two discs. Flow takes place from right to left. At pressure levels less than the adjustment-screw setting, the path is obstructed, as shown by the position of the orifice discs in Figure 3-25(C).

In hydraulic systems, pressure-relief valves are used to dump the output of a positive displacement pump back into the reservoir when the pressure rises to a dangerous level. In this pressure-control application, the relief valve serves as a safety control device.

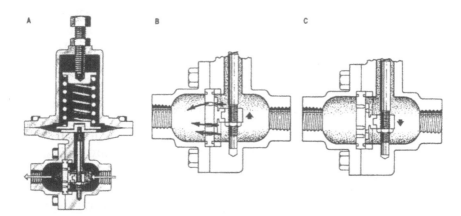

Figure 3-25. An orifice-disc pressure relief valve: (A) Cutaway view of valve; (B) Area of orifice disc showing path open; (C) Area of orifice disc showing path closed, (Courtesy Jordan Valve)

In pneumatic systems, pressure-relief valves are used to meter small amounts of air or to bypass specific parts of the system when activated. In this case, excess air is not released into the atmosphere. The output port of a reducing valve may be altered in size to achieve metering control.

Pressure control may also be used to establish specific operating sequences. Valves that achieve this type of control direct pressure in some predetermined sequence at certain pressure levels. Control of this type is achieved by a relief-valve type of construction. When the main system pressure overcomes the valve setting, it shifts to a different port. Sequencing valves are usually included in automated system operations. These valves are commonly actuated by system controllers.

Directional Control

Directional control is achieved by devices designed to start, stop, or reverse system flow without causing an appreciable change in pressure or flow rate. One-, two-, three-, and four-way valves are some of the more common directional control devices. These devices may be actuated by pressure, mechanical energy, electricity, or manual operation.

The control action of a directional valve can be achieved in a variety of different ways. One-way valves, for example, may operate on the *seated-ball principle* of the relief valve. One-way valves are commonly called *check valves* in most applications. These valves will permit flow in only one direction.

Figure 3-26 shows the operation of a check valve. When pressure is applied to the inlet port, it drives the ball away from its seat. When this occurs, the flow path becomes open and fluid will pass through the valve. In the reverse direction, pressure will force the ball into its seat. This action will prevent flow in this direction. Valves of this type are often used to permit free flow around some controls when the flow direction is reversed.

Two-way valves are simple directional control devices that have input and output connections placed in series within the transmission line. Valves of this type are designed primarily to achieve full control. They permit flow or shut it off. This type of control is achieved by placing gates, plugs, discs, spools, and numerous other objects in the line in such a manner that normal flow is obstructed.

A *ball-type* directional control valve has a ball that can be rotated manually by the outside control handle. When the handle is in line with the feed line connections, flow occurs. Turning the handle to a right angle positions the ball to stop the flow. This type of valve is used primarily for

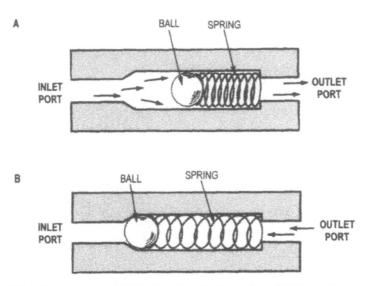

Figure 3-26. Operating principle of a check valve: (A) Free flow condition; (B) Closed flow condition.

high-pressure, manual control applications in industry, and is only one representative of a large number of two-way control valves.

Three-way valves are primarily designed to permit the operator to shift to two different sources of pressure, or to direct pressure to alternate devices. Generally, this type of valve is used to alter cylinder operation or to control hydraulic or pneumatic motors. Valves of this type can be actuated manually, automatically, by pilot pressure, or by electricity. Basic designs include shifting spools, poppets, and shear-seal plates. Figure 3-27 illustrates the structures of these basic types.

Figure 3-28 shows the basic operation of a spool-type, three-way directional control valve. When the manual control shaft is pushed toward the right, it causes the spools to shift right. Flow is then through ports P and A. When the shaft is pulled to the left, the spools shift to the left. This permits flow through ports P and E or T. The label P is a designation of pressure; A indicates actuating port; and E or T indicate exhaust or tank. The letter designations are assigned by the American National Standard Institute.

As a general rule, three-way valves are designed for only two-position operation. Some valves do, however, have a neutral or off position. This additional position simply increases the control capabilities of the valve.

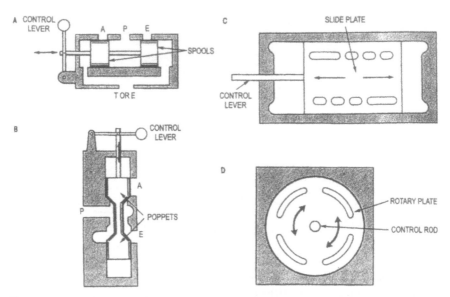

Figure 3-27 Basic types of three-way valves: (A) Spool type; (B) Poppet type; (C) Sliding-plate shear-seal type; (D) Rotary-plate shear-seal type.

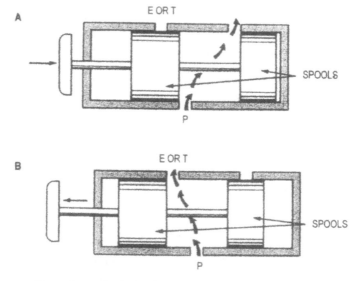

Figure 3-28. Operating principle of a spool-type three-way valve: (A) Flow path with spools to right; (B) Flow path with spools to left.

Four-way valves are used in control applications to start, stop, or reverse the direction of flow. In its simplest form, this type of valve has four working connections, but is also manufactured with two, three, five, and six different position combinations. Four-way valves are commonly used to control forward and *reverse actuation* of a double-acting cylinder, or to *reverse the* rotational direction of a fluid motor.

Figure 3-29 shows the basic operation of a four-way, spool-type valve. In position 1, the pressure feed line (P) is neutralized, or in the off position. Position 2 shows the flow direction from P to A with an exhaust from B to E2. Position 3 shows flow from P to B with exhaust from A to E1. With this type of control, flow direction can easily be reversed by manually shifting spool location. Actuation of this valve can be achieved mechanically, manually, electrically, or with pilot pressure.

Flow Control

Flow control refers to those components designed to alter the volume or flow rate of hydraulic or pneumatic systems. The rate at which air or hydraulic fluid is delivered to the load of a system determines its operational speed. Motor speed, for example, is directly dependent on the flow rate of the applied fluid. By altering this rate, a motor can be made to operate at a wide variety of speeds.

Cylinder actuating speed is also controlled by fluid flow devices. To alter the linear motion of a cylinder, fluid may be controlled at the input feed line, the return line, or a combination of both. The term metering is often used to describe this function.

Figure 3-30 shows the operation of a flow control valve. Construction of this valve permits controlled flow from P to F. The direction is adjusted by the needle valve. The letters P and F refer to the pressure and free-flow connections of the valve. Flow from F to P forces the ball of the check valve to move away from its seat, which produces uncontrolled flow. An arrow on the valve refers to the controlled flow direction.

Controllers

Controllers are components that are used to alter a particular system process through some type of automatic operation. Figure 3-31 shows an example of a pneumatic system that is altered by a controller. Air from the header line is applied to the controller through a supply line and alters the final control element to some decided value that has been determined by a specific setpoint adjustment. A change in header pressure is monitored

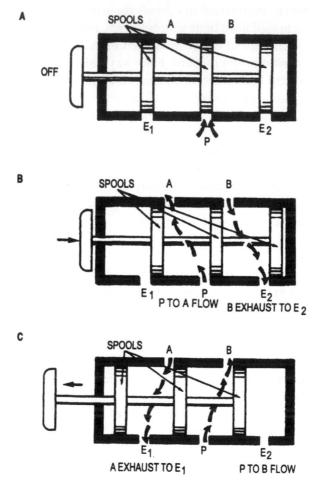

Figure 3-29. Operating principle of a spool-type four-way valve: (A) Position 1; (B) Position 2; (C) Position 3.

and compared with the assigned setpoint value, and the controller alters the final control element in such a way that it meets the demands of the system. All this action is achieved automatically.

Industrial controllers are normally designed to respond to system changes in pressure, flow, temperature, differential pressure, vacuum, and absolute pressure. They will accept these signals when operated remotely, with a transmitter, or they may be attached locally to a system to monitor direct signal applications. Setpoint levels and process-variable values are usually displayed on an indicator mounted in the controller housing.

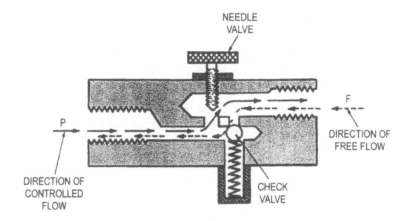

Figure 3-30. Operating principle of a flow control valve.

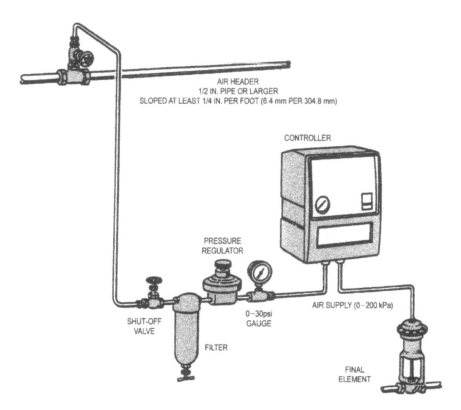

Figure 3-31. A pneumatic controller application. (Courtesy ABE Kent Taylor)

Some controllers, by comparison, are nonindicating and achieve control operations without any display of system values. Controllers may also be used to indicate values on a *strip chart* or *circular chart* when permanent system recorders are needed. A boiler recording controller that monitors the ratio of air and steam flow is shown in Figure 3-32. This type of controller is used to alter the combustion level of a water-tube boiler system. Air pressure or electrical energy is commonly used to operate most industrial controllers today.

Load Devices

The primary purpose of a pressure power system is to produce some form of useful work. Up to this point, our efforts have centered around the source, transmission path, and control. The power output of a system refers to those parts that are designed to do the actual work. The term *load*

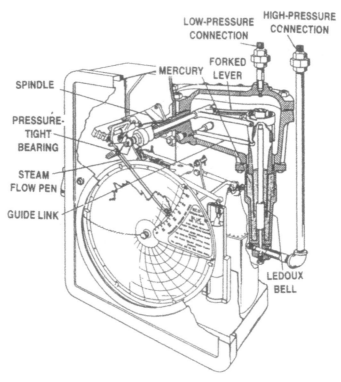

Figure 3-32. A boiler recorder controller.
(Courtesy Bailey Control Systems)

is often used to broadly describe all components that consume power or do work. The term *actuator* is therefore a more meaningful term when referring to specific load components.

In industry today, pressure systems are designed primarily to produce work in the form of mechanical motion. In this regard, special actuators are used to produce either linear or rotary motion. Hydraulic, pneumatic, and steam systems produce linear motion through the action of a cylinder as the actuator. Rotary actuators are used to produce a twisting or turning motion to do certain types of work. The basic operating principles that apply to these actuators are similar in nearly all respects, and they will be discussed together in the following section.

Linear Actuators/Cylinders

In industry, cylinders are used to develop the force needed to lift, compress, hold, or position objects during different manufacturing processes. In order to produce linear motion, hydraulic fluid, air, or steam is forced under pressure into a cylindrical chamber. A piston placed within the chamber is free to move as a result of the pressure applied to it. The area of the piston determines the amount of force it will develop from an applied pressure. The area of a round piston can be determined by the formula

$$A = \pi D^2 / 4$$

where

  A = area of the piston in in^2 or m^2
  π = a constant (3.14159)
  D = diameter of the piston in inches or meters

The force developed by this piston is then determined by the formula

$$F = PA$$

where

  F = force developed in pounds or newtons
  P = applied fluid pressure in psi or pascals
  A = area of the piston in in^3 or m^2

Single-acting Cylinders

When a cylinder has only one input port, it is described as a *single-acting cylinder*. Figure 3-33 shows that when fluid is forced under pressure into the actuating port, it causes the piston to move. In this case, weight is lifted by the fluid force driving the piston. The combined load of this system is the weight being lifted, the friction between the piston and the cylinder walls, and the heat developed by fluid friction.

Returning the piston to its original position can be achieved by first stopping the forward flow of fluid, as with a shut-off valve. Second, the fluid under the piston must be released. The weight of the load would then cause the piston to retract. Hydraulic jacks and automobile lifts typically employ single-acting cylinders of this type.

Double-Acting Cylinders

Double-acting cylinders, as the name implies, have power action in two directions. Two ports are needed with this type of cylinder. Fluid is applied to one port and expelled from the other port during its extending operation. Retracting the piston is achieved by reversing the fluid flow. A four-way valve is commonly used to control the motion of this cylinder. Industrial applications are found in such things as punch presses, rolling mills, machine-tool clamps, paper cutters, robots, and production-line actuators.

Figure 3-34 shows the basic operation of a double-acting cylinder. To initially extend the piston rod, fluid must be applied to the right side of

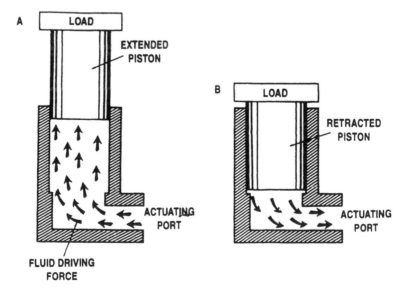

Figure 3-33. The operation of a single-acting cylinder: (A) Piston extended by fluid driving force; (B) Piston retracted by release of fluid driving force.

the piston and removed from the left side. This action forces the piston to move to the left. Switching the fluid flow causes the piston to move to the right. As a general rule, the retracting force is less than the extending force. In this case, the area of the piston is smaller due to the connection of the piston rod.

The double-acting cylinder just discussed is of the *differential* type. The linear action of the piston is determined by the pressure difference on each side of the piston, and the piston rod extends only from one side of the cylinder. A non differential type of cylinder would have rods extending from both ends of the piston. Cylinders of this type can provide an equal force in either direction.

A double-acting pneumatic cylinder with separate ports can be described as an *air motor*. It produces continuous reciprocating motion when air is alternately applied to the inlet ports. Cylinders of this type are very popular.

Rotary Actuators

Rotary actuators are fluid-power devices designed to produce a limited amount of rotary motion in either direction. Figure 3-35 illustrates two of the most common types. Fluid applied to port A causes the ro-

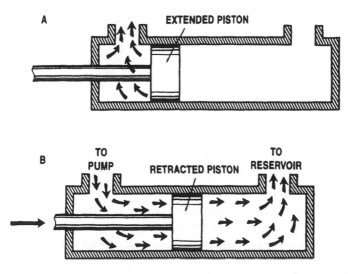

Figure 3-34. Operation of a double-acting cylinder: (A) Fluid driving piston to the left; (B) Fluid refracting piston to the right.

tor to move a certain distance in a clockwise direction. Counterclockwise rotation is achieved by applying fluid to port B and expelling fluid from port A. The single-vane rotor can be made to turn approximately 280° in either direction. The double-vane rotor has twice the turning power, but can only turn approximately 100° in either direction. Actuators of this type are used in industry to lift or lower, open or close, and in indexing operations. Rotary actuators are commonly used in continuous reciprocating operations, such as that of a punch press.

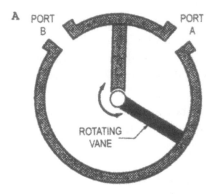

Motors

Motors are designed to convert the force of a moving fluid into rotary motion. As a general rule, fluid motors and pumps are similar in appearance and operation. In a fluid motor, the power of a moving fluid is used to produce rotary motion by driving vanes, gears, or pistons. The pump must be driven by a rotary force to produce fluid flow.

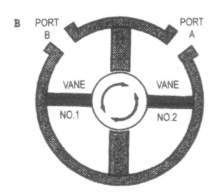

An unbalanced-vane fluid motor is designed with either four or eight vanes. Rotary power output of a fluid motor can be increased with more vanes. This particular motor can be used either as a motor or as a pump. Gear pumps can also be used as motors. Gear motors are capable of operat-

Figure 3-35. Rotary actuator operation: (A) Single-vane type; (B) Double-vane type.

ing at speeds up to 5,000 revolutions per minute (rpm). Both internal and external gear motors are available today.

Fluid motors are generally classified according to the type of *fluid displacement* they employ. Gear, vane, and piston motors usually have a *fixed* displacement characteristic. This type of motor accepts a certain amount of fluid and moves it with each revolution. The operating speed of the motor depends entirely on the amount of fluid supplied by the source. *Variable* displacement motors, by comparison, are designed so that the amount of fluid circulated during each revolution can be changed. The piston type of motor is in this classification. The length of its stroke 5 altered to produce the variable displace lit characteristic. The speed of this motor can therefore be changed by an outside physic adjustment. Operating speeds of up to 3,000 rpm are typical.

The rotary motion produced by a fluid motor is a form of power designed to do work. Its turning capability is a measure of *torque*, which is equal to the developed force multiplied by the radius of the rotating arm. Mathematically, this is expressed by the formula

$$\text{torque (in-lbs)} = \frac{\text{pressure (psi)} \times \text{displacement (in}^3/\text{revolution}}{2\,\pi}$$

The output power developed by a fluid motor is commonly expressed as horsepower. Mathematically, horsepower can be determined by the formula

$$\text{horsepower (hp)} = \frac{\text{torque (in-lbs)} \times \text{speed (rpms} \times 2\,\pi}{33{,}000 \text{ ft-lb/mm}}$$

Motor performance information can normally be obtained from the manufacturer to make these calculations.

Indicators

Indicators that measure pressure play an important role in the overall evaluation of a system. Regulators and pneumatic receiver tanks often employ pressure indicators or gauges as permanent fixtures. A wide range of pressures must be measured today. Negative pressure (vacuums) as small as 2×10^{-5} lb/in^2 (1.379×10^{-1} Pa) to positive pressures as high as 1×10^6 lb/in^2 (6.895 GPa) must be measured in industrial equipment. This wide range of measurement requires a number of different indicating devices.

Elastic deformation pressure elements are used as industrial pressure indicators. These indicators employ an element that physically changes shape when different pressures are applied. *Spiral* and *helix coil* elements have a tendency to uncoil when pressure is applied. The *Bourdon tube* element tends to straighten when pressure is applied to it. The physical change produced by these different elements can be used to move an indicator hand on a scale, or a stylus on a paper chart recorder.

For a pressure indicator employing a Bourdon element, when pressure is applied to the element, it tends to straighten, causing the indicating hand to move. Indicators of this type can be purchased to measure pressure in a number of ranges from 15 to 1,000 psi (103.425 to 6,895 kPa).

A typical pressure-indicating recorder uses a spiral element. Pressures from 14 to 4,000 psi (96.53 to 27,580 kPa) can be measured and recorded on this type of indicator. Instruments of this type are called *direct-reading* pressure indicators. This means that the output is taken directly from the pressure element. These indicators are mounted at a fixed location.

Resistance Transducers

A common *resistance transducer* is designed to respond to the physical changes in pressure that are applied to a strain gauge. The resulting change in resistance is used to cause an ac bridge to become unbalanced. Indicator values due to bridge imbalance are then used to register precise pressure values.

The strain gauge of a resistance transducer is generally a simple grid of fine wire (see Figure 3-36). The resistance of this grid changes directly with length and inversely with its cross-sectional area when the grid becomes deformed. The actual resistance of a strain gauge element is expressed by the formula

$$R = KL/A$$

where

    K = specific resistance constant for wire type
    L = length of wire
    A = cross-sectional area

Only small amounts of distortion are needed to change the resistance of a strain gauge through its total range. This type of element may be formed into a variety of different structures. The most significant change in resistance occurs when a distorting force causes the grid wise to be elongated horizontally. As a general rule, strain gauges are used to detect small changes in pressure.

Figure 3-37 shows an ac resistance bridge circuit attached to a strain gauge. The strain gauge serves as one of the resistor legs of the bridge, while the temperature-compensating gauge serves as the alternate resistive leg. Through this type of circuit construction, variations in tem-

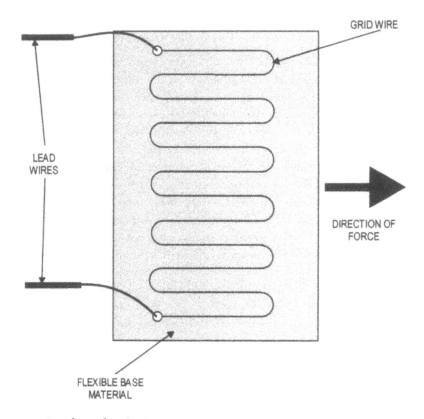

GRID WIRE

LEAD
WIRES

DIRECTION OF
FORCE

FLEXIBLE BASE
MATERIAL

L = length of wire   A = cross-sectional area

Figure 3-36. A bonded type of resistance strain gauge.

perature are cancelled so that only strain resistance is measured by the bridge.

When pressure is applied to the bellows element of the indicator, it causes the beam to flex upward. This causes a corresponding change in strain-gauge resistance. The bridge immediately indicates an imbalance at output points A and B. This signal voltage is then amplified and applied to the slide-wire servomotor. The motor moves the slider contact of the slide-wire resistor to a position that will cause the bridge to return to its balanced condition. The physical position of the slider is then used to register a pressure value on the graduated scale. The desired response speed of this type of indicator is largely determined by the selected amplifier and balancing servomotor. High-speed response adds significantly to the cost of this type of indicator.

SUMMARY

Pressure is one of the more important process variables used in the manufacturing of industrial products. Pressure systems employ some type of fluid, gas, steam, or air as a means of controlling the flow of energy or power. Hydraulics, pneumatics, steam or vapor, and static-head pressure systems fall into this classification.

The source of a pressure system is mainly responsible for the transformation of energy to the eventual result. Flowing fluids develop pressure by being forced through components that have some type of surface resistance. The source of a pneumatic system is usually a motor-driven compressor. Static-head pressure is developed as a result of liquid placed in an elevated tank. Steam pressure is the result of boiler ac ton which causes a liquid to be changed into a vapor due to the application of heat.

The transmission path of a pressure system simply provides a path for the system medium to follow. Both thermal and mechanical energy are transferred through the system during its operational cycle.

Control of a pressure system ranges from a simple on-off operation to a variety of variable system changes. Open-loop or closed-loop control can be used to achieve this operation.

The load of a pressure system achieves some type of work function. Specific loads and composite hidden loads are typical of pressure systems.

Indicators are primarily used to monitor system operation and in maintenance procedures. In general, most indicators respond to elastic deformation principles.

Pascal's law is the basis of pressure-system operation. This law states that pressure applied to a confined fluid is transmitted throughout the fluid and acts on each surface at a right angle to the surface.

Force is defined as something that modifies or produces motion. Weight expressions are used to denote force values. A force applied to a specific unit area is called pressure. When an applied force causes an object to move, work is being done. Power refers to the length of time in which work is being accomplished.

A unique difference between hydraulic and pneumatic systems is in the compressibility of fluids. Air is compressed to increase its pressure. Hydraulic fluid does not compress under normal operating conditions. Its pressure is developed by circulating fluid through the system resistance.

The pump of a hydraulic system is designed to provide the system with fluid flow that will develop pressure. This device accepts fluid at an inlet port, forces it to move through a confined area, and expels it at an outlet port. Hydraulic pumps operate continuously, while air compressors work on a demand basis. Positive-displacement pumps move certain volumes of fluid during each revolution. Nonpositive-displacement pumps have no set amount of fluid that will pass during one revolution. Reciprocating action, gears, and vanes are used in positive-displacement pumps. Centrifugal force is used to produce nonpositive pump operation.

Conditioning is needed to prolong component life in fluid systems. Filters, strainers, and heat exchangers are some of the common hydraulic conditioning components. Regulators, filters, and lubricators are used to condition air systems.

Compressors are designed to squeeze air into smaller volumes and store it in a receiving tank. Compressed air can be released immediately or used according to the demands of the system. Positive-displacement compressors employ a bellows assembly, vanes in a rotor, reciprocating pistons, diaphragms, lobed rotors, or helical screws in their operation.

The transmission path of a fluid system is either flexible or rigid metal tubing. Rigid tubing is used in applications that are stationary and free from vibration. Flexible tubing is selected according to pressure, temperature range, and exposure resistance.

Fluid system control is designed to alter pressure, flow direction, and volume of flow. Pressure control is achieved by valves that relieve pressure, reduce its volume, bypass components, and operate them in a sequence. Direction control is designed to start, stop, or reverse system flow. One-, two-, three-, and four-way valves are some of the common types of directional control devices in operation today. Flow control is designed to alter the volume of fluid flow or flow rate.

The load of a fluid system that does useful work is commonly called an actuator. Linear motion is achieved by an actuating cylinder. Single-acting and double-acting cylinders are both used in industrial applications today. Applying pressure to one side of a piston causes it to move linearly. Applying pressure to the other side of the piston causes it to refract in a similar action.

Rotary actuators are designed to produce a limited amount of rotary motion in either direction. Single-vane motors are used to achieve this function.

Fluid motors are designed to produce continuous rotary motion. The power of a moving fluid or air flow causes vanes, gears, or pistons to move in a rotary direction. The power output of a fluid motor is commonly measured in horsepower.

Chapter 4

Thermal Systems

OBJECTIVES

Upon completion of this chapter, you will be able to

1. Explain the operation of a basic thermal system.
2. Describe some of the thermal systems used in industry.
3. Describe how control is achieved in a thermal system.
4. Discuss the response of thermal controllers.
5. Describe how on-off control is achieved.
6. Identify different modes of control of a thermal system.
7. Describe testing and evaluation of a thermal system.
8. Explain the role of electronic instrumentation in a thermal system. 9. Describe the role of thermal-system sensors and transducers.
10. Explain how infrared radiation is used to determine temperature values.

KEY TERMS

Absolute zero—The temperature at which molecular motion that constitutes heat ceases. This occurs at –273°C. –459°F, 0°R, or a Kelvin.

Agitator—An assembly that causes the circulation of air or liquid in a convection furnace, heating chamber, vat, or container.

Arc—A condition that occurs when electric current flows between two electrodes and produces intense heat.

Blast furnace—An area, constructed of fireproof material, that produces heat when large volumes or blasts of air are forced into a chamber.

Boiler—A chamber where fuel is burned to produce heat and ultimately change water into steam or hot water.

Bridge—An electrical circuit used to measure an unknown value; its components form two current paths that are commonly connected together at the center.

Conduction—The process by which heat or electricity is transferred through particles of a material. Convection The transfer of heat by currents that move in liquid and gas.

Crucible—A container used for melting a substance that requires a high degree of heat.

Expansion principle—The explanation for why liquid or metal is enlarged when heat is applied.

Fined system—Liquid, gas, or vapor, sealed in a closed element, that expands when subjected to heat.

Galvonometer—An electromagnetic instrument used to measure small values of current.

Inductance—The property of an electrical circuit that opposes a change in current due to energy stored in a magnetic field.

Inductive heating—A method of heating in which metal is placed inside a coil of wire and high frequency ac voltage is applied to the coil.

Kelvin (K)—An absolute temperature scale based on 100 Celsius units between the ice point and boiling water. $0°C = 273.15°K$.

Kinetic theory—A set of ideas that deal with the effects of forces on the motion of material bodies or the movement of particles in a substance.

Mode of control—A method that adjusts the operation of a system in order to restore it to a desired level.

Multivibrator—An oscillator that produces a repeating square or rectangular waveform.

Negative temperature coefficient—A condition where a value drops with an increase in temperature or increases with a decrease in temperature.

Overshoot—The amount by which a changing process variable exceeds the desired value as changes occur in an operating system.

Pyrometer—Another name for a temperature indicator.

Radiation—The transmission of energy from a source, such as a hot surface, by the emission of electromagnetic waves.

Radiation pyrometry—A method of measuring the temperature of an object by the amount of thermal energy radiating from its surface.

Rankin (R)—An absolute temperature scale based on the Fahrenheit scale of 180 units between the ice point and the boiling point of water. $459.67°R = 0°F$.

Resistance—An opposition to the flow of electrical current in a circuit; its unit of measurement is the ohm.

Resistance heating—A process that develops heat due to electric current passing through a resistive element. When current is forced to pass through a conductor, heat is produced as a result of the resistance.

Resistance temperature detector (RTD)—A sensor used to determine the temperature of an environment due to value changes in a resistive element.

Resolution sensitivity—The specific amount of temperature a controller needs to initiate a change.

Response time—The time it takes a system to react to change and produce a variation in its output.

Sensing element—A device that is used to detect a change in the operation of a system. Temperature sensors are thermistors, thermocouples, and RIDs.

Silicon controlled rectifier (SCR)—A solid-state electronic switching device that is used to control the operation of a load.

Slidewire—A variable electrical resistance that has a contacting slider which permits adjustment of its value.

Temperature—The relative hotness or coldness of a body as determined by its ability to transfer heat to its surroundings or another body.

Thermal gradient—A measure of temperature values, beginning at the source and moving toward the load.

Thermal lag—A delay in heat distribution that occurs between the source and the load. Thermistor A temperature-sensitive resistor. Thermocouple A junction of two dissimilar metals that when heated will produce a voltage that is proportional to the applied temperature. Thermometer A temperature measuring instrument. Thermopile An energy detector that consists of thermocouples connected in a series configuration.'

Thermoresistor—A device that responds to heat by producing a change in resistance. Thermistors and RIDs are thermoresistors.

Time-proportioning control—An electrical control procedure that changes the number of cycles of electrical power delivered to a load in a given period of time.

Transducer—A device that changes an input signal of one type into an output signal of a different type that is proportional to the original input.

Undershoot—A momentary control condition that occurs when the value of a process drops below some predetermined setpoint value.

Zero-voltage switching—A time-proportioning control procedure that
causes the power to be turned on and off at the zero crossover point
of an ac sine wave.

INTRODUCTION

Of all the manufacturing processes utilized by industry, temperature
measurement and manipulation are by far the most common. In fact, over
50 percent of all measured variables in industry tend to involve some form
of temperature measurement or change. With improved technology, the
trend will obviously be toward even better and more improved measurement and control techniques.

Systems that utilize thermal processing must produce or generate
heat, provide a path for it to follow, initiate some type of control action,
utilize energy to accomplish a specific work function, and ultimately record temperature variations through precise measurement techniques.
These functions are a major concern of all industrial personnel involved in
process control applications. A basic understanding of the thermal system,
its applications, and measuring instrumentation is of great importance to
someone pursuing this field as a career.

THERMAL PROCESS SYSTEMS

Manufacturing processes that respond in some way to temperature
are classified as *thermal systems.* In industrial applications, thermal systems
are used to control a wide range of processes, from complex manufacturing operations to single-function systems that control only one process.
The main concern here is directed toward thermal-system basics that can
be applied to any application regardless of its size or complexity.

A thermal system must have a primary energy source, a transmission path, control, a load, and possibly one or more indicators in order to
function properly. These parts, as stated earlier, are the basic elements of
all functioning systems. To distinguish this type of system from others,
further classification is required. In general, a thermal process system is
defined as any system that responds in some way to changes in temperature. These systems may generate heat for molding or forming operations,
for food processing and packaging, or for fabrication operations, or they

may respond to temperature changes that are used to control another process.

The energy source of a thermal system also requires some additional classification in order to distinguish it from other systems. Energy, for example, is present in three basic forms: *heat, light,* and *mechanical motion.* It is also available in the form of *electricity, chemical action,* and *nuclear energy.* These six forms of energy are all related to some extent. This relationship is based on the fact that energy of one form can be readily changed or transferred into one of the other existing forms. The energy source of a thermal system usually operates by changing energy from one form into something of a different type. In this regard, electricity and chemical action generally serve as the primary energy source for most industrial thermal systems. In addition, the sun can be used as a primary source of thermal energy. A great deal of research has been done to make better use of the sun as a primary source of energy. Thermal energy may also be obtained by friction, compression, and mechanical action, but energy in any of these forms is difficult to harness as a primary source for an operating system.

The transmission path of a thermal system is unique when compared with other systems. The material of the path may be anything that is either solid, liquid, or gas. If one end of a solid metal bar is placed in an open-flame heat source, the other end of the bar will soon become hot. The process by which heat is transferred from the heated end to the cold end is called *conduction.* According to the *kinetic theory of energy,* conduction is the transfer of heat through a collision of *molecules.* Essentially, heat from the source causes molecules at one end of the bar to move more rapidly. Increased molecule velocity ultimately causes collisions with neighboring molecules, which in turn cause them to move faster. The process continues until the molecular motion has increased throughout the bar. Of solid materials, metals are the best thermal conductors, and nonmetals generally serve as insulators of heat. Figure 4-1 shows an example of the conduction principle of heat transmission.

Liquids and gases are heated by the *convection* process. When a container of liquid is placed on a heat source, at first only the liquid at the bottom of the container receives heat. Since most liquids are poor conductors of heat, little heat is transferred to other parts of the liquid, yet as the bottom layer of the liquid begins to expand, it becomes less dense than the cooler liquid above. The warmer liquid therefore moves to the top of the container and the colder liquid moves to the bottom. In this manner, dif-

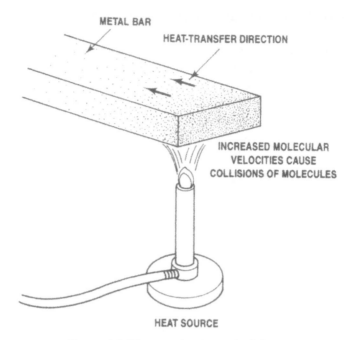

METAL BAR

HEAT-TRANSFER DIRECTION

INCREASED MOLECULAR
VELOCITIES CAUSE
COLLISIONS OF MOLECULES

HEAT SOURCE

Figure 4-1. Heat-conduction principle.

ferent layers of liquid begin to move. The moving liquid is called *circulating currents*. Convection, therefore, is the transfer of heat when liquids or gases are heated. Figure 4-2 shows an example of the convection principle in a forced-air heating system.

Heat may also be transmitted from a source through the radiation of *electromagnetic waves* where matter does not exist. An example of this is the heat that reaches the earth from the sun. In effect, thermal energy is given off or radiated away from a heat source through *infrared rays*. Any object possessing heat gives off these rays. In theory, discrete bundles or *quanta* of energy move away from the thermal source in wavelike patterns at the speed of light. Energy released from atoms having the greatest mass has the shortest wavelength and the highest frequency. Heat produced by radiation has limited application in industrial thermal systems. Figure 4-3 shows an example of heat transfer by the radiation principle.

The control function of a thermal system alters the flow path of heat between the primary energy source and the load device. Specific devices called *controllers* are usually responsible for this operation. These devices attempt to maintain system temperature at a desired level. *Two-position* control causes on and off conditions to be performed. *Proportioning* control

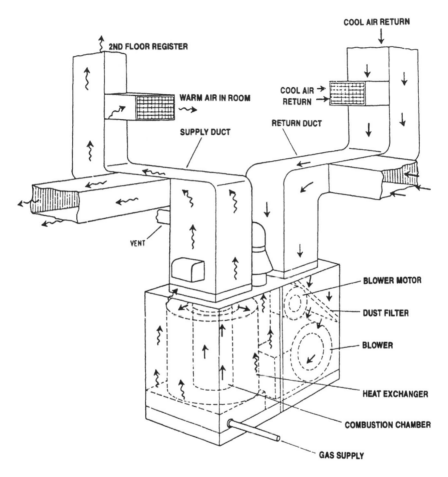

Figure 4-2. Forced-air gas furnace *employing the convection principle.* **(Courtesy American** *Gas* **Association, Inc.)**

maintains less pronounced variations in temperature. On-off and proportioning control are the most common methods of pure control. System controllers respond by sensing level changes between the actual temperature of the system and a predetermined setpoint value. The control function of a thermal system is limited because of the nature of the energy being manipulated.

The load of a thermal system represents the work function or material that is to be maintained at a specified temperature. Since heat flows away from a body of higher temperature and toward a substance of lower temperature, the load must be directly dependent on the source. The ther-

Figure 4-3. Radiation principle.

mal demand of the load may be of a steady nature for a prolonged period, variable, or cyclic. The application of the system determines the characteristic nature of the load.

Thermal indicators represent the part of the system that is responsible for measurement. This operation may be performed at a number of different locations, depending on the demands of the system. Typically, temperature indicators are attached to the output or system load, the controller, or the source. Periodic monitoring of the indicators ensures operating efficiency within the designed limits of the system.

Temperature measurements are made in a number of ways. Non-electrical indicators, such as *thermometers*, are designed to respond to temperature variations caused by changes in liquid volume and gas or vapor pressure. This also includes the dimensional changes of solid materials. Electronic indicators respond by measuring the temperature of solids, liquids, or gases by direct contact through *thermoelectric transducers*. This can be done indirectly through *infrared radiation detectors*. The range and span of the temperature being measured, accuracy, and speed of response are the primary factors to consider when selecting an indicator for a specific application.

INDUSTRIAL THERMAL SOURCES

The heat of an industrial thermal system is produced by a change in the state of matter. Heat produced as the result of a fire is an example. Burning changes the state of coal, wood, or whatever fuel is being used. In theory, the fuel is united with oxygen, and both fuel and oxygen are changed to another material. Carbon in the burning fuel also unites with oxygen in the air and produces a different type of matter called *carbon dioxide*. The union of carbon and oxygen is responsible for liberating heat.

There are many examples of heat being produced by uniting certain substances with air. In the human body, oxygen in the air that we breathe combines with our food to produce body heat. Iron exposed to the oxygen of outside air produces rust and a measurable amount of heat, but the amount of heat produced in this case is not significant because the combining process is slow.

Heat is also produced when a liquid or gas changes form. The transition of water to ice is a prime example of this principle. In a similar manner, heat is given off when steam or gas is condensed into water. Steam heating systems operate on this principle.

INDUSTRIAL FOSSIL-FUEL HEAT SOURCES

The most common fossil-fuel heat source is the *furnace*. This type of device is built of metal, brick, or a combination of these fireproof materials. After the structure has been built, fuel is burned in the center or *fire chamber* to produce heat. Furnaces are used to develop heat for comfort, heat treating, or to melt materials for different manufacturing purposes. This source is designed to provide the greatest amount of heat from the fuel being used and to concentrate it where it does the most effective work.

Warm-air furnaces are designed to heat air and distribute it through ducts to different rooms or parts of a building to provide a comfortable temperature. Forced warm-air furnaces employ a blower or electric fan to move greater volumes of air through the system. This type of system permits the furnace to be located a greater distance away from the area being heated by the system. The fuel for these furnaces may be coal, coke, fuel oil, or natural gas. Three representative comfort-heating furnaces are shown in Figure 4-4.

Fossil fuel is used as a source of energy for large industrial furnaces used in heat-treating operations. It is also used in the production of iron, steel, bricks, cement, glass, and many other materials. These furnaces differ from the warm-air type because they produce extremely high temperatures. Again, coal, coke, fuel oil, and natural gas are typical fossil fuels used as the primary energy source. *Metal-refining furnaces* usually force large quantities or blasts of air into the fire chamber to increase the heat already produced. This type of heat source is commonly called a *blast furnace*. Figure 4-5 shows a representative iron-ore blast furnace.

Small, gas-fired, heat-treating furnaces of the batch type are commonly found in many industrial operations. These units are used in applications that require precise control of temperature, time, and atmosphere.

Steam and hot-water heating systems represent another type of heating system that finds widespread usage in industry. Systems of this type may be used exclusively for facility heating purposes, or for facility heat-

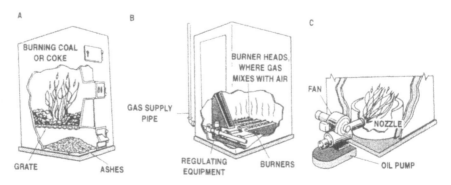

Figure 4-4. Comfort heating furnaces: (A) Hand-fired coal or coke furnace; (B) Gas-burning furnace; (C) Oil-burning furnace.

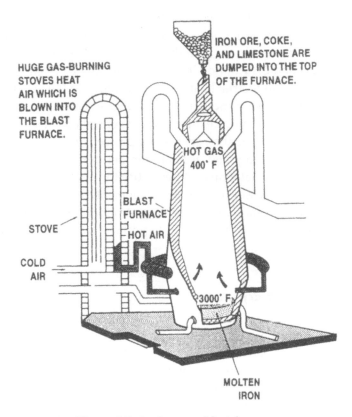

HUGE GAS-BURNING
STOVES HEAT
AIR WHICH IS
BLOWN INTO
THE BLAST
FURNACE.

IRON ORE, COKE,
AND LIMESTONE ARE
DUMPED INTO THE TOP
OF THE FURNACE.

HOT GAS
400° F

BLAST
FURNACE

STOVE

HOT AIR

COLD
AIR

3000° F

MOLTEN
IRON

Figure 4-5. An iron-ore blast furnace.

ing and manufacturing process applications. This type of system is generally considered to be the most dependable. It also provides the most uniform distribution of heat throughout the system.

The essential parts of a steam or hot-water system are the boiler, distribution pipes, and unit heaters or radiators. The boiler is where fuel is burned to produce heat and ultimately change water into steam or hot water. In a steam system, steam is circulated at a temperature of 170°F to 200°F, and the distribution lines are kept free from air. The lack of air keeps the system at less than normal barometric pressure. As a result, steam is produced at a lower temperature. Temperatures in the range of 170°F can be used effectively to produce steam in this type of system.

The radiators (load) of a steam or hot-water system give off heat by conduction, convection, and radiation. Conduction is quite small, with convection and radiation being the most significant methods of heat transfer. The radiator is located where heat is to be supplied to the system, and the temperature of the area where the radiator is located determines the load placed on the radiator. Steam or hot-water systems usually have many radiators located throughout the system. The load is a composite of all the radiators and distribution lines connected to the system.

Steam-heated industrial heat-treating furnaces of the batch type are frequently found in many industrial applications. These units may contain their own boiler or may be attached to a central distribution system. Units of this type combine the benefits of a protective steam atmosphere, forced circulation heating, and accurate temperature control in a single unit. They may operate at temperatures of 0°F to 1500°F. A *thermocouple* is used frequently as a sensor for the control element. Proportioning control is also widely used, because this type of control permits the unit to come up to temperature quickly and smoothly without overshoot, while holding for the duration of the cycle.

ELECTRICAL HEAT SOURCES

Electrically energized heat sources have a number of advantages over nearly all other industrial heating methods. In furnaces, electricity can be used to produce temperatures that range from 3500°F to 5000°F. In addition, electric heat does not produce gases that may have a harmful effect in metal processing. Temperature is more accurately controlled, and the entire process is more efficient and cleaner to operate.

Three common types of electrical heat sources used in industrial applications today are the arc, resistance, and *induction furnaces.* Each general type uses electricity as the primary energy source and produces heat as the end result or output of the load device.

A simplified diagram of an *electric arc furnace* is shown in Figure 4-6. High-current electricity is applied across the carbon electrode and the graphite crucible. This causes an electric arc to be produced. The arc occurs when the electrode touches metal. After an arc has been started, the electrode is withdrawn. Temperatures of 3500°F are typical of electric arc furnaces.

An *electric resistance furnace* is very similar in operation to the oven of a home-type electric range or comfort-heating unit. When current passes through a conductor, heat is produced as a result of *conductor resistance.* These special-resistance heating elements are then placed around an insulated chamber, where small batches of material are placed for heat treating. One type is a *nitrating* heat-treating furnace. This type furnace provides forced convection, so that all surfaces of the work are exposed to the same treatment. It provides rapid heating and uniform distribution of heat and atmosphere inside of the chamber. The controller of this furnace responds to a thermocouple located inside of the chamber. A time-rate adjusting control alters the electrical energy proportionally so that the heat conforms to the demands of the load. Control of this type is smooth without overshooting, and holds the temperature at precise levels.

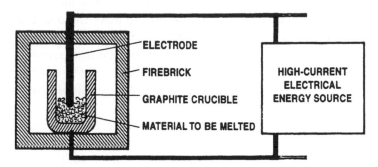

Figure 4-6. Basic diagram of an electric arc furnace.

Induction heating is accomplished by sending high-frequency alternating current through a wire coil. Metal placed inside of the coil develops intense levels of heat very rapidly. Suppose that an iron bar is placed inside of the coil of an induction heater. The iron bar becomes magnetized as alternating current flows through the coil. This causes the iron molecules to change polarity with each directional change of current flow. Each molecule develops an internal friction that causes the iron bar to become very hot. Induction heating units develop intense levels of heat in confined areas in very short periods of operating time.

If nonmagnetic metals are placed in the coil of an induction heater, they will not become hot due to the resistance of the material. For nonmetallic materials, heating is achieved by placing the material in a carbon crucible. The crucible is then placed in the coil. Crucible heat produced by induction is transferred to the nonmetallic substance by conduction. Metal can be heated in an insulated container without altering the temperature of the container. Induction heating may also be localized in a very small area. For example, the teeth of a gear wheel may be heated red-hot for tempering purposes, while the body of the gear remains cool and retains its toughness. If this were attempted by other heating processes, the entire gear would be heated. Induction heating can be achieved very quickly with rather significant levels of accuracy.

TEMPERATURE SYSTEM CONTROLLERS

The controller function of a thermal system is primarily achieved by a combination of components placed in a unit known as a *controller*. The function of this unit is to sense the temperature of the system and to determine the amount of heat flow needed to meet the demands of the operating setpoint of the system. In achieving this function, several factors affect the accuracy of the controller. These include *temperature gradients, thermal lag, component location*, and *controller operation-mode* selection. *Mean temperature consistency* and *system bandwidth are* primarily affected by these factors.

Thermal Gradients

Measurement of different temperature values, starting at the heat source and moving toward the load, indicates that a decided drop in temperature occurs near the load end of a system. This variation in temperature is commonly called the *thermal gradient of* the system. Thermal gradients occur in all temperature systems and it is quite important to recognize where they exist. Since heat is effectively transferred from the source to the load in only one direction, thermal gradients are inevitable and necessary. System control can be effective only when thermal gradients are taken into account. It therefore becomes imperative that all measurements affecting the control function be made as near as possible to the area being influenced by the controller. If these measurements are taken near the source end of the system, the readings are somewhat higher than those that appear at the load. Ideally, the controller sensing element should be attached to the load or placed in the actual work area of the system. The setpoint of the controller must take into account the thermal gradient and be adjusted or compensated for relative to the controller's location in the system.

A number of things can be done to minimize the gradients that occur in a thermal system. These include balancing heater capacity against heat demand, proper sensing-element location, sensing-element control range, and general system insulation against heat loss. In practice, it is desirable to minimize the thermal gradients in order to improve the accuracy of the controller.

Thermal Lag

Thermal lag is another inevitable condition that is present to some extent in all temperature systems. Thermal lag is primarily a delay in heat distribution that occurs between the source and the load. The distance between the source and the load and the resistance to heat flow are the primary factors that influence this kind of delay. Thermal lag, *dead time*, and *transport lag* are all used to describe this condition of operation.

Accurate system control is largely dependent on thermal lag. When the delay factor is low, controller action closely follows system setpoint adjustments. Long delay periods, by comparison, tend to cause temperature overshoot. This represents a period when there is more heat delivered to the load than is actually needed to recover from the temperature drop. In addition, the sensing element may not be able to respond quickly enough to deliver the necessary heat needed by the load. This causes the temperature to drop below the setpoint level. The term *undershoot* usually describes this condition. Both overshoot and undershoot cause the width of the control range to be expanded quite significantly during normal operation.

In practice, thermal lag cannot be entirely eliminated. With proper system design, however, a large amount of delay can be minimized and some degree of compensation can be made for the lags that still exist. Proper selection of the transmission path material between the source and the load is a prime factor in system design. Typically, solids, liquids, or gases are all used in the transmission path. As a general rule, however, material selection is primarily dictated by the application of the system. When close control is desired, several transmission path factors must be considered. These are described in the following statements:

1. When liquids, air, or gas are used in the transmission path, they should be in a continuous state of agitation.
2. All metal used in the transmission path should have a high level of thermal conductivity.
3. Thermal path insulating materials should possess a low conductivity.
4. Conditions in the transmission path that cause air or liquids to become stagnant should be avoided.

In general, thermal lag can produce some misleading information about the performance of a rapidly changing temperature system. In some systems, the effects of excessive lag can be so great that the sensing element at the load may be calling for more heat when the system is just beginning to respond to heat from a previous change. Systems with fast-response controllers and slower-responding indicators, such as mercury thermometers, usually have inherent delay problems of this type.

System-component Placement

In an operating thermal system, component placement has a great deal to do with the effectiveness of the control function. If the heat source, the sensing element, and the load could be grouped together in a compact central area, there would be very little problem with control. A short heat path from the source to the load would enable the sensing element to respond quickly to any and all system changes. This would minimize overshoot, undershoot, thermal lag, and thermal gradients. In most industrial thermal-system applications, it is rather difficult to achieve intimate component-grouping arrangements. The size of the system and the remote location of the source and load create problems. In practice, there is no single answer to system-component placement. Designers generally try to arrive at a compromise that will permit the best level of control for a particular system.

A general rule to consider when selecting the location of thermal-system components is based on the nature of the control characteristic. For systems where the desired heat demand is steady, the sensing element should be positioned closer to the source. When system demand is of a variable nature, the sensing element should be oriented closer to the load area.

In liquid and gas systems, where the demand for heat is of a rather steady nature, the sensing element should always be placed above the heat source. This is done to minimize the bandwidth of the control range. The transfer of heat in this type of system is primarily achieved by the convection process. Figure 4-7 shows a realistic way to place the sensing element in this type of system. The agitator of this system is used to distribute convection currents and to minimize thermal gradients and thermal lag.

Thermal-system Controller Selection

Selection of an appropriate controller for a specific thermal system involves a number of important considerations. Essentially, controller performance should match the application of the system in order to ensure

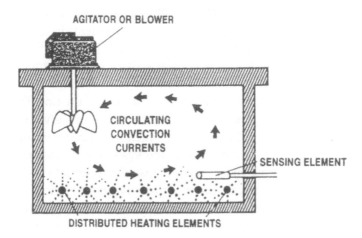

AGITATOR OR BLOWER

CIRCULATING
CONVECTION
CURRENTS

SENSING ELEMENT

DISTRIBUTED HEATING ELEMENTS

Figure 4-7. A Sensing element placement in steady heat systems.

that it achieves the desired level of control. In addition, the selection process should take into account temperature operating range, response time, resolution sensitivity, mode of control, and sensor type. When these factors are satisfied, the controller should be compatible with the system and its application.

The temperature operating range of a controller refers to the upper and lower temperature extremes over which a controller will respond. This range of operation is primarily determined by the type of sensing element employed and the mechanical or electrical operation of the controller. Typical *electrical* sensors respond to temperature by generating a voltage, changing resistance, comparing colors, or responding to infrared radiation. *Liquid-filled* sensors, by comparison, respond to temperature variations by producing changes in liquid volume, changes in gas or vapor pressure, or dimension changes of solid materials.

In practice, the temperature ranges of controllers vary a great deal among different manufacturers. Some representative ranges are −100°F to 600°F, 200°F to 1500°F, and −20°F to 275°F (−3°C to 315°C, 93°C to 815°C, and −28°C to 135°C) for thermostat and thermoswitch sensing elements. In liquid-filled systems, representative ranges may be 350°F to 650°F (176°C to 343°C) or somewhere within a −450°F to 1400°F (−267°C to 760°C) limit. *Thermocouple* systems generally respond to a very wide range of temperature, with extremes ranging from −400°F to 3200°F (−240°C to 1760°C). *Thermoresistive* systems cover a range of −430°F to 1800°F (−256°C to 982°C). Because of nonlinearity problems over a wide range, thermistor systems have a narrower temperature operating range. This range can be extended, however, by using alternate sensing elements in a system. *Infrared* sensors respond to temperatures in the range of –20°F to 5400°F (–30°C to 3000°C). This type of sensor is not placed in direct contact with the heat source.

The *response time* of a controller is a measure of the time that it takes the sensing element to generate a signal that will ultimately initiate a system state change. To a large extent, this factor is also determined by the type of sensing element employed by the system. Response time is normally based on the time that it takes the sensor to initiate a 63.2 percent change in value over its calibrated output. Typical electrical sensors are of the *bimetal-strip switch* or *thermostat* type, or are *thermocouples, thermoresistor detectors,* and *thermistors*. Nonelectrical sensors are of the gas- or fluid-filled type. The response time of the sensing element is a major factor in the selection of a controller. When high-speed response is needed, thermistor and thermoresistor detector elements are applicable. Response times ranging from 0.035 to 5 seconds (s) are typical. The response time of a thermocouple is next, with a range of 0.04 to 7.5 s. Thermostats and thermoswitches are somewhat slower, with response time starting at 1 s. Filled sensing elements overlap some of the electrical sensors, with a response time of 0.5 to 10s or more.

To demonstrate the response time of a sensor, we will describe how a thermistor responds electrically. A thermistor changes its resistance value according to the temperature of the environment where it is placed. Response time is generally expressed in *time constants*. The time constant of a thermistor is the time required, in seconds, for it to change 63 percent of its resistance value when subjected to a new or different environment. In this regard, a thermistor taken from an environment of 72°F and placed in an environment of 172°F will change to a temperature of 135°F in one time constant. This is based on the fact that 172°F minus 72°F produces a span of 100°F, 63 percent of 100 is 63, and 72 plus 63 equals 135. This means that a thermistor goes through a 63 percent change in temperature and resistance in one time constant. In practice, five time constants are required for a thermistor to completely reflect a new environmental temperature change.

In general, response time is not very significant in systems where the temperature remains fairly constant for long periods. In systems where temperature changes occur frequently and rapidly, however, response time becomes a very important consideration of the system.

Resolution sensitivity of a controller normally refers to the specific amount of temperature change needed by a particular controller to initiate a state change. Typical expressions of sensitivity are given as a specified number of degrees or as some percentage of the controller's total operating range of scale. In practice, sensitivity is a good measure of the controller's temperature bandwidth. The process of changing good controller sensitivity into accurate control calls for a number of careful design procedures. As a rule, this means that good sensitivity is somewhat more costly to achieve. For most applications, controller sensitivities of 2°F to 5°F (1.1°C to 2.7°C) are more than adequate when the system is properly installed. Some representative sensitivities for thermostats and thermoswitches are 0°F to 1°F, 1°F to 5°F, 5°F to 10°F, and 2°F to 8°F (0°C to 0.5°C, 0.5°C to 2.7°C, 2.7°C to 5.5°C, and 1.1°C to 4.4°C). In liquid-filled controllers, a representative sensitivity is 2°F to 8°F (1.1°C to 4.4°C). Thermistor and thermocouple elements are by far the most sensitive of all detectors, with 1°F or 0.2 percent of the temperature span being typical values.

Modes of Control

Mode of control describes the method by which a controller adjusts system temperature in order to restore it to a desired level. In practice, thermal systems employ the two-position mode (on-off), proportioning control, or PID control. In the two-position mode of control, the source is simply turned off when the load temperature has exceeded the setpoint value. When the temperature of the load drops below the setpoint value, the source is turned on again to return it to the desired level. Figure 4-8, shows a typical temperature response of a two-position controller. As noted, there is a great deal of overshoot-undershoot oscillation in this mode of control.

The *proportioning* mode of control provides a means of variable temperature adjustment. This reduces the overshoot-undershoot problems of the two-position mode of operation. In electrical thermal systems, proportioning control is achieved simply by altering the amount of power applied to the heating element. One type of proportioning control is called *time-rate* control. This type of control is used in applications that require precise control of a process temperature. Time proportioning operates in the same ways as on-off control when the process temperature is outside of the proportional band. When the temperature approaches the setpoint value and enters the proportional band, the load device is switched on and off at an established time. At the lower limit of the band, the on time is

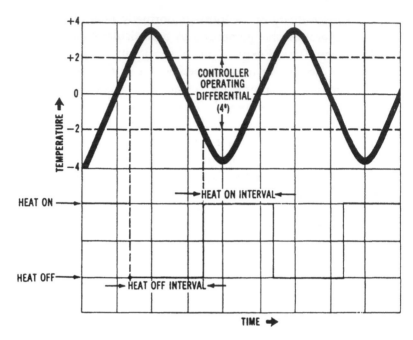

Figure 4-8. Response of a two-position controller.

greater than the off time. As the temperature approaches the setpoint, the ratio of on to off time changes, and the amount of on time decreases as the off time increases. This causes a change in the effective power delivered to the heating element work load. Essentially, it has a throttling-back effect that results in less temperature overshoot. This on and off action continues until the times are equal. When this occurs, it indicates that the system has reached a balanced state and the temperature is at a point just below the setpoint value.

When a time-proportioning controller reaches its balanced state or becomes stabilized, there is a resulting droop or offset in the temperature value. In the time-proportioning operational profile in Figure 4-9, the offset is slightly below the setpoint value. This condition will continue as long as there are no abrupt changes in the work load. If an offset cannot be tolerated, there are ways to compensate for it by combining proportional control with another mode of operation. Integral control, for example, can be added to proportional control to achieve PI control.

Integral or reset control is an operation in which the rate of change in the output is proportional to the input. Integral control can be achieved manually by an operator. This adjustment brings the temperature in

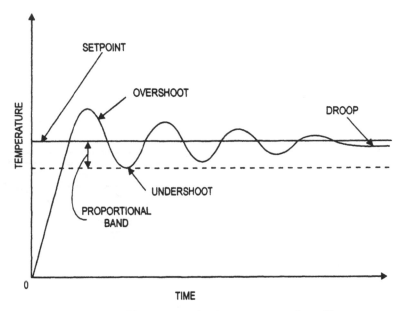

Figure 4-9. Time proportioning operational profile.

agreement with the setpoint value. When manual control is used, it causes temperature agreement within a narrow span of the setpoint value for which it is adjusted. If there is a drastic change in temperature, the agreement between the setpoint and process temperatures will be lost. Manual reset adjustments generally require continuous monitoring of the output temperature. Figure 4-10 shows a time-versus-temperature profile of proportional control with manual reset.

The automatic reset function of integral control allows the controller circuitry to automatically compensate for an offset before it exists. An integrator circuit automatically compensates for the difference between setpoint and operating temperature. This operation automatically drives the temperature up to the setpoint value. Figure 4-11 shows a time-versus-temperature profile of a proportional plus integral controller. Note that the offset condition does not exist in the operational profile.

All of the temperature control procedures that we have discussed up to this point have shown a temperature overshoot problem. This is the point where the temperature rises well above the setpoint value in normal operation. In some temperature control operations, this condition cannot be tolerated. It can be prevented or reduced by adding derivative or rate control to a proportional controller. This type of controller then becomes a proportional plus derivative instrument.

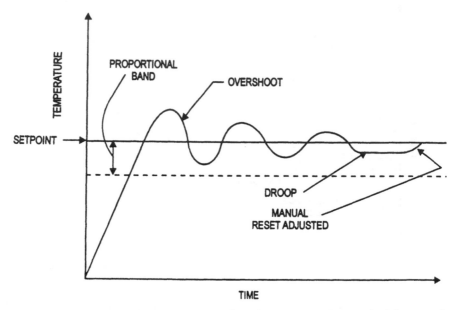

Figure 10. Time-versus-temperature profile of proportional control with manual reset.

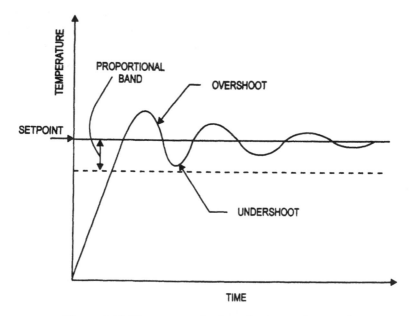

Figure 4-11. Time proportioning plus integral control.

Derivative control is an anticipatory function. It measures the rate of increase in temperature and forces the operation into a proportioning action on an accelerated basis to reduce the increase. This kind of circuit operation prevents a large degree of overshoot during start-up operation and reduces overshoot when disturbances tend to drive the temperature up or down. Derivative action is a control operation in which the output is proportional to the rate change of the input.

Temperature controllers that combine proportioning, integral, and derivative operations are called PID instruments. This type of controller is used for difficult processes which result in frequent disturbances, and applications where precision temperature regulation is required. A time-versus-temperature profile of the operation of a PID controller is shown in Figure 4-12. Note that the temperature has no droop and a reduced amount of overshoot at the start of its operation. Electronically, this type of control can be accomplished with standard integrated circuits.

Load Control

The final control element of an electrical thermal system is directed by a signal developed by the controller. The control element is simply turned on and off at some desired rate or adjusted to some value that maintains

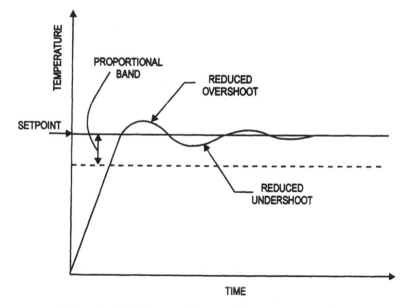

Figure 4-12. PID controller time/temperature profile.

the temperature of the system at a desired level. The final control element dictates the control action achieved by the output. Electronically, output control is accomplished by *electromagnetic relays, solid-state relays,* and *solid-state switching devices.*

On-off control and time-rate control can be accomplished by electromagnetic relays. An electromagnetic relay has an energizing coil with electrical contacts that respond to the action of the coil. This device has moving parts and is susceptible to vibration and mechanical failure. The operational life of a relay can be extended by reducing the amount of load controlled by the contacts. The relay does, however, produce a positive break in the load circuit, which is an important operational characteristic. Relays can be mounted in any position, are easy to install and service, can have a number of normally open and normally closed contacts, and are fairly inexpensive.

A *solid-state relay* (SSR) is commonly used in controller operations to alter the status or an electrical load. It has no moving parts and provides electrical isolation between the controller output and the electrical energy source. It is also resistant to mechanical shock and vibration, has a fast response time, and does not have contact bounce problems. The switching device of an SSR can be a *transistor,* a *silicon-controlled rectifier,* or a *triac.* All of the load current being controlled must pass through the switching device. As a rule, it is equipped with a metal strip or *sink* for heat dissipation. A solid-state relay can effectively be used to replace the electromagnetic relay as a power control device. Its disadvantages include the inability to provide a positive circuit break, initial cost, ambient-temperature damage susceptibility, transient-line voltage damage, and failure when subjected to overrated conditions.

Solid-state switching devices are widely used to achieve load control in the operation of a controller. These devices include transistors, silicon-controlled rectifiers, and triacs. Mechanical failure is reduced because no moving parts are involved in the control procedure. Solid-state switching is also responsible for variable control of the load. This permits various proportioning control operations to be accomplished. One type of proportioning control is achieved by regulating the number of cycles of electrical power supplied by the source to the load in a given period of time (see the time-rate temperature control procedure of Figure 4-13). Systems that employ this type of control produce a very low level of electrical interference because the power is turned on and off at the zero crossing point. Some controller manufacturers refer to this type of control as *zero-voltage switch-*

ing (ZVS). A single ZVS integrated circuit is available for this operation.

An alternate method of achieving electrical proportioning control employs the *phase-angle firing principle*. In this method, a triac or dual SCRs are used to turn on the electrical power to the system heating *element* for only a portion of the cycle, as shown in Figure 4-14. Phase-angle firing, as a rule, generates some pronounced transient spikes in the ac power line. A system employing this type of control must have some form of transient-spike suppression. Phase-angle firing improves operational efficiency and permits precise control of the load device.

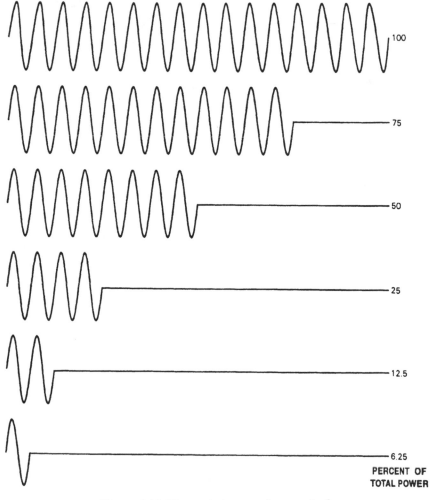

Figure 4-13. Time-rate temperature control.

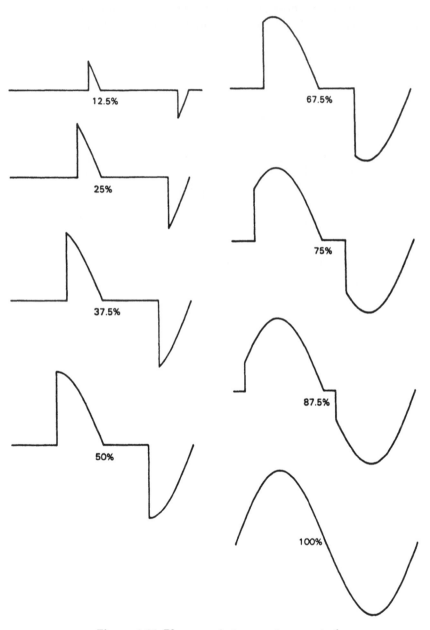

Figure 4-14. Phase-angle temperature control.

Filled Systems

In addition to the electrical control process, proportioning control can be accomplished by filled systems. Changes in sensor output can be used to change the wiper arm of a potentiometer. In this action, a motorized valve can be adjusted to different temperature levels. Through this

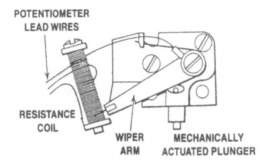

Figure 4-15. Proportioning potentiometer.

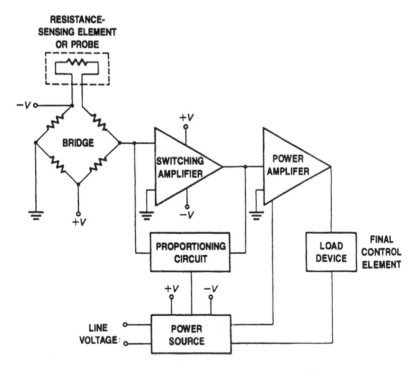

Figure 4-16. Electrical temperature controller.

control process, valve positions can be altered to any position from closed to 100 percent open, depending on the deviation of the setpoint. Figure 4-15 shows a proportioning potentiometer, and Figure 4-16 shows an electrical controller circuit that can be used for either two-position or proportioning control.

Sensors

Sensor-element operation is an extremely important consideration when selecting a thermal system controller for a specific application. Such things as response time, temperature operating range, resolution sensitivity, and repeatability are dependent on the sensor element. In addition, the physical size of the sensor has a great deal to do with component location and installation design procedures.

Filled sensing elements are mainly designed to respond to changes in temperature by producing a physical change in a pressure- or volume-sensitive component. Controllers of this type employ a sensing element or *bulb,* a capillary tube, and a pressure- or volume-sensitive component such as a Bourdon tube, bellows, or diaphragm.

The sensing element or bulb of a filled system contains a fluid or gas that changes its volume or pressure with temperature. The pressure- or volume-sensitive part of the system responds to these changes by delivering a motion or physical change that is applied to the control element. Comparisons of the setpoint value and sensor output then determine the action of the controller.

Class I filled systems employ liquids other than mercury. An inert hydrocarbon, such as xylene, is very common. Sensors of this type have an expansion rate that is six times greater than that of mercury. A characteristic of this sensor is its small size.

Class II pressure elements have the filling medium in both liquid and gaseous forms. The combination of the two interface in the sensing bulb to produce a vapor. There are four different combinations of liquid-gas elements: labeled IIA, IIB, IIC, and IID.

Class III refers to gas-filled systems. Nitrogen is used for conditions up to 800°F, and helium is used for extremely low temperatures. This type of sensor is made rather large and somewhat longer than the others to reduce temperature interference along the capillary.

Class IV filled systems employ mercury as the medium. Mercury has a rapid response time and is very accurate. It also provides a great deal of physical power to actuate the pressure-sensing element. At high temperatures, pressures may reach 1,200 psi, while dropping to 400 psi at the lower end of the temperature range. Mercury-filled systems are commonly used in industrial applications.

Thermostats

In general, *thermostats* and *thermoswitch* sensors are slower in response time compared with other sensing elements. For example, thermostats must produce switching action by physically distorting a bimetal strip with the application of heat. As a rule, some degree of response lime is needed to cause the bimetal material element to react. Standard response times range from 1 to 10 s or more. A rapid-responding thermoswitch has a high-expansion outside shell and a low-expansion internal strut assembly. The temperature at which the contacts make and break can be adjusted by the temperature-adjusting screw.

A thermostat with a bimetal disc assembly is shown in Figure 4-17. This sensing element has *positive reinforced snap action*. The action is known for its repeatability and reliability. Thermostats of this type are calibrated at the factory and housed in a tamperproof enclosure.

Thermocouples

A *thermocouple* sensing element is composed of two dissimilar metal wires connected in isothermal contact. The isothermal connection may be achieved by welding, fusing, or twisting the two wires together.

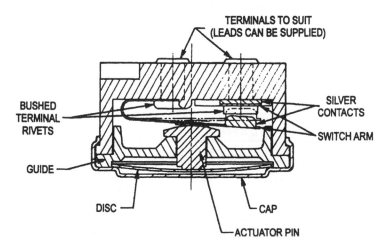

Figure 4-17. Bimetal disc thermostat: Cutaway drawing. (Courtesy Philips Technoiogies/Airpax Protection Group)

When a thermocouple is connected into a circuit, as in Figure 4-18, it serves as the heat-sensing junction (T_H). If one junction is maintained at a reference temperature (T_C), the magnitude of the resulting current will be proportional to the temperature at the measured junction (T_H).

In practice, the output of a thermocouple sensor is measured in millivolts. In some applications, the millivolt output of a thermocouple is transposed into temperature for indication on a calibrated scale. In controller operation, thermocouple output voltage is compared with the setpoint adjustment voltage to determine the action of the controller. Some type of amplification is usually required to make this comparison. Today, opAmps are used exclusively to achieve this amplification.

Figure 4-19 shows a thermocouple input amplifier with cold-junction compensation. This circuit has a 10-mV/°C output. Gain is achieved by adjustment of the feedback network composed of resistors R_7 and R_8. The LM335 is a precision-calibrated temperature IC. It responds as a two-terminal zener diode, whose breakdown voltage is directly proportional to absolute temperature at +10 mV/K. This device permits cold-junction temperature compensation of the thermocouple. The LM329B is a precision temperature compensated 6.9-V zener reference diode. The LM308A is an opAmp that responds well to low input currents and has a low offset voltage. In a controller, the output of this circuit is connected to one input of a voltage comparator. The alternate input comes from the setpoint circuit. The values of these two voltages are compared and determine the output of the circuit. A variety of modifications can be made to permit this circuit to be used in a controller.

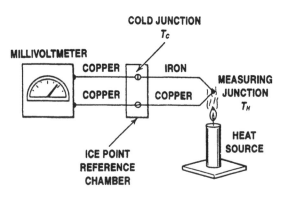

Figure 4-18. Thermocouple circuit.

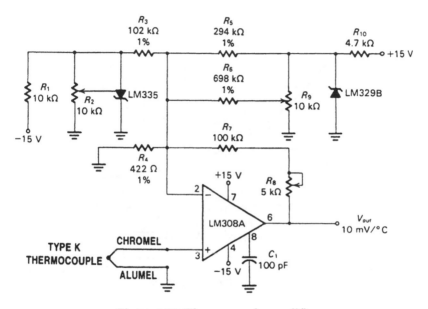

Figure 4-19. Thermocouple amplifier.

Thermoresistive Sensors

Thermoresistive sensing-element operation is based on the property of certain metals to change resistance with changes in temperature. These changes in resistance can then be used to alter the current in an electric circuit. They may also be used to develop a *changing voltage* that can be amplified to a level that will permit control. Of all the materials used in thermoresistive elements, platinum tends to have the most desirable characteristics. As shown in Figure 4-20 the resistance of platinum is linear for temperatures from 9°C to 800°C. In addition, the melting point of platinum is high; platinum does not change physically for temperatures below 1200°C. The tensile strength of platinum is 18,000 psi and its resistance is 60.0Ω/circular mil at 0°C. Industrial thermoresistive elements are called *resistance temperature detectors*. These elements have a response time of 0.035 to 5s for temperatures of −100°F to +1000°F, with a 0.1 percent tolerance and a ±0.01 percent repeatability.

The resistance change of an RTD is quite small under normal operating conditions. As a rule, it can be used to sense temperature changes directly as a variation in voltage across a current-driven resistor, or to sense output-resistance changes in a bridge circuit. Normally, the developed voltage is used to control a load device. OpAmps are used exclusively

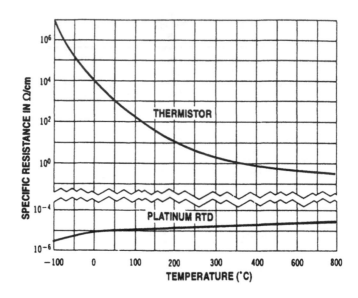

Figure 4-20. A resistance/temperature comparison of thermistors and platinum thermoresistive detectors.

today in controllers that respond to RTD sensors.

Figure 4-21 shows the circuitry of an adjustable reference opAmp temperature-sensing circuit. The RTD of this circuit is connected in the feedback path of the opAmp. The resistance of the RTD varies from 100 to 200Ω over the temperature range. The gain of the opAmp is altered by these changes. The AD584 is an adjustable multireference IC that is set for 6.2-V output at the emitter of the 2N2219 transistor. Potentiometer R_2 is used to alter this reference voltage. Potentiometers R_4 and R_6 adjust the span and offset of the opAmp. The span adjustment is made first, by altering R_2 to produce a 1.8-V output when 266°C is applied. The offset adjustment is made by altering the value of R_4 so that the output is 0 V at 0°C. The scale of the circuit is somewhat arbitrary. In general, the output voltage range is primarily based on the device being fed by the circuit. This circuit permits temperatures of 0°C to 266°C to be measured as voltages of 0 to 1.8 V.

Thermistors are an additional type of thermoresistive sensing element of the semiconductor or solid-state type. As the name implies, a thermistor is a temperature-sensitive resistor.

The differences between a thermistor and a platinum thermoresistive element are shown in Figure 4-20. The resistance of the thermistor

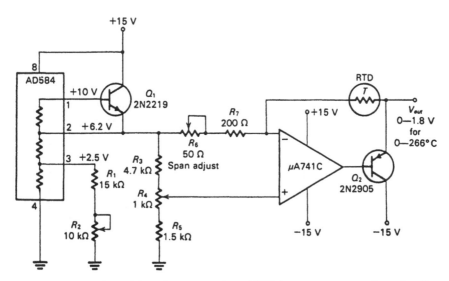

Figure 4-21. Adjustable reference opAmp RTD temperature-sensing circuit.

drops in value as temperature increases. This characteristic is known as a *negative temperature coefficient*. The response time of a thermistor compares favorably with that of a thermoresistive element. Device resistance of a thermistor can be selected from a number of values that range from several $M\Omega$ at $-100°C$ to less than 1Ω at $400°C$.

The use of a thermistor as a control element to drive a load device generally necessitates some degree of amplification. As a rule, the amount of amplification needed to achieve control is based on the load device being driven by the controller. OpAmps are commonly used in industrial controllers today. The controller can be used to achieve on-off, proportional, or proportional with integral and derivative control. The thermistor's amplified change in voltage as a result of a change in temperature is used to vary the load device according to the type of controller mode that is needed.

Figure 4-22 shows the circuitry of a simplified on-off temperature controller. The thermistor is used in a bridge circuit. Balancing the bridge is the equivalent of adjusting the setpoint of the circuit to some desired temperature. A change in thermistor temperature causes the bridge to go out of balance and produce a corresponding change in voltage at the input of the opAmp. This voltage is then amplified and applied to the load device. If the load, such as a heating element, is controlled by a relay or silicon controlled rectifier (SCR), on-off control can be achieved. When a

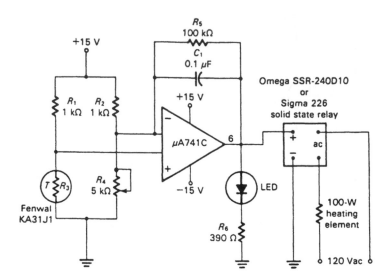

Figure 4-22. On-off temperature controller.

power transistor or IC is used to alter conduction of the load, some degree of variability is achieved. This permits proportional or PID control.

TEMPERATURE INSTRUMENTATION

Instrumentation is a generic term for devices or procedures used in the measurement and evaluation of industrial process applications. *Temperature instrumentation* is more specialized because it applies to those things that are used to test and evaluate the temperature of a system. Precise measurement of temperature is a key factor in nearly all manufacturing operations, regardless of the product involved. In thermal systems, instrumentation can be applied to any discrete part of the system. It is essential that a person working with this type of system (or any process control application that responds to temperature) has an understanding of the basic principles of instrumentation.

A general way of classifying all temperature-measuring instruments is by the temperature operating range of each device. Figure 4-23 shows an instrument operating range comparison chart. Note that several instruments may be capable of measuring temperatures over the same given range, but not all instruments are suited for a given temperature measurement. Selection of a specific instrument must also take into account such

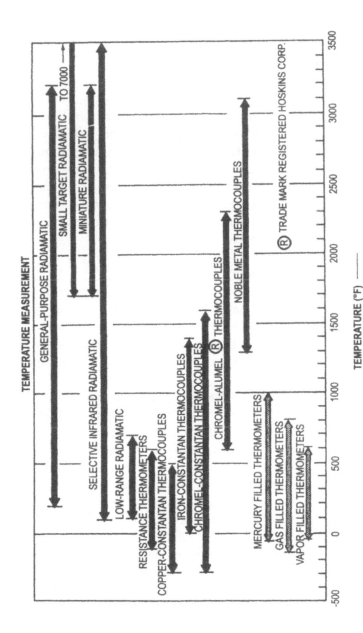

Figure 4-23. Temperature instrument measuring range. (Courtesy Honeywell, Inc.)

things as response time, accuracy, life expectancy, and the basic method of operation. In general, the measuring instrument should be compatible with the capabilities of the controller, process, and system.

NONELECTRICAL INSTRUMENTATION

Temperature measurement is divided into two general classifications: *nonelectrical instruments* and *electrical* or *electronic instruments*. Nonelectrical instrumentation was developed first and probably has more applications today than electric or electronic instrumentation. Nonelectrical instrumentation essentially exemplifies the expansion principle. Changes in temperature cause dimensional changes in solid materials, or cause gas, vapor, or liquid to expand. As a result of this action, changes in temperature can be transformed into physical or mechanical changes that can be read on a calibrated scale.

Filled-system Thermometers

The principle of *volumetric liquid expansion* with increasing temperature is one of the oldest methods employed for the determination of temperatures. The *glass stem thermometer*, which uses this principle, was developed by G.D. Fahrenheit in the eighteenth century. In the general range of –500°F to 1000°F, this type of indicator is about as important as any other method today.

A *filled-system thermometer* is a glass tube, with a small hole bored lengthwise and a bulb reservoir at one end. This construction is shown in Figure 4-24. When the reservoir is filled with liquid, the remaining air is removed from the tube and the top of the tube is sealed. This creates a closed thermal expansion system.

In a sealed container that has a fixed but larger volume area, liquid will expand according to ambient temperature. The expansion of pure mercury is 0.01 percent/°F, and is linear from its approximate freezing point of –38°F to its boiling point of 1000°F.

Industrial applications of the thermometer are common. Some thermometers are permanently attached to a system, while others are portable. In both cases, the thermometer is limited to direct reading applications that are not adaptable to recording or automatic control situations. With some degree of modification, it is possible to adapt the thermometer principle to perform these operations. *Filled temperature-measuring*

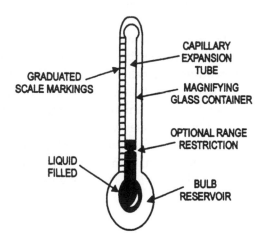

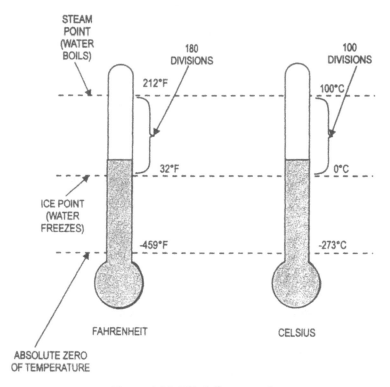

Figure 4-24. Filled thermometer.

instruments are the result of this modification.

A filled temperature-measuring instrument is similar to the sensing element, capillary tube, and pressure-volume element of a filled-system controller. In a temperature-measuring instrument, the end result of its operation is a usable reading on a calibrated scale. The term temperature recorder is often used to describe this type of device in industry.

Typical industrial filled-system thermometers normally employ a circular chart as a calibrated scale, with a recording pen attached to the pressure element. Figure 4-25 shows an enlargement of the spiral Bourdon element of a recording instrument. Instruments of this type are available for class I, II, or Ill operation. These have ranges from –450°F to 1450°F.

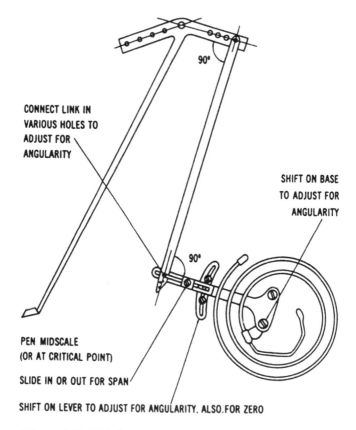

CONNECT LINK IN VARIOUS HOLES TO ADJUST FOR ANGULARITY

SHIFT ON BASE TO ADJUST FOR ANGULARITY

PEN MIDSCALE (OR AT CRITICAL POINT)

SLIDE IN OR OUT FOR SPAN

SHIFT ON LEVER TO ADJUST FOR ANGULARITY. ALSO.FOR ZERO

Figure 4-25. Filled system pressure measuring element.

Bimetallic Thermometers

Bimetallic thermometers are nonelectrical temperature-measuring instruments that find about the same amount of acceptance in industry as the glass bulb thermometer. Bimetallic thermometers make use of the coefficient of the linear expansion principle. This principle states that a solid material will change dimensions when its temperature is changed. If two different kinds of metal strips are bonded together and heated, the resulting strip will bend in the direction of the metal with the lower expansion rate. The amount of resulting deflection that occurs is proportional to the square of the length, and the total change in temperature is inversely related to material thickness.

To take full advantage of the expansion principle, the element of a bimetal thermometer should be rather long. In practice, the element is formed into a flat spiral or a single helix. The outside end of the element is then mounted to a structure and an indicating hand is attached to the inside of the loose end. An increase in temperature causes the element to wind up, which produces a clockwise deflection. Movement of the indicating hand registers a change in temperature on a calibrated scale.

Figure 4-26 shows an example of a spiral-element bimetal thermometer. The sensing element, in this case, is exposed for quick and accurate response to surface temperature. Where applications demand better protection; the element may be sheltered by a cover.

ELECTRICAL AND ELECTRONIC INSTRUMENTATION

Electrical instrumentation is different from nonelectrical measuring techniques in several ways. First, electricity must be supplied by an auxiliary source to make the system function. Second, a *transducer* is employed to change temperature variations into electrical signals. This type of instrumentation also has the advantage of small-mass sensing elements from the measuring area. In addition, electronic instrumentation lends itself well to portable applications, which increases the versatility of the measuring equipment.

Electrical and electronic instrumentation is classified according to the fundamental operating principle of the sensing element. This includes resistance changes, voltage generation, radiation, and optical comparisons. The sensing element or transducer of these instruments must have the ability to distinguish between temperature changes and judge the amount of heat in an object.

Temperature-measuring transducers are similar to those used in the sensor element of a controller. In fact, sensor elements are often used interchangeably in controllers and temperature recorders, and these functions are frequently combined to achieve both control and temperature indications in a recording type of controller. The operation of some of the important instruments commercially available and used by industry will now be discussed.

Thermoresistant Instrumentation

Thermoresistant instrumentation is based on the property of certain metals to change resistance when subjected to heat. In general, all metals possess

Figure 4-26. A bimetal thermometer with a spiral element: (A) Back view; (B) Side view; (C) Dual Fahrenheit and Celsius scales.

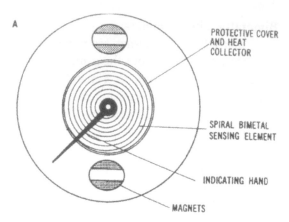

A

PROTECTIVE COVER AND HEAT COLLECTOR

SPIRAL BIMETAL SENSING ELEMENT

INDICATING HAND

MAGNETS

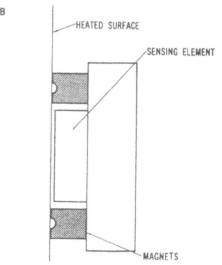

B

HEATED SURFACE

SENSING ELEMENT

MAGNETS

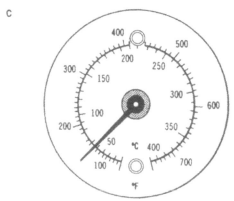

C

this characteristic to an extent, but certain considerations affect which metal is used. The most significant of these considerations is the purity of the metal and its ability to be formed into a fine wire. In addition, the metal should respond to rapid changes in temperature, have a repeatable temperature coefficient, respond over a linear resistance range, and possess a high resistance to temperature change ratio.

The sensor of a thermoresistant instrument is a long piece of wire, in the shape of a coil, wound around a ceramic core. The entire assembly is then enclosed in a protective sheath. Connection to the wire coil is made by passing leads through the ceramic core. This forms a stress-relief junction. For a typical resistance thermal detector element, the active length of the sensor is small compared with the rest of the assembly. Thermoresistive sensing elements are constructed from a number of common metals and alloys, including platinum, nickel, tungsten, and copper. Vanadium, rhodium, silver, iron, and tantalum are used occasionally. Platinum is best suited for industrial applications because it is readily available in an almost pure state and can be easily formed into a wire. In addition, it has linear resistance over a very useful range. The accuracy of platinum is so good that it serves as the international standard for measuring temperatures between –297.35°F and 1102°F.

The display or *readout* of a thermoresistant instrument is obtained by connecting the sensing-element output to a bridge circuit, Two-, three-, and four-wire bridge-circuit configurations are in use today. A simple two-wire bridge circuit is shown in Figure 4-27. Note that the RTD element is connected to the bridge as R_X.

The resistance relationship of the bridge circuit in Figure 4-27, when it is balanced, is shown in this formula:

$$R_2/R_3 = R_1/R_X$$

When balance occurs, the current through R_2 and R_3 equals that through R_1 and R_X (the RTD element). A zero deflection of the meter shows this condition.

To determine temperature, the resistance of the RTD sensor must first be determined. In practice, R_3 is adjusted to produce a balance indication on the galvanometer after the RTD has changed value. When it is balanced, resistance of the RTD is then determined by the formula:

$$RTD = R_3/R_2 \times R_1$$

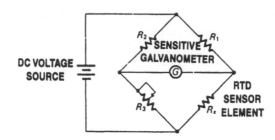

Figure 4-27. A Two-wire bridge temperature-measuring circuit.

When the value of the RTD element is determined, temperature can be determined by this value on a resistance-temperature graph.

Typical temperature bridge circuits often have R_3 calibrated in temperature values instead of resistance. As a result, when the circuit is balanced, the temperature is read directly from a calibrated dial or on a scale.

An *automatic, self-balancing, Wheatstone bridge resistance thermometer circuit* is shown in Figure 4-28. With this type of circuit, any condition of imbalance caused by heat applied to the resistance element is recognized and corrected by the slide-wire resistor balancing motor. An indication of this change is then recorded on a suitable scale or chart.

The operation of a self-balanced bridge thermometer is based on the dc voltage developed across the bridge circuit as a result of the resistance element. Any dc voltage appearing at A-A is changed into ac voltage by action of the opAmp chopper circuit. By input transformer action, this ac is stepped up and applied to a voltage amplifier at B-B. The output of the voltage amplifier at C-C is then used to drive the power amplifier. Power-amplifier output at D-D then controls the rotation and direction of the balancing motor. Mechanical connection of the motor and slide-wire potentiometer is made through E-E. As a result of this action, an increase or decrease in element resistance is automatically transposed into a physical change in slide-wire resistance, which nulls the bridge. This action can also be used to drive a recording stylus or deflect a meter, thus indicating the temperature of the resistance element. Temperature recorders of this type are commonly used to monitor continuous temperature values in process control applications.

An addition to the thermoresistant measuring-device family is the *portable digital readout* type of instrument. An instrument of this type will measure temperatures over a range of –200°F to 1000°F (–128°C to 537.7°C). The resolution of this instrument with a platinum RTD probe

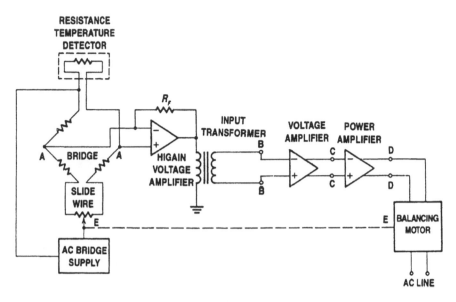

Figure 4-28. Automatic self-balancing thermometer.

is 0.1°F (0.05°C). Similar models have ranges of –50°C to 500°C, 0°F to 2000°F, and 0°C to 1370°C. Placing the sensor probe on a surface or in liquid, powder, air, or gas causes an instant digital reading to be obtained.

A simplified block diagram of a representative digital thermometer is shown in Figure 4-29. The operation of this circuit centers around an *analog-to-digital (A/D) converter*. This part of the circuit is designed to change temperature (analog information) into a digital signal. A variety of different conversion methods can be used. Typically, the A/D converter is a *voltage-to-frequency* conversion process: A change in temperature causes a change in voltage, which is then translated into a frequency. The frequency appears as a series of pulses that are representative of binary information. These data are then counted, decoded, and applied to the readout as a display of temperature. The display response time is very rapid. Some models of this thermometer will trigger and hold momentarily at the maximum sensing temperature. The versatility of this device and its operational simplicity and accuracy make it a very popular industrial temperature-measuring instrument.

Thermocouple Indicators

Thermocouple temperature indicators respond to the electrical properties of metal. When two dissimilar metals are heated at a common con-

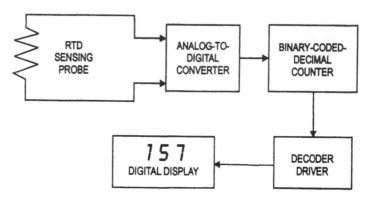

Figure 4-29. Block diagram of a digital thermometer.

nection point, a dc voltage is generated. The resulting voltage measured across the free ends of the thermocouple wires can be used to indicate the temperature applied to the measuring junction. Figure 4-30 shows a simplified circuit diagram of a thermocouple temperature indicator. In industrial applications, this type of instrument is commonly called a *millivolt pyrometer* or simply a *pyrometer.*

The basic components of a pyrometer include a *d'Arsonval* type of galvanometer, a thermocouple sensing element, and a compensating re-

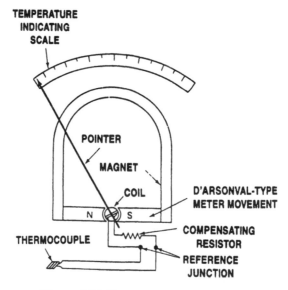

Figure 4-30. Temperature pyrometer.

sistor. When heat is applied to the thermocouple measuring junction, a resulting voltage appears at the free ends, or reference junction. This voltage causes a corresponding current in a series circuit formed by the meter coil, compensating resistor, and thermocouple. The current through the meter coil produces an electromagnetic field which is in opposition to the permanent magnetic field of the horseshoe-shaped magnet surrounding the coil. An interaction between the generated electromagnetic field and the permanent magnetic field causes the meter coil to move or deflect. An indicating pointer attached to the coil displays the amount of deflection in degrees or millivolts.

Several different types of thermocouples are available today for industrial applications. The combination of metals must possess a reasonably linear temperature-millivolt relationship to be of value in this type of measurement. Figure 4-31 shows the temperature-millivolt characteristics of several common types of thermocouples.

An indicating recorder circuit diagram for a continuous-balance thermocouple temperature-measuring system is shown in Figure 4-32. Note that this circuit is similar to the self-balancing Wheatstone bridge resistance thermometer in Figure 4-28. The primary difference between the two circuits is in the thermocouple connection to the bridge, but the final operation of these circuits is essentially the same. The conversion stage of the circuit in Figure 4-32 uses an electromagnetic unit instead of an

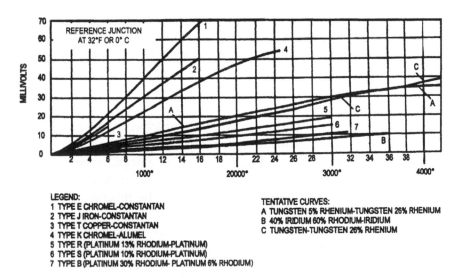

LEGEND:
1 TYPE E CHROMEL-CONSTANTAN
2 TYPE J IRON-CONSTANTAN
3 TYPE T COPPER-CONSTANTAN
4 TYPE K CHROMEL-ALUMEL
5 TYPE R (PLATINUM 13% RHODIUM-PLATINUM)
6 TYPE S (PLATINUM 10% RHODIUM-PLATINUM)
7 TYPE B (PLATINUM 30% RHODIUM- PLATINUM 6% RHODIUM)

TENTATIVE CURVES:
A TUNGSTEN 5% RHENIUM-TUNGSTEN 26% RHENIUM
B 40% IRIDIUM 60% RHODIUM-IRIDIUM
C TUNGSTEN-TUNGSTEN 26% RHENIUM

Figure 4-31. Temperature-millivolt characteristic of typical thermocouples.

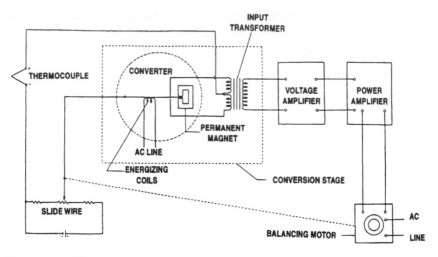

Figure 4-32. Thermocouple indicating temperature recorder for continuous balance. (Courtesy Honeywell, Inc.)

opAmp. This type of conversion has been available for a number of years, and is reliable and widely used in temperature recorders.

When more accurate thermocouple measurements are desired, it is common to use *cold-junction compensation.* In this situation, a reference junction is placed between the thermocouple and the meter or readout device. The reference junction is then maintained at a constant temperature of 0°C (32°F). The only variables that remain are the cold-junction temperature and the resistance of the reference-junction readout lead wires. Special cold-junction compensators and *ice-point cells* simplify this measuring technique.

Thermistor Instrumentation

A *thermistor* is one of the simplest and most versatile temperature-measuring components available. This component, being a solid-state device, differs from its RTD counterpart by having a negative temperature coefficient of resistance. As a result, increases in temperature cause a corresponding decrease in resistance. This effect is the reverse of that in a metal that has a positive temperature coefficient.

The resistance of a thermistor is primarily controlled by the temperature of its environment. When using a specific thermistor, it is possible to predict how it will respond to a change in temperature. The reference-temperature resistance and the temperature of the environment in which the thermistor is placed must be known in order to determine the response. Predictions of this type can be achieved by using the manufacturer's resistance-temperature tables.

The process of measuring temperature with a thermistor simply involves monitoring corresponding circuit changes in current or voltage. The circuit in Figure 4-33 shows a dc energy source, a variable resistor, a thermistor, and a *microammeter*. Any temperature change that takes place around the thermistor will produce a change in current. With the meter calibrated in temperature values, direct readings can be obtained. In this type of circuit, the thermistor may be located a long distance from other circuit components without adversely affecting accuracy. Additional copper wire, for example, adds only a small amount of circuit resistance to a normally high-resistance circuit.

The variable resistor (R_1) provides calibration for the thermistor circuit in Figure 4-33 The range of the microammeter has a great deal to do with the setting of R_1. It is generally advisable to make a multipoint calibration for a circuit of this type; a typical thermistor does not ordinarily have a wide range of linearity. It is also important to have the voltage source stabilized when the circuit is used for long periods of time; this will assure proper calibration. The source voltage should be kept at a minimum value to reduce the self-heating effect of the thermistor.

Thermistors are generally used in a bridge type of circuit configuration. A bridge has improved sensitivity over the series type of circuit. Bridges can be energized by either an ac or dc power source. They have a voltmeter or ammeter indicator and four resistance arms. The sensitivity

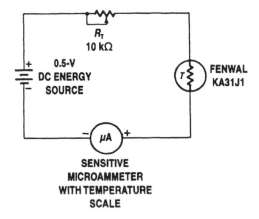

Figure 4-33. Thermistor temperature-measuring circuit.

of the indicator determines the temperature-range capabilities of the circuit. In some cases, a full-scale deflection of the meter may correspond to a reading of only it.

Figure 4-34 shows a simple four-arm bridge in which the thermistor forms one of the arms. A microammeter or millivoltmeter is used as an indicator. The variable resistor is used for balancing the bridge. The bridge may be balanced to null at any temperature within the operating range of the thermistor. When the bridge is nulled, the following formula applies:

$$R_1/R_3 = R_2/R_4$$

The value of R_2 could then be determined by transposing the formula so that:

$$R_2 = R_1 \times R_4/R_3$$

If R_3 and R_4 are of equal value, R_4/R_3 equals 1. This means that the setting of R_2 directly indicates the resistance of the thermistor. Thermistor resistance for any temperature can be accurately determined if a precision adjustable resistor is used for R_2. A *decade resistance* box or calibrated slide-wire resistor is normally used for this purpose. The null resistance can then be converted to temperature units, in degrees. In a commercially prepared thermistor bridge, the resistance setting of R_2 may be graduated

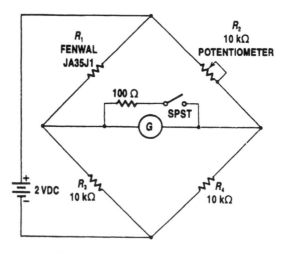

Figure 4-34. Thermistor bridge.

directly in degrees. Temperature values are taken directly from the setting of the resistor.

When two thermistors are used in a bridge, differential comparisons can be made. If the two thermistors are placed in different arms of the bridge, a greater circuit imbalance occurs with each change in temperature; so if matched thermistors are used, it is possible to detect temperature changes as small as 0.0005°C. Figure 4-35 shows a two-thermistor bridge circuit.

When a thermistor is used to drive a device that achieves some form of measurement, it generally necessitates amplification. In this case, the thermistor is being used to get data, such as a change in temperature, to achieve control of a circuit or load device. A simplified thermistor control circuit is shown in Figure 4-36. When the circuit is energized by the switch, resistor R_1 must be adjusted to produce a null indication on the milliammeter. This calibrates the circuit to ambient temperature. Grasping the thermistor by hand should cause an imbalance and produce a current reading on the milliammeter; removing the hand will cause the circuit reading to return to a null state. As a rule, any changes in thermistor resistance will cause a response in the output of this circuit. Resistor R_2 could be exchanged for the dc input of a solid-state relay and provide control of a substantial load.

Thermistor temperature-measuring circuits are inherently sensitive, stable, and fast responding. They require rather simple circuitry, lead

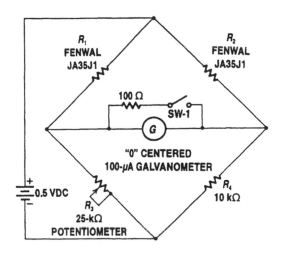

Figure 4-35. Differential thermistor bridge.

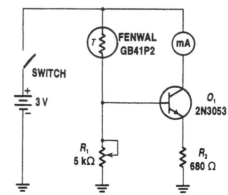

**Figure 4-36. Thermistor con-
trol circuit.**

length is not a significant problem, and device polarity does not effectively alter circuit operation. In addition, a thermistor does not require reference temperatures or cold-junction compensation, and it is rather inexpensive. Its disadvantages are nonlinearity over wide spans and instability for temperatures in excess of 200°C. The ability of a thermistor to produce changes in resistance that are almost entirely a function of temperature makes it a vital measuring device.

RADIATION PYROMETERS

Radiation pyrometry refers to a method of measuring the temperature of an object by the amount of thermal energy radiated from its surface, Through this method of measurement, temperatures can be determined without direct contact with the object. A special type of optical system is employed that collects visible and infrared energy and focuses it on a detector element, which changes this concentrated energy into an electrical signal. The signal is then amplified and applied to a readout or display element, which indicates temperature by meter deflection, a chart recording mechanism, a digital display, or a computer terminal. Figure 4-37 shows a diagram of a radiation pyrometer system.

The energy detector of a radiation pyrometer frequently employs a device known as a *thermopile.* Technically, a thermopile is a number of discrete thermocouples connected together in series. Thermal energy is focused through an optical lens to the center of the thermopile. The composite output of this device is a dc voltage that is directly proportional to

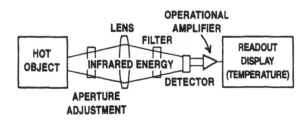

Figure 4-37. Simplified radiation pyrometer temperature-measuring system.

the amount of thermal energy falling on the surface.

A radiation pyrometer can measure temperatures between 70°F and 3500°F (21°C to 1926°C) with a target as small as 6mm. A diagram of such a unit is shown in Figure 4-38. A special *photon detector element* is used in this unit to change thermal energy into dc electrical energy. The signal is interrupted 1,380 times per second by a chopper, which changes the dc into an ac signal. This signal is then amplified by a high-gain operational amplifier and applied to a *demodulator*. Another chopper signal, with pulses in synchronism with the original signal, is applied to the demodulator input. The demodulator output is a dc signal that is applied to a meter or readout assembly. Instruments of this type are ideally suited to measure temperature in furnace atmospheres, of moving objects in rolling mills, or those beyond the range of a conventional temperature-measuring instrument.

Infrared Thermometry

A number of unique advances in optoelectronic technology have resulted in a variety of *infrared (IR) thermometers* for industrial and scientific use. This type of instrument permits noncontact measurement of hot surfaces, moving devices, and inaccessible areas at remote locations. It will measure high and low temperatures at a short distance and can be adjusted for long-distance targets. Measurement has an accuracy of within 1 percent for temperatures up to 5400°F (3000°C), with a response time of 250 ms. Most units of this type have a laser-sighting option that makes it easy to pinpoint small targets at a distance in low-light conditions.

The operational theory of an infrared thermometer is an optoelectronic principle. Energy is emitted by all objects having a temperature greater than absolute zero. The emission of energy increases as the object gets hotter. This permits the measurement of temperature by determining the amount of emitted energy from the surface of an object. The emitted energy is radiation in the infrared portion of the electromagnetic spec-

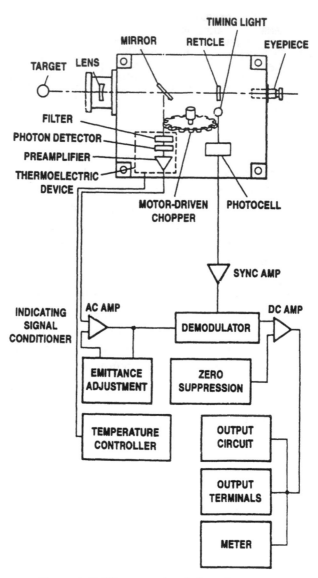

Figure 4-38. Diagram of an infrared pyrometer.

trum. A hand-held version of a radiation pyrometer with a digital readout is frequently used. This device typically uses a special *vacuum-deposited ultrastable IR sensor* of the *complementary-metal-oxide-semiconductor* (CMOS) integrated-circuit family. The instrument is primarily designed to measure temperatures up to 3200°F (1760°C) through flames and heavy smoke. It often responds to energy in the range of 3.5 to 4.1 microns.

This narrow response range minimizes errors caused by reflectance from walls and flames. The temperature readings of the instrument are updated three times per second. The instrument measures temperatures of objects up to 150 ft away and will respond to a target diameter of only 1 in from a distance of 10 ft. The lightweight pistol-style thermometer is a general purpose instrument that responds to molten glass, ceramics, metals, slag, and annealing materials.

SUMMARY

Of all the manufacturing processes used in industry today, temperature is by far the most common. Over 50 percent of the measurements made in industry-related fields involve some form of temperature measuring.

In a thermal system, the energy source operates by changing energy of one form into something different. Chemical energy and electricity are typical primary energy sources for heat. The transmission path of a thermal system may be liquid, solid, or gas. Energy transfer takes place through conduction, convection, or radiation. Control of a thermal system is designed to alter the flow path of heat between the source and the load device. Controllers are usually responsible for this system operation. The load of a thermal system receives heat from the source and performs a work function. System indicators are primarily responsible for measuring temperature.

Industrial thermal-system sources usually employ fossil fuel or electricity to produce heat. Fossil-fuel systems combine fuel and oxygen to produce heat. Electrical energy is used to produce heat in the 3500°F to 5000°F range. Arc, resistance, and induction heating are produced by electricity.

Controllers are used in heat systems to achieve the control function. A controller senses system temperature and decides on the amount of heat needed to meet the demands of the operating setpoint. Controller accuracy is determined by temperature gradients, thermal lag, component location, and controller selection.

Temperature instrumentation refers to things which are primarily designed to test and evaluate system temperature. Nonelectrical instruments respond to the expansion principle when a change in temperature occurs. A filled element has liquid, gas, or vapor sealed in a closed element; and when heat is applied, the sealed material expands accordingly. Filled glass tube thermometers respond to this principle. A filled-system thermometer employs a sensing element, a pressure-responding element, and a chart or readout device. Bimetallic thermometers are nonelectrical instruments in which dissimilar materials expand at different rates and cause the element to bend or deform.

Electronic instrumentation requires electricity in order to produce indications of temperature. Thermocouple indicators respond to dc voltage generated by heat. Voltage values, in the millivolt range, are translated into temperature values which are displayed by meters, strip recorders, or digital readouts.

Thermistors are solid-state elements that have a negative temperature coefficient. An increase in thermistor temperature causes a decrease in thermistor resistance. Bridge circuits are normally used to produce readout signals.

Radiation pyrometers measure the temperature of an object by sensing the amount of thermal energy radiating from its surface. An optical system focuses infrared energy on a detector element which changes it to an electrical signal. The signal ultimately applies itself to a readout device. Radiation pyrometers are wed in general purpose applications up to 1000°F and in hot metal applications up to 3500°F. A unique feature of these instruments is that they can measure the temperature of a moving object without touching the object.

Chapter 5

Level Determining Systems

OBJECTIVES

Upon completion of this chapter, you will be able to
1. Explain how the systems concept applies to level determination.
2. Show how buoyant force is used to determine level.
3. Define the capacitance effect and show how it is used to determine level.
4. Identify some common level sensors.
5. Describe how radiation is used to determine level.
6. Explain how ultrasonic signals are used to determine level.
7. Describe how pressure is used to determine level.
8. Show how photoelectric level control is achieved.
9. Define a number of common terms used in association with level control.
10. Show how weight is used to achieve level control.

KEY TERMS

Acoustics—The science of sound passing through liquids, gases, and solids.

Analog—Naturally occurring system variables that have an unlimited set of values having negligible separation between adjacent values.

Archimedes' principle—The reaction that occurs when an object is immersed in a liquid.

Bleed nozzle—A nozzle or valve in a pneumatic system that releases spent air from the system.

Capacitor—Two or more metal plates separated by an insulating material.

Density—The mass per unit volume of a substance.

Density change principle—The effect that ultrasonic waves have when transmitted through materials that are of the same density.

Digital—A representation of naturally occurring variables by a limited number of values that have a discrete separation with no interim values.

Domain theory—A theory of magnetism which states that groups of atoms, produced by the movement of electrons, align themselves in groups (domains) in magnetic material.

Gamma rays—Radiation that comes from changes in the nucleus energy of atoms; usually occurs as a result of nuclear collision or radioactive changes.

Geiger-Mueller tube—A gas-filled tube that detects the presence of radioactive substances by means of ionizing particles that set up momentary pulsations in the gas.

Head   The height or depth that a liquid extends above the point where atmospheric pressure is at its lowest.

Ion—An atom that has gained or lost electrons, permitting it to take on a positive or negative charge.

Level—The amount of material in a bin, tank, or hopper.

Linear variable differential transformer (LVDT)—A transformer with a movable core that produces a value change in its output voltage.

Load cell—A transducer that determines the weight of large storage tanks or containers.

Oscillator—A self-excited active electronic device and circuit whose output voltage is a periodic function of time.

Photocell—A device that releases electrons when subjected to light energy.

Photoconductive cell—A device whose internal resistance changes with the application of light energy.

Purge bubbling A pressure measurement procedure that indicates the pressure needed to force gas or air into a liquid at a point beneath the surface of the liquid.

Radar—A system for locating distant objects by the use of radio waves.

Radiation—The transfer of energy by electromagnetic waves, or the emission of particles or waves by radioactive bodies.

Resonant frequency—The natural frequency of a transmitter, crystal, or an industor-capacitor circuit.

Sensor—A device that responds to a physical stimulus and converts it into an output signal.

Specific gravity—A ratio between the density of a liquid to that of an equal volume of water, or the density of a gas to that of air.

Tare—The ratio of dead weight to full or capacity weight.

Transducer—A device that receives information in one form and converts it into another quantity.

Ultrasonic—Vibrations or frequencies above the range of human hearing.

Volume—The space occupied by a given amount of material.

INTRODUCTION

The measurement of liquid and solid material levels in tanks or vessels is a process that applies to nearly all industry, no matter what product is being manufactured. The level of material in a tank, bin, hopper, or other container indicates the amount of material available for accomplishing a particular process. This information is used to determine the length of time in which a production run can occur. The information is also used to calculate the amount of material that is consumed during a particular manufacturing process.

Level determination, like other process applications, requires measurement before any control function can be implemented. Level measurement is somewhat unique because it applies to all kinds of material. Liquids, solids, or, possibly, an interface between two different materials must be detected and measured before some type of control can be achieved. The control function of a level system is usually not as complex as that needed by other process systems. Control is generally used to turn a pump on or off, depending on the demands of the system level. Control may be more complicated where continuous level monitoring is needed. Because of the importance of level measurement and control as a process, these two functions will serve as the control concept for this presentation.

THE SYSTEMS CONCEPT

The system framework that was used to describe other processes is unique when applied to level determination systems. In a strict sense, level determination is often considered to be a control function that could apply to any one of several different processes. When viewed as a system, it must incorporate a source, path, control, load, and at least one indicator. The system can respond to such things as milk, gasoline, oil, solvent, plastic granules, or pieces of coal. The primary purpose of the system is level determination, while the actual process being manipulated is somewhat incidental.

In order to demonstrate the uniqueness of level determination as a system, refer to the solvent distribution system in Figure 5-1. In this system, the source is a bulk storage tank situated at a convenient location in the industrial facility. The tank serves as a reservoir that stores a large quantity of solvent. As a system source, the reservoir represents potential energy that can be released whenever the demand for it arises. When released, solvent flows through a path of connecting pipes or tubes, and control of solvent flow is automatically achieved by the controller. The process tank ultimately represents the load of the system. Work is achieved by the load when it releases solvent to the output network distribution path during operation. Indicators may be attached to any part of the system and are used to measure and evaluate system operation.

Level determination systems have a number of unusual features when compared with other process systems. As a general rule, level determination is a static process. System pressure, for example, maybe the result of storage-tank height and size instead of an operating pump. Head pressure, in this case, can be quite high.

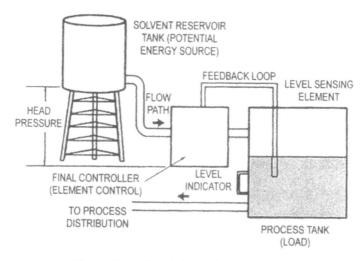

Figure 5-1. A level determination system.

The load part of a level system is also unusual, because the material may be dormant for long periods. The nature of the material being processed has a great deal to do with some of the other unusual features of this type of system. Solid materials, for example, pose many problems that are not prevalent in liquids.

In this investigation of level determination systems, attention will be focused particularly on the function of control. Through this function, a large part of system automation is achieved. Such things as measurement, evaluation, system controllers, and final control-element operation are included in this study. In general, level determination systems represent a major part of all process considerations.

LEVEL DETERMINATION EQUIPMENT

The quantity of material used by industry that requires some form of level determination is so great that it makes this process an essential manufacturing operation. There are two classifications of equipment utilized in level control operations. The first is systems that control liquid level. The second group is made up of those that control solid material. In some cases, there is overlapping of the basic operating principles used in this equipment. In practice, solid level determination tends to lean toward weight, pressure, ultrasonics, and radiation techniques. Liquid levels are determined by these techniques plus several others. Because of this diversity, this discussion will be directed toward all level measuring and control techniques. When something applies to only one type of level measurement, this will be indicated.

Equipment used to determine the level of a storage tank is based on the operating principle of the sensing element. This element is responsible for detecting a change in level and generating a signal that is used to correct the problem. In general, level systems will respond to either mechanical, pneumatic, electrical, radiation, or ultrasonic information.

The selection of a specific method of control is often narrowed down by basic considerations. These include level range, the type of material involved, operating pressure, accuracy, cost range, construction techniques, and temperature. After a control method has been selected for a specific application, it is essential that the person responsible for its operation or maintenance be familiar with its limitations.

LEVEL MEASURING TECHNIQUES

The first and most significant part of level determination is measuring technique. Level must be measured and evaluated before it can be decided whether a control action is needed. In practice, level is either measured directly or inferred by some indirect action. The direct method uses an actual physical change in the material itself to obtain a measurement. An inferential method, by comparison, employs some outside variable, such as pressure, to indicate changes in level. Typically, mechanical mechanisms, pressure gauges, and diaphragm elements. are used to achieve this type of measurement, and mechanical, pneumatic, and electrical controllers are energized by these measurement techniques.

The Buoyant-force Principle

Buoyant-force level sensors respond to Archimedes' principle, which was discovered around 250 B.C.: *A body placed in water is buoyed up by a force that is equal to the weight of the water it displaces.* The body, in this case, is called a float. In effect, the float is a liquid-level to mechanical-motion transducer.

Float Level Sensors

In general, the float-mechanism element rides on the surface of the liquid in a tank as a means of detecting the liquid's level. If the float passes a predetermined switching point, a mechanical action signal is generated. In practice, float level changes are commonly used to control electrical-switch contacts or pneumatic-valve operation. This action is of the *digital type:* It turns a liquid control valve on when the level is low or off when the tank is full.

Figure 5-2 shows a cutaway view of a *permanent-magnet* float-actuated liquid-level switch. In this unit, one permanent magnet is attached to the float, which rises and falls with changes in liquid level. A second permanent magnet is mounted in the switch head. The magnets are positioned so that like poles face each other through a nonmagnetic diaphragm. Changes in liquid level will cause the float to follow accordingly. When the float magnet is positioned close to the head magnet, repulsion occurs. This causes the switch to have a snap action, which can he used to turn a pump on or off.

A permanent-magnet float-actuated liquid-level switch with pneumatic control is shown in Figure 5-3. A rise in liquid level will cause the float to move in an upward direction. This action causes the float mag-net to repel the head magnet downward. As a result, air supplied to inlet valve A passes into outlet C and is coupled to a control valve or pressure switch that stops the flow into the tank.

A decrease in liquid level causes the float to move to the position indi-cated in the diagram. In this position, inlet supply air is shut off. This causes vent B to open and exhaust control-valve or pressure-switch air into the atmosphere. Through this operation, there is no significant loss in air other than that exhausted during the changeover action. The maximum supply pressure that can be handled by this type of controller is 100 psi.

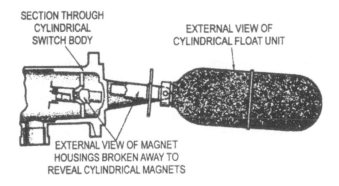

SECTION THROUGH
CYLINDRICAL
SWITCH BODY

EXTERNAL VIEW OF
CYLINDRICAL FLOAT UNIT

EXTERNAL VIEW OF MAGNET
HOUSINGS BROKEN AWAY TO
REVEAL CYLINDRICAL MAGNETS

Figure 5-2. Permanent-magnet float-actuated liquid-level switch. (Courtesy Bestobell/Columbia Controls)

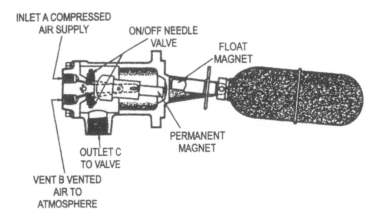

INLET A COMPRESSED
AIR SUPPLY

ON/OFF NEEDLE
VALVE

FLOAT
MAGNET

PERMANENT
MAGNET

OUTLET C
TO VALVE

VENT B VENTED
AIR TO
ATMOSPHERE

Figure 5-3. Permanent-magnet pneumatic liquid-level control. (Courtesy Bestobell/Columbia Controls)

A *unifloat* liquid-level controller that operates on the magnetic-float principle is typical. In this controller, a single float travels up and down the entire length of a guide tube with the rise and fall of liquid. A magnet guide tube of any practical length can be attached to the unit. Magnetically latched *reed switches* are placed at desired levels in the guide tube. As the magnet passes a given control level, it magnetically actuates a reed switch which energizes externally mounted relays. With this type of unit, level adjustments can be easily achieved by altering the position of a common support wire *in* the guide tube. This type of controller is versatile, reliable, can be adjusted, and is quickly installed. It is used in a broad range of applications requiring accurate multiple-function/multiple-level control for nearly any type of liquid.

Displacer Level Sensors

The *displacer* level sensor is different from the float type of level sensor. Displacer elements, for example, respond to a force that is equal to the weight of the liquid displaced by the sensing element. Sensors of this type employ a probe that is submerged in the liquid.

Figure 5-A shows an illustration of the displacer principle. In Figure 5-4(A), the buoyant force F_B is quite small and the displacer-weight force F_W is large. This condition causes the connecting cable-spring assembly to be fully loaded by the weight of the displacer.

In Figure 5-4(B), the action is reversed. Buoyant force is increased due to the rise in liquid level and the displacer weight is reduced. As a result of this level change, the connecting cable-spring assembly has reduced loading. Less deflection of the indicator hand demonstrates the physical change that has taken place.

The physical change in force that occurs in the displacer element can be used to indicate liquid-level changes or to actuate a controller. *Displacer controllers* respond to a two-state or digital switching action. Some advantages of the displacer over the float type of controller element are an extended range of level control, immunity to surface turbulence, adjustable level spans, and reduced influence of fluid density.

A displacer-actuated, pneumatic liquid-level controller is shown in Figure 5-5. Force developed by the displacer of this unit is transmitted to the pneumatic relay by a constantly engaged magnetic field. The transmitted force is dependent on the amount of liquid displaced. The resulting output is an increasing or decreasing linear signal.

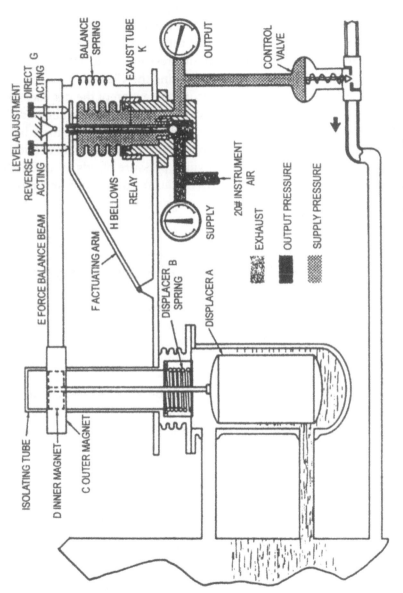

Figure 5-5. Displacer-actuated pneumatic liquid-level controller. (Courtesy Varec Div, Emerson Electric Co.)

In operation, an increase in liquid level unloads the displacer spring. This change in force is then transmitted to inner magnet D and through the magnetic field to outer magnet C. Force balance beam E continues the transmission process by applying force to actuating arm F, which compresses bellows H and causes the exhaust tube to open the air inlet port. This immediately causes an increase in air pressure to the control valve, which restricts liquid flow to the tank by seating the valve.

When a decrease in liquid level occurs, the action of the displacer is reversed. Displacer spring tension is increased, which forces magnets U and C downward. This action is again transmitted through the force balance beam to the top of the bellows. In this case, it allows the exhaust tube to lift from the ball. As a result, control-valve pressure bleeds off through the exhaust. This action causes the valve seat to open, admitting more water into the tank.

Capacitance Level Sensors

A capacitor is two or more metal plates separated by a dielectric or insulating material. A capacitor probe made of two metal conductors can be used to detect liquid-level changes. The liquid serves as the dielectric that changes the value of the capacitor probe as it changes level. The liquid must be nonconductive or have a low dielectric constant.

Figure 5-6 shows a cross-sectional view of a capacitor probe used to determine the liquid level of materials with a low dielectric constant. When liquid enters the probe, the capacitance increases. The change, in this case, occurs because the dielectric constant of the liquid is greater than air. A probe exposed to air with a dielectric constant of 1.0 could have a normal capacitance of 10 pF. When the same probe is completely immersed in a liquid, such as chlorine, with a dielectric constant of 2.0, its capacitance changes to 200 pF. If the probe is only half immersed in the same liquid, it will have a capacitance of 150 pF. Changes in capacitance can be equated to linear changes in liquid level. A *capacitance bridge* excited by frequencies of 500 kHz or 1.5 Mhz is normally used to determine capacitance, which is then translated into feet or meters.

When conducting liquid levels that are to be measured by the capacitance method, there is a decided change in the installation. A single insulated conductor probe serves as one plate of the capacitor, while the metal tank serves as the other plate. This change is shown in Figure 5-7. Level changes in the conducting material inside the tank again produce a change in capacitance, which still increases linearly with level.

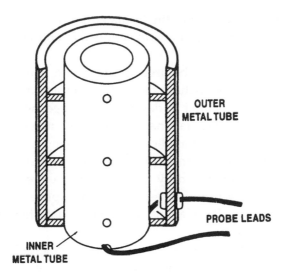

Figure 5-6. Partial cutaway view of a capacitor level-sensing probe.

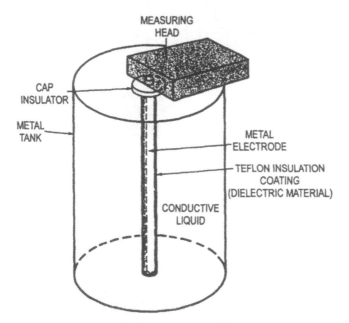

Figure 5-7. Single-probe capacitance sensor.

An interesting feature of the insulated-probe type of installation is that it can be used to measure both conductive and nonconductive liquids.

Capacitance can also be used to determine liquid levels without touching the liquid. Installations of this type employ the *proximity* probe in Figure 5-8, which is used to detect level changes where the liquid is above 800°F or where it contaminates or coats an immersed probe.

In a proximity-sensor application, level changes produce a capacitance change between the sensor plate and the liquid surface. In practice, capacitance changes detected by the sensor probe are used to produce changes in frequency values. These values are expressed in level indications of feet or meters. Installations of this type usually have a digital readout as a level indicator.

The advantages and disadvantages of capacitance level determination should be obvious after the preceding discussion. On the plus side, there are no moving parts and the installation is simple. The proximity

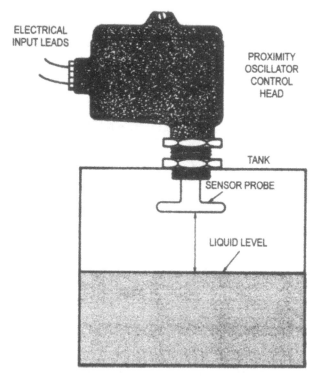

Figure 5-8. Proximity capacitance detector.

unit is excellent for hot liquids because it does not make contact with the liquid. The disadvantages are that accuracy is affected by material characteristics and probe coating. Also, some rather sophisticated equipment is needed to detect changes in capacitance. This method of level determination is somewhat expensive and should only be used where its selection can be adequately justified.

Electrical-conductivity Level Determination

When a liquid is a fair electrical conductor, it can be used in an *electrical-conductivity* level determination system. In practice, when the liquid of this system has a resistance of less than 20 mΩ/cm, a sufficient amount of current can be made to pass through it to actuate a sensitive relay without amplification. Through amplification, liquid resistivities of 20 MΩ/cm can be used to control liquid levels. Systems of this type employ one or more fixed level-sensing electrodes mounted on a suitable holder in a tank or vessel containing liquids. When the level of the liquid reaches the electrode, an electric current passes between the electrode and the metal tank or a second *electrode*. This current is used to actuate a pump relay or control valve.

To see the operation of an electrical-conductivity system, refer to the relay-actuated solenoid valve circuit in Figure 5-9. In Figure 5-9(A), one tank is at a low-level position. When this occurs, the liquid has dropped below the longer electrode and the circuit path between the longer electrode and the metal tank or ground has been broken. With no current, the relay drops out and the filling solenoid valve is energized. The tank is then refilled.

When the liquid level of the tank reaches its upper limit, as in Figure 5-9(B), the solenoid valve is shut off by the relay. The high-level control point of the system is established by the shorter electrode. When the liquid reaches this level, conduction between the two electrodes takes place and the relay is again energized. This, in turn, shuts off the solenoid valve and the filling operation stops. The circuit remains in this state until the lower-level point is exceeded, which again starts the filling operation.

A pump-down control application of the electrical conductivity type is shown in Figure 5-10. In this application, assume that the solenoid filling circuit in Figure 5-9 is used to fill the tank to its upper level. The tank will stop its pump-down operation when it reaches the low-level limit. The filling operation is left out in this application to simplify the pump-down circuit.

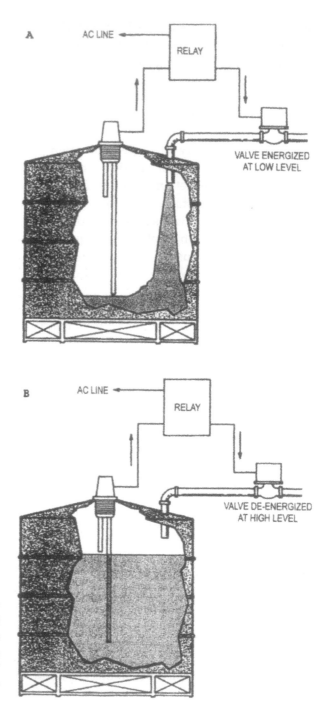

Figure 5-9. Electrical-conductivity level control system: (A) Low level; (B) High level. (Courtesy B-W Controls, Inc.)

Note the low-level condition in Figure 5-10(A) as a starting point. At this level, the pump-down operation has stopped. When ac line voltage is applied to the primary coil of the relay, a magnetic flux is established in the relay coil. Voltage is induced in the secondary coil, which is attached to the electrode circuit. No current flows in the electrode circuit because points 8 and 9 remain open. The pump-down motor is likewise off because of contacts 6 and 7. The circuit remains in this state until the tank is refilled.

When the tank is filled, as indicated in Figure 5-10(B), the pump-down operation begins. Liquid-level contact between the pump start and pump stop electrodes completes the secondary circuit of the relay. Current in the secondary coil sets up a bucking action, which diverts the magnetic field to the lower part of the core. The relay armature is immediately attracted to the lower part of the core, and contact points 8 and 9 are closed. This latches the relay through the liquid, the ground, and the stop electrode. The pump-down motor starter circuit is completed at the same time through contacts 6 and 7. Tank pump-down continues until the liquid drops below the pump stop electrode. When this occurs, the latch circuit is broken and the relay armature drops out. The pump-down operation stops until the tank is refilled.

Liquid-level control by conductivity has numerous applications in food, beverage, dairy, drug, and chemical process industries. Its advantages are low installation cost, simplicity, no moving parts, and its ability to detect even moist bulk solids. Disadvantages include possible sparking, electrolytic corrosion of electrodes, and electrode coating with some materials. In general, this method of level determination is suited to a wide range of industrial process control applications.

Radiation Level Determination

A fairly recent addition to level determination equipment utilizes the emission of gamma rays from a radioactive source, a detector, and a transmitter/indicator. Several different radiation level determination installations are shown in Figure 5-11.

The source of a radiation level determination system may be mounted outside of a tank at a single point, in an external-strip source lead, or mounted inside thick-walled tanks. The level of a tank determines the amount of radiation from the source. Note that there is no electrical connection between the source and the detector element. If the radiation source of a system is placed at the bottom of a tank and the detector at the top, detector signal levels increase when the liquid drops to a low level.

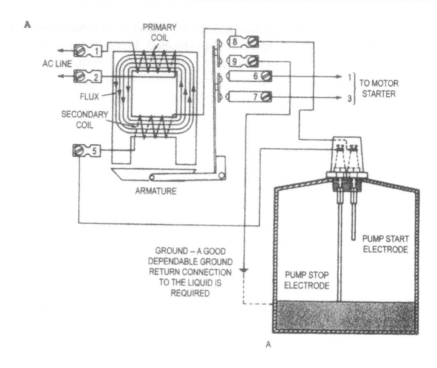

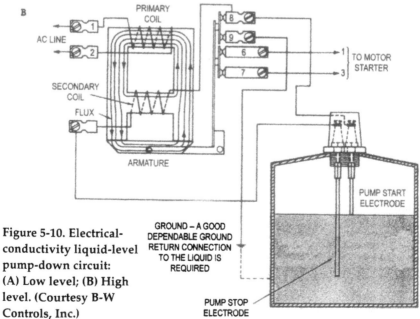

Figure 5-10. Electrical-conductivity liquid-level pump-down circuit: (A) Low level; (B) High level. (Courtesy B-W Controls, Inc.)

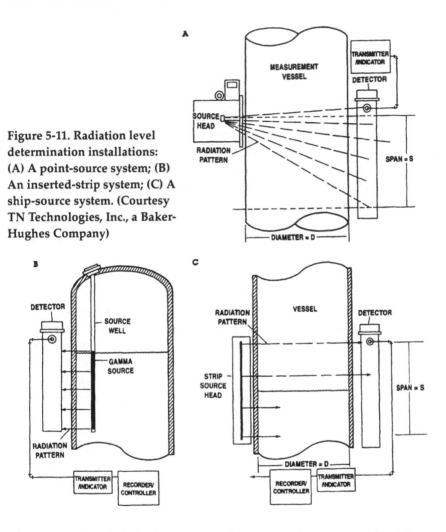

Figure 5-11. Radiation level determination installations: (A) A point-source system; (B) An inserted-strip system; (C) A ship-source system. (Courtesy TN Technologies, Inc., a Baker-Hughes Company)

The output signal of the detector can then be used to actuate a liquid control valve or start a pump motor. Strip-type sources mounted on a tank permit continuous (analog) level monitoring to be achieved. In this type of installation, liquid-level variations cover up different parts of the source. At low levels of liquid, less of the source is covered, causing the detector to have a higher output. At high levels, source radiation from the strip is smaller. If the span length of the source is made larger than the control range desired, the detector output will be linear.

The head of a radiation system is normally constructed of steel and filled with lead to provide radiation shielding. Lockable shutters are usually provided so that radiation can be blocked out during shipping, installation, and maintenance. Radium, cesium 137, or cobalt 60 are the common sources of radiation. The size, weight, and material selection of the source depends on the application. The radiation material selected, and its strength, is determined by the diameter of the tank, wall thickness, and the construction material. In general, radium is used where a small source is required. Cesium 137 is a very common source because of its low cost and long life expectancy. It decays at the rate of 2 to 3 percent per year. Cobalt 60 is used only in thick-walled installations where high penetration is needed. Radium, being a naturally occurring *radioisotope*, does not require Atomic Energy Commission licensing, while cesium 137 and cobalt 60 do, so the higher cost of a radium source with a half life of 1,400 years is often justified over the other two sources. All manufactured radiation source heads must meet or exceed the safety requirements of the U.S. Nuclear Regulatory Commission.

The detector or sensor of a radiation level system can be a *Geiger-Mueller tube*, an *ion chamber, or a scintillation counter*. In practice, the ion chamber seems to be the most commonly used. Detectors of this type employ a small chamber that houses an active form of a solid material, liquid, or gas. When gamma rays are applied to the chamber, they collide with the active material and produce ionization. The amount of ionization that takes place is then used as a measure of the radiation entering the chamber.

Figure 5-12 shows a thawing of a gaseous-diode ionization chamber. When ionization occurs by radiation exposure, the gas molecules in the chamber are changed into positive and negative ions. Ions and free electrons are then collected by the respective electrodes of the diode and cause a resulting current in the external circuit. At low voltage levels, the resulting current is extremely small and tends to arrive in small bursts or pulses at irregular intervals. The RC circuit placed in series with the chamber helps to maintain the output at a fairly constant level. The time constant of this circuit is normally several seconds, in order to develop a representative current. Voltage developed across the resistor can be amplified and used to drive an indicating device, actuate a motor, or turn on a signal alarm to indicate an improper level.

Level control by nuclear radiation has numerous applications in process tanks, reactors, bins, pipes, hoppers, and other types of vessels. Because of the high reliability, low radiation requirements, and the ability to operate at high temperatures, these systems are particularly useful in the

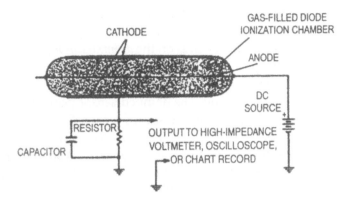

Figure 5-12. Gaseous-diode ionization chamber.

most demanding level measurements of liquids, slurries, and solids. Since the components of the system are usually external to the vessel, chemical or physical characteristics of the process material do not adversely affect system performance.

Applications of radiation-type level systems are very common in industry. This type of system has high reliability, is mostly immune to outside variables, and is not extremely expensive compared with other systems. Its main disadvantage is the misconception that radiation systems are dangerous and expensive to install and maintain. In the future, this should be dispelled, and radiation types of level systems will find widespread usage in process control applications.

Ultrasonic Level Systems

Ultrasonic level determination systems make use of a form of sound energy that lies beyond the range of human hearing. These systems utilize a combination of *acoustics* and electricity. Acoustics is the science of sound and makes use of the basic principles of sound passing through liquids, gases, and solids. The electrical function of ultrasonics deals with the techniques of generating high-frequency electrical energy.

Ultrasonic waves behave the same as audible sound waves except that they are inaudible. Mechanical vibrations move molecules of the medium into alternate states of expansion and compression. Energy in these waves can then be transferred through solids or liquids, depending on the application. Level determination of the ultrasonic type can be applied to both liquid and solid materials.

An ultrasonic level system essentially employs a transmitter or ultrasonic source and a receiver. Depending on the design of the system, these components may be independently mounted or housed in a common assembly. Operation primarily depends on the *damped-sensor principle* or the *density-change principle.*

There are numerous variations of the two basic operating principles of ultrasonic level determination. The damped-sensor principle is simple because it requires only an ultrasonic source sensor. Figure 5-13 shows sensors attached to four different tank locations. The sensor probe, in all cases, has a piezoelectric crystal head (see following section). With high-frequency voltage applied, the crystal head vibrates at an ultrasonic frequency. In position A, the head is in the vapor zone of the tank and vibrates normally. In a switching application, this sensor would stop vibrating or be damped when the liquid reaches it. This effect could be used to trigger an output signal, stop a pump, or activate an alarm. Installation B shows an exterior-mounted sensor head. When the liquid level rises to or above this location, it damps the vibration. Levels below the sensor do not change its natural vibration. Probes C and D are side-mounted sensors with exterior connections. The sensor D is damped and sensor C is vibrating. For all practical purposes, the damped-sensor principle is best

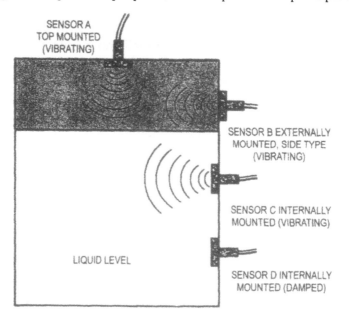

Figure 5-13. Mounted ultrasonic sensors. (Courtesy TN Technologies. Inc., a Baker-Hughes Company)

suited for liquid-level control applications. The liquid, in general, should be relatively clean.

The density-change principle takes advantage of the fact that ultrasonic waves behave the same when transmitted through materials that are of the same density. When a wave reaches a pronounced change in density, it is reflected. This principle is used in radar and sonar systems and is technically called *echo ranging*. This type of system is more complicated than the damped-sensor principle because it requires a transmitter and a receiver in order to function. There are systems with independent transmitter and receiver units, or these may be combined in a single sensor. Piezo-electric crystals are employed in both the transmitter and the receiver.

Figure 5-14 shows three combination transmitter/receiver sensors. The gap separating the transmitter and the receiver is an important part of the unit. In this case, there is no transfer of the ultrasonic signal unless the gap is filled. Once the gap is filled, there is a transfer of energy from transmitter to receiver. An output signal from the receiver is then amplified and ultimately applied to the final control element of the system. Relays, pump motors, alarms, recorders, indicators, or pneumatic-actuated valves can be controlled by the output signal.

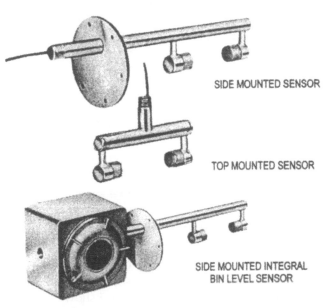

SIDE MOUNTED SENSOR

TOP MOUNTED SENSOR

SIDE MOUNTED INTEGRAL
BIN LEVEL SENSOR

Figure 5-14. Ultrasonic transmitter/receiver probes.

Piezoelectric Transducers. The primary reason why piezoelectric transducers vibrate when energy is applied is somewhat complex. The most accepted theory among scientists, however, applies to the action of discrete areas called *crystal domains*. When no electrical voltage is applied to the crystal, these domains are aligned in a random manner. As a result, no net mechanical change appears across the crystal. When voltage is applied, the domains tend to align in a specific order according to the polarity of the applied energy. This action causes a shock wave or vibration to be generated within the crystal. The wave then causes a violent change in the shape of the crystal. Repetitive polarity changes from an ac energy source cause these vibrations to occur at a corresponding rate.

The physical size, shape, and material of a crystal are primarily responsible for establishing its natural vibrating frequency. In a transmitter, the natural crystal vibrating frequency is called the *resonant frequency*. The resonant frequency of a crystal represents the frequency at which the most violent physical changes occur when voltage is applied. The natural frequency of the system is determined by the structural cutting of the crystal.

A receiving crystal is essentially the same as a transmitting crystal. Its operation is dependent on a signal sent out by the transmitter. When vibrations from the transmitter crystal strike the receiver, vibrations occur in the receiver. These vibrations, in turn, are used to generate a very small ac signal. This signal is then amplified and used to actuate the final control element of the system. The transmit/receive principle is similar to the sympathetic vibrations that are set up in two tuning forks of the same frequency when one fork is placed into vibration.

Sludge-level Control Systems

Ultrasonic level control systems are often used to determine the amount of sludge in a clarifier or settling tank. This application calls for a special sensor probe. An ultrasonic sludge probe, similar to the top-mounted sensor in Figure 5-14 can be used for this operation. This probe employs a combination transmit and receive sensor with a specially designed sludge gap.

When the sensor is placed in a tank, the transmitter sends out a signal that is picked up by the receiver. In thin clear liquids, there is less signal attenuation than in thicker liquids holding solids. Attenuation increases with the sludge buildup. This change in signal strength can be detected and used to control a sludge pump. Figure 5-15 shows a block diagram of this type of sludge-level control system.

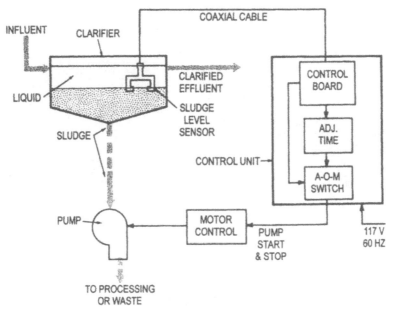

Figure 5-15. Block diagram of a sludge-level control system.

Ultrasonic Bin-level Systems

Ultrasonic signals are frequently used to determine the levels of dry products and solids. In this application, an ultrasonic signal is sent out by the transmitter and picked up by the receiver. When the beam is interrupted in some way by the presence of the product, it initiates a signal change. The sensor element of this system can be mounted on silo walls as thick as 6 in. Figure 5-16 shows some representative sensors that are used in ultrasonic bin-level systems. This type of installation calls for only a sensor and a control unit. The gap in the sensor is approximately 4 in (to cm). This gap is much wider than in sensors used to determine liquid levels. Units with this type of gap are somewhat immune to dust and dry-product coating, and can withstand up to 50 psi of bin pressure.

In general, the ultrasonic technique of level measuring has proved to be a reliable and effective method of control. Its advantages include no moving parts, inexpensive installation and maintenance costs, and the possibility of noncontact detection. Accuracy for a large tank is ±1 percent or within 2 cm. Both two-state switching and continuous control are possible with ultrasonic level measuring units. Measurements are unaffected by variations in material composition, density, and thermal or electrical

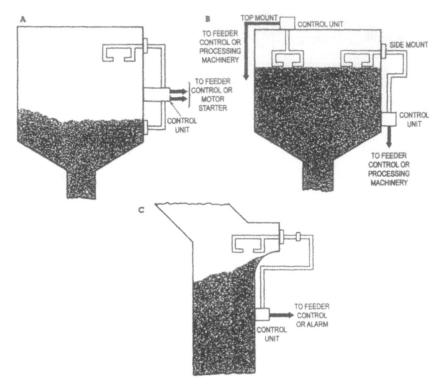

Figure 5-16. Ultrasonic bin-level applications: (A) Automatic high-low sensing; (B) High-level sensing; (C) Jam control.

conductivity. Ultrasonic units, in general, have great flexibility and numerous applications.

Bubbler Level Systems

The *bubbler* or *purge* method of measurement is probably the oldest and one of the simplest methods of liquid-level determination. In this type of system, measurement is based on the fact that the pressure of liquid at a reference point is directly proportional to the height of the liquid above this point. A *dip tube is* installed vertically so that its open end is placed near the bottom of the tank being measured. A regulated air supply applied to the dip tube is adjusted so that it produces air bubbles from its open end. The amount of air pressure needed to produce this action corresponds to the head pressure of the liquid at the bottom of the tank (see Figure 5-17).

To make level measurements by the purge system, the air supply regulator must be adjusted so that its pressure is slightly greater than the

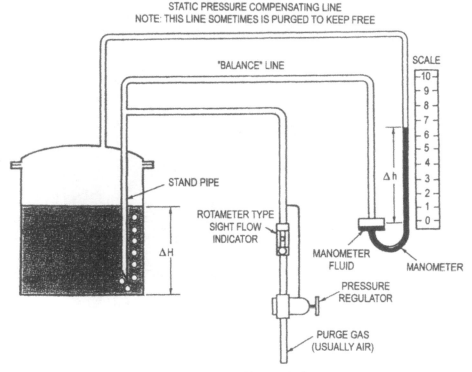

Figure 5-17. Bubbler-level/purged-air system.

liquid head pressure. The air pressure gauge is normally calibrated in feet, inches, or meters. The height of the liquid (in ft) is equal to the pressure (in psi) divided by the density of 1 ft^3 of the liquid.

Purge systems can also be used to measure the level of pressurized tanks. In systems of this type, the process becomes somewhat more complicated. The actual liquid level now becomes a difference measurement between head pressure and tank pressure. An equalizing line must be added to the top of the tank, and level is measured by a *differential pressure gauge*.

Figure 5-18 shows a diagram of a level bubbler unit with some innovations over the older, more conventional unit. When a pressure pipe is inserted into the liquid and air or inert gas is supplied to line B, it purges the pipe. Pressure in the pipe will only increase until all of the liquid is evacuated. Any additional air will bubble out freely through the opening at the bottom of the pipe. The head pressure that remains returns to the well through line A. This pressure raises the indicating fluid column to the depth of the liquid.

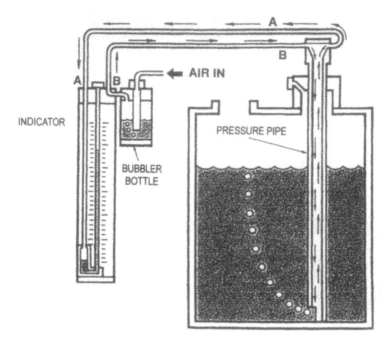

Figure 5-18. Level bubbler diagram.

A diagram of a digital-indicator purged system that is attached to a closed-atmosphere tank is shown in Figure 5-19. The tank of this system is not vented to the atmosphere. This type of system uses three lines. Lines A and B have the same function as in the two-line system in Figure 5-19. The third line, labeled C, is used to separate the pressure or vacuum above the liquid in the tank from the head pressure. A transducer is added to this assembly to convert pressure into an electrical signal that is sent to the digital display.

Pressure-sensitive Level Determination

There are several indirect level measuring processes that respond to changes in pressure. A pressure gauge attached to the zero reference level of the tank can be used to determine liquid levels. Tank-level height can then be determined by dividing pressure by the density of the liquid involved. A change in level produces a corresponding change in bottom-tank pressure. In practice, the pressure gauge is normally calibrated in ft or m instead of psi. This is shown in Figure 5-20.

When corrosive or high-temperature liquids are to be measured, they should not come into direct contact with the pressure gauge. In level mea-

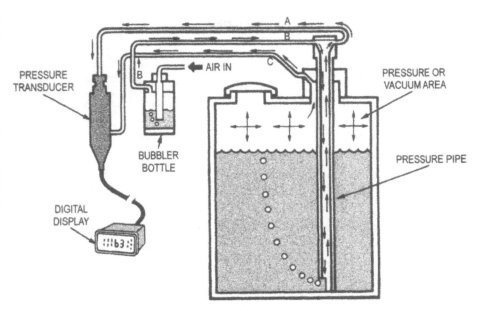

Figure 5-19. Closed-atmosphere digital liquid-level measuring system.

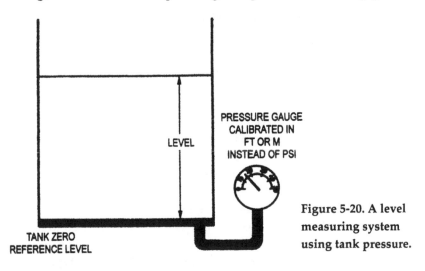

Figure 5-20. A level measuring system using tank pressure.

suring applications of this type, *diaphragm boxes (air traps)* are commonly used to reduce this problem. Figure 5-21. shows an air box or pressure trap placed in a tank. As the level of the liquid rises, the pressure of the air trapped inside of the box increases accordingly. Trapped air, in this case, forms a barrier to the liquid entering the box. As a result of this condition, only air pressure is applied to the indicating gauge. The pressure indicator

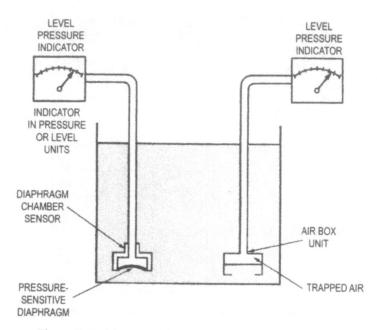

Figure 5-21. Air-trap and diaphragm level indicators.

can be calibrated in either pressure or liquid-level units.

A diaphragm box, shown in Figure 5-21, achieves liquid-level control through the transmission of air pressure by the same principle as the trapped air box. In this device, air is trapped inside of a diaphragm-enclosed chamber. As the level of liquid rises, pressure applied to the bottom of the flexible diaphragm increases proportionally. Pressure, in this case. acts upon air that is isolated from the liquid. Pressure-sensitive level determining systems of this type are normally called *filled systems.*

Pressure-switching Level Control Systems

In some level systems, pressure of a certain value is used to turn on a switch that energizes a pump motor or actuates a control valve. *Pressure-actuated* electric switches are frequently used to achieve this type of level control.

Pressure switches can be actuated by an air trap, by a diaphragm chamber, or by attaching the sensor directly to the tank. A pressure switch, can be used to achieve level control. Air pressure applied to the unit moves a diaphragm which actuates a microswitch in the assembly. Pressure switches of this type may be purchased for a specific actuating pressure. In some cases, these switches are adjustable to a limited extent. Both liquid and solid material levels can actuate pressure switches.

Weight-level Determination

Another significant indirect method of determining the level of a material is by changes in weight: An increase or decrease in level will cause a corresponding change in container weight. The tank or vessel may be weighed, mechanically or electrically, to determine level. The particle size of the material must be uniform and the moisture content must be constant so that changes in weight can be attributed entirely to changes in level.

In small container filling applications, weight control is common. In this application, the container is placed automatically on a mechanical or electronic scale and a certain amount of material weight is equated to a filled container. The end result of this operation is a filled container of a specific weight.

Large containers and, in some cases, storage tanks are also filled by the weight process. This process can be applied to both liquid and uniformly shaped solid materials. Electrical *strain gauges* and hydraulic or pneumatic *load cells* are used to achieve this sensing operation.

A strain gauge, as shown in Figure 5-22 is a resistive transducer that changes value when the surface to which it is attached is subjected to a stress. Physically, a strain gauge is constructed of a fine metal wire about 0.0001 in wide, mounted on an insulating strip. The metal used has high elasticity so that it will easily change dimensions. When subjected to a stress, the metal stretches. Therefore, the cross-sectional area of the metal is reduced and the length is increased. The resistance of the metal can be expressed mathematically as

$$R = p \times l/A$$

where

R = resistance of the conductor
p = resistivity constant
l = length of the conductor
A = cross-sectional area of the conductor

Therefore, as the material of the strain gauge is stretched, its resistance will increase because of the change in cross-sectional area and length.

Figure 5-23 shows a strain gauge attached to the elastic support

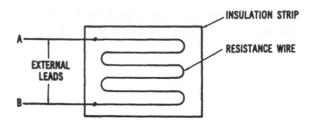

Figure 5-22. A simplification of the strain gauge.

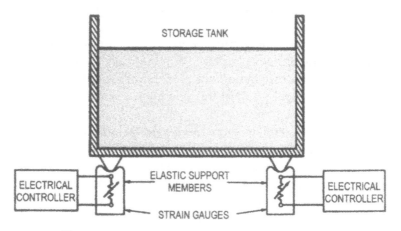

Figure 5-23. Strain gauges attached to a storage tank.

members of a large storage tank. When a dry material of uniform size fills the tank, the elastic supports change shape. Changes in weight are then transposed into resistance values which can be used to represent different tank levels.

In applications where dry materials of a uniform size are loaded into a tank, there may be limes when nonuniform levels occur. To reduce this problem, matched strain gauges are attached to each elastic support member. By averaging individual resistor values or determining the total resistance of a parallel strain-gauge circuit, the level can be determined.

Another electrical transducer that may be used to determine the level of a container according to changes in weight is the *linear-variable differential transformer* (LVDT). The operating principle of this device is illustrated in Figure 5-24. A movable metal core is placed within an enclosure that has three windings wrapped around it. The center winding, or primary winding, is connected to the ac source. The two outer windings have voltage

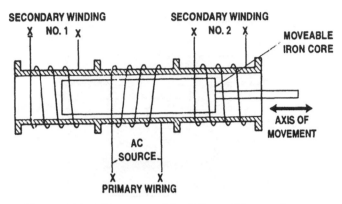

Figure 5-24. A linear variable-differential transformer.

induced from the primary winding. When the movable core is placed in the center of the enclosure, the voltages induced into the two outer windings are equal. Any movement of the core, in either direction, will cause one induced voltage to increase and the other to decrease. It is possible to measure the difference in voltage induced in the two outer windings in terms of the amount of movement of the core. The variation in flux linkage due to the movement of the metal core is responsible for the change in induced voltages. A linear movement that is caused by something such as a change in container weight can be converted into an electrical signal. As a rule, small container level control applications utilize the LVDT while larger tanks respond to strain-gauge sensors.

Load-cell Weight Transducers

Load cells are used as transducers to determine the weight of large storage tanks in level control applications. A hydraulic load cell is shown in Figure 5-25. In this device, weight applied to the loading head produces a change in internal fluid pressure. The hydraulic output of the load cell may be connected to a Bourdon tube controller or to a pressure gauge. The load cell is essentially part of a fluid-filled type of control system.

Hydraulic load cells are self-contained units that do not require an outside source of energy to operate. They tend to be stable under varying temperature conditions, provided that temperature extremes do not alter the operation of the diaphragm material. Pressures of 800 to 1,000 psi are typical for the cell shown in Figure 25. When higher-pressure cells are needed, or when they are to be subjected to temperature extremes, special all-metal load cells should be used.

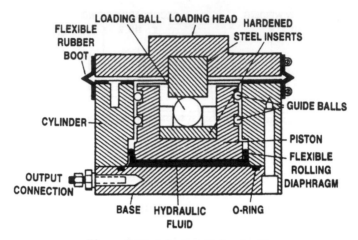

Figure 5-25. A hydraulic load cell.

Pneumatic load cells are also used in level control applications that respond to container weight. Cells of this type respond to a total weight that is somewhat less than that used by comparable hydraulic units. These units provide a high degree of accuracy, which is important in small container level-filling applications.

Pneumatic level control systems using load cells are well suited for food and drug filling applications. They are inherently explosion-proof and rather insensitive to extreme changes in temperature. In the event of a cell rupture or line leakage, no contaminating fluid would be injected into the filling medium. Figure 5-26 shows a cross-sectional view of a typical pneumatic load cell.

A pneumatic load-cell system requires a carefully regulated clean air supply from an outside source in order to operate. This device then responds to the force-balance principle. With no load applied, air supplied to the net load weight chamber will force the bleed nozzle up and open. This exhausts the air into the atmosphere. The output pressure goes to zero or to its lowest level. With the weight of the load applied, both diaphragms are forced to move down accordingly. This action forces the bleed nozzle to close a bit, less supply air is exhausted, and the output pressure in the net load weight chamber increases. Through this type of action, changes in load weight are reflected as variations in output air pressure.

The remote tare-control valve of the load cell is used to compensate for empty container weight and full capacity weight. Tare, in this case, is a ratio of dead weight to full capacity weight. In designing a system of this

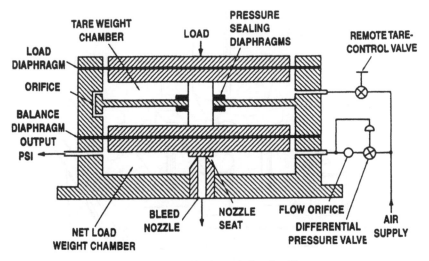

Figure 5-26. Pneumatic load cell.

type, it is desirable to keep the tare ratio as low as possible so as to avoid accuracy problems. Tare ratios in excess of 2 to 1 should be avoided in order to produce a satisfactory output indication.

The output of a pneumatic load cell is normally attached to a fluid-filled controller. Changes in weight are used to develop variations in pressure that are ultimately used to energize the system controller. Final control of the system is some type of filling control valve or electrically actuated filling pump.

Photoelectric Level Control

Photoelectric level control systems are frequently used to fill small containers in automatic process applications. In Figure 5-27, containers passing along a conveyor line are positioned under a material dispenser. When the container moves to the correct position, the actuator opens the dispenser unit, which initiates the filling operation. When the filling material reaches the level of the light source, the light beam is interrupted. With no light striking its surface, the detector actuates a relay in the dispenser. This stops the filling operation and permits the next container to align itself in the proper location. Level control applications of this type are found in food and beverage industries.

A photoelectric level control system essentially employs a light source and a detector or sensor. Light energy passing from the source to the detector makes the system complete. As a result of this condition, a

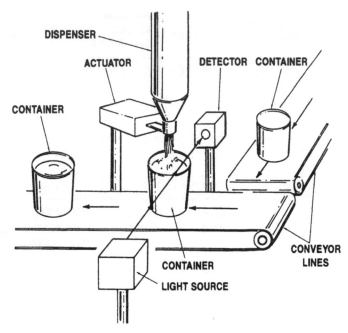

Figure 5-27. A photoelectric container-filling level control operation.

final control element or load device is either actuated or turned off according to the design of the system.

Figure 5-28 shows an example of a simplified photoelectric liquid-level control circuit. In this circuit, light is diffused when the liquid level exceeds the position of the detector. The detector, in this case, is a photoconductive cell or light-dependent resistor. High-intensity light reaching the photocell causes it to become low resistant. Reduced light levels or diffused light causes the cell to be high resistant.

The operation of the detector and the relay control element is dependent on the resulting current from the source and the photoconductive cell. In circuit operation when the container level is above the detector/source position, the light to the photocell becomes diffused and detector resistance increases immediately. This, in turn, decreases circuit current and turns off the relay. Container filling action stops at this time. When the container level drops below the detector, intense light shines on the photoconductive cell. This action causes the cell to be low resistant and energizes the relay to start the filling operation. Systems of this type respond to digital or on-off changes and maintain the level of a container at a predetermined value.

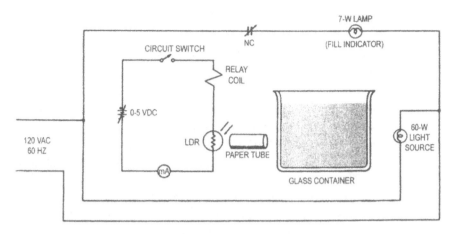

Figure 5-28. A photoelectric level control circuit.

SUMMARY

Level determination has numerous applications in industry no matter what product is being manufactured. In particular, this type of system is responsible for filling a tank, bin, hopper, or container to a certain level. When the predetermined level is reached, the system is turned off. Measurement and control are the primary system functions of concern in level determination operations.

The buoyant-force principle is used in liquid-level sensors and displacers. A float mechanism rides on the surface of a liquid and generates a mechanical action as a result of level changes. Typically, float-level operations are digital. Displacer elements respond to a force that is equal to the weight of the liquid displaced by the sensing element. Pneumatic systems are frequently actuated by displacer elements today.

The capacitance effect is used to determine level through two probes inserted into a nonconducting material, or single-probe elements are used, with the metal of a storage tank serving as a second conductor. Conducting and nonconducting material levels can be determined by the capacitance effect. The output may be either two-state or continuous. In addition, capacitance can be used to determine levels by the proximity process.

Electrical conductivity is used to determine level by employing one or more probes. When the level of a material reaches the electrode, an electric current passes between the metal tank or to a second electrode. The resulting current can be used to actuate a pump relay or control valve.

Radiation level determination utilizes the emission of gamma rays from a radioactive source. There is no direct electrical connection between the radiation source and the sensor. If the radiation source is located at the bottom of a tank and the detector at the top, detector signal levels increase when the liquid drops to a low level. Radium, cesium 137, and cobalt 60 are the common sources of radiation. The size, weight, and material selection of the source depends on the application. Detectors are either Geiger-Mueller tubes or ionization chambers. Ionization chambers produce an electrical output when radiation enters a diode chamber.

Ultrasonic level determination makes use of inaudible signals radiated from a source. The damped-sensor principle stops generating a signal when the sensor is covered by the material being controlled. The density-change principle responds to signals that are transmitted through the material. This type of system employs both a transmitter and receiver. Both liquid and solid material levels may be determined by ultrasonic units.

The bubbler or purge system is an indirect method of level determination that responds to level head pressure. Air applied to a dip tube will escape or bubble from the tube when its pressure is slightly greater than the head pressure. Tube pressure can be readily calibrated as tank level. These systems are also used to determine the level of a pressurized tank.

Several other pressure-sensitive level determination processes are used in process control, including pressure gauges, diaphragms, and air boxes. Essentially, pressure is expressed as tank-level height. Pressure-actuated switches represent an important part of level control.

Weight level systems are used to determine the level of liquids and solids of the same particle size or moisture content. Small container applications of this principle are common in industry. Electrical strain gauges, linear-variable differential transformers, and hydraulic or pneumatic load cells are typical weight transducers. Tare is an important ratio between dead or empty container weight and full capacity weight.

Photoelectric level control circuits are frequently used to determine the level of small containers. Essentially, this type of system has a light source and a detector or sensor element. When a light beam between the source and the detector is broken, it can be used to produce level control. Photoconductive cells and light-dependent resistors are frequently used to detect light levels for control applications. Systems of this type generally produce two-point (on-off) control.

Chapter 6

Flow Process Systems

OBJECTIVES

Upon completion of this chapter, you will be able to
1. Explain how the systems concept can be applied to flow.
2. List two general types of flow measuring instruments.
3. Describe the purpose and operational principle of the flowmeter.
4. Explain the operation of a variable-area flowmeter.
5. Describe the operation of a magnetic flowmeter.
6. Identify and explain how a velocity flowmeter operates.
7. Clarify positive displacement when applied to a flowmeter.
8. Define the operating principle of an ultrasonic flowmeter.
9. Explain the difference between mass and turbine flowmeters.
10. Define the Coriolis principle.

KEY TERMS

Absolute pressure—Gauge pressure plus atmospheric pressure.

Beta ratio—A ratio that compares inside pipe diameter to orifice diameter; further determines the flow velocity and differential pressure of the system.

Bourdon tube—A thin, springy, cylindrically shaped tube with one end sealed and one end open; used to measure pressure.

Bernoulli's principle—A principle stating that the higher the speed of a flowing stream of fluid or gas, the lower the resulting pressure.

Centrifugal force—Force on an object moving in a circular path which is exerted outward from the center of rotation.

Centripetal force—Force on an object moving in a circular path which is exerted inward toward the center of rotation.

Coriolis force—Force that occurs because of atmospheric pressure acting on flow relative to the Earth's rotation.

Concentric—Having a common center.

Differential pressure—The difference in pressure between two distinct sources measured with respect to each other.

Doppler theory—An effect in which there is a measurable change of sound (or light) frequency as a function of the relative velocity of the source to the receiver.

Eccentric—Not having the same center.

Faraday's electromagnetic induction law—A law stating that the amount of voltage induced in a conductor is proportional to the velocity with which it moves at right angles through a magnetic field.

Flange—One section of pipe that is joined to another with a large rim for strength.

Flowmeter—A device used for measuring the flow or quantity of a moving fluid.

Flow nozzles—A projecting spout that offers a restriction in the flow path.
Flow rate—A measurement that determines the amount of flow that moves past a point in a given unit of time.

gpm—Gallons per minute.

Head—Pressure resulting from gravitational forces on liquids.

Impact instrument—An element that flow must strike in order for a measurement to occur.

Totalizing flow—A measurement that records or tabulates the total amount of flow that occurs during a period of operation.

Turbulent flow—Flow that occurs when a value exceeds a Reynolds number of 4,000.

Velocity—The speed or rate of occurrence, measured in ft/s or in/s, for an action to take place.

Vena contracta—The smallest cross-section of fluid that occurs downstream from a freely discharging aperture.

Vena contracta area—An area of the flow stream that is the diameter of the pipe behind the orifice.

Vena contracta taps Locating points of minimum pressure downstream from the restriction.

Venturi tube—A length of tube with a varying cross-sectional area. Flow through the tube causes a pressure drop according to the velocity of the flow.

Volumetric measurement—A measurement of the volume or capacity of an object.

Vortex—A circular motion of fluid that tends to form a cavity in the center of the circle and draw more liquid toward the vacuum that occurs.

Vortex shedding A phenomenon that takes place when an obstruction bar produces a number of predictable swirls downstream from its placement.

Vortices Circular movement, swirls, or whirlpools that occur in a flowing stream.

INTRODUCTION

Nearly all of the products manufactured by industry are influenced in some way by the flow of materials. This type of processing ranges from the simple flow of fuel or gas in a heat-treating furnace to the control of large amounts of oil or gas passing through a pipeline. The measurement and control of flow variables is one of the most important process control applications of industry.

Flow processing is simple when compared with other process systems. The process is used to perform a measurement function or to achieve some type of regulation operation. Any complexity related to this process appears in the diversity of the measuring techniques used by the system. The sensing elements of a flowmeter are either of the inferential type or they deduce an output by direct displacement of quantities. The output signal of the sensing element may be either mechanical or electrical, depending on the type of sensor employed. Flow sensors and instrumentation will serve as the basis of this chapter.

THE SYSTEMS CONCEPT

The systems concept that has been used with other processes can also be applied to flow metering. Flow is often considered to be a control operation or measurement technique that could apply to anyone of several different processes. When something is viewed as a system, there must be a source, path, control, load, and one or more indicators. The flow process can involve chemicals, solvents, paint, water, gas, or anyone of an infinite variety of materials. The primary function of this type of system is control of a flowing medium or its measurement. The actual material being manipulated is somewhat incidental.

To see the uniqueness of flow as a process system, look at the chemical distribution system in Figure 6-1. In this system, the source is a bulk storage tank situated at a convenient location in the industrial facility. The tank serves as the source and represents a large reservoir of the chemical. As a source, the tank represents potential energy that can be released when the demand arises. When the chemical is released, it flows through a path of connection tubes or pipes. Chemical flow is then measured by a recording indicator. In addition, flow is detected by a sensor whose output is used to control a metering-valve mixing operation. Here, control is a function of flow. The output of the mixing operation serves as the load of the system. Work done as a result of the volume of chemical flow is based upon the demand of the system.

Flow processing systems generally have a number of unusual features when compared to other process systems. Liquids, for example, are subjected to resistance due to friction as they flow through a pipe or hose. If the flow velocity of the fluid is low, it will move parallel to the sides of the pipe. In this case, it is said to have a *laminar, viscous,* or *streamline* type of flow. If the flow rate is increased beyond certain critical values, *eddy currents* begin to form and the flow pattern becomes *turbulent.* Flow is also dependent on pipe size, material density, velocity, and viscosity. Solid material flow, in comparison, causes a number of problems that are not prevalent in liquid and gas flow.

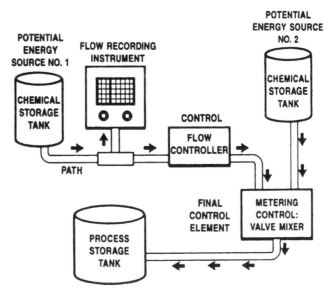

Figure 6-1. A chemical processing plant.

A large part of the automation that occurs as a result of system flow is dependent on instrumentation and measurement techniques. The sensing element of a flow system serves as a signal source in this operation. A variety of measuring techniques and sensing procedures are used in flow systems. These procedures and methods are essential in the study of flow processing systems.

FLOW INSTRUMENTATION EQUIPMENT

In industry today, flow equipment plays a significant role in manufacturing. A high percentage of industrial materials are subjected to a type of flow processing during manufacturing. Flow has, therefore, become an essential division of industrial processing.

One way of classifying flow process equipment is to place it in two general divisions. *Flow-rate instrumentation* is the first class. This type of equipment is used to determine the amount of fluid that moves past a given point at a particular instant. An oil-pipeline instrument measures flow in gallons per minute (gpm). *Totalizing flow measurement* is the other classification. Instrumentation of this type responds to the total flow that moves past a given point during a specified period. A gasoline pump generally has a totalizing flowmeter.

Totalizing and flow-rate instrumentation have a wide range of applications in industry. In general, the material that flows in the system is very important. Liquids, gases, and solid particulates suspended in liquid are typical materials that respond to flow instrumentation. Some degree of instrument modification must take place in order for an instrument to respond to more than one type of material flow.

Another important way of classifying flow instrumentation is through groupings according to the principle of operation and design. Methods include *differential-pressure producers, variable-area instruments, velocity techniques,* and *obstructionless flowmeters,* as well as magnetic, ultrasonic, and positive-displacement instruments.

The selection of a particular type of instrument according to its principle of operation is based on the material being measured, flow volume, accuracy, and the type of control desired. In this classification of instruments, there are many methods of determining flow. Strangely enough, few of these instruments have found widespread acceptance in industry because of limitations and cost. In this discussion of flowmeter principles, the limitations and the advantages will be presented.

DIFFERENTIAL-PRESSURE PRODUCERS

Differential-pressure producing instruments are used extensively in a number of industrial flowmeters. Essentially, this type of instrument takes advantage of the fact that a change in pressure will occur when liquids or gases are forced to flow through a restriction placed in a flow path. The resulting difference in pressure or *head* is used to indicate different values of flow. In practice, flow or head pressure is measured before and after a restriction. The pressure differential is then changed into an indication of flow rate.

Figure 6-2 shows a simplification of the differential-pressure principle. Flowmeters that employ this principle are used primarily to measure clean fluids. This type of device is rather low in cost, requires little maintenance, usually has no moving parts, and is easily installed in a system. The

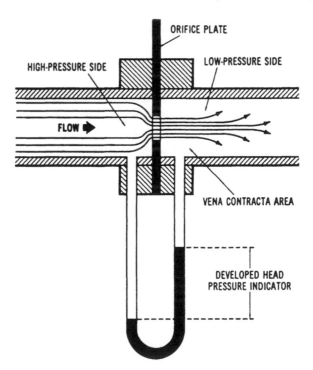

Figure 6-2. A simplification of the head flow principle.

restriction part of this instrument is commonly called the *primary element*. The *secondary element* of the instrument is used to indicate a difference in pressure before and after the restriction. The pressure test points are called the *vena contracta taps*. These taps are located one pipe diameter upstream and at the point of minimum pressure downstream from the restriction.

Differential-pressure Theory

Differential-pressure flowmeters respond to a fundamental law known as *Bernoulli's principle*. This principle states the following: *The higher the speed of a flowing stream of a fluid or gas, the lower the resulting pressure.* A decrease in flow speed likewise causes a corresponding increase in pressure. In practice, flow moves faster through the narrow portions of a pipe than through the wider portions. This means that a restriction placed in the flow path will cause a change in flow speed. As a result of this action, a decided pressure differential will develop before and after the restriction. A measurement of this pressure differential can be used to establish flow rate.

Bernoulli's principle applies to flowmeters that alter the velocity of the flow stream in a predetermined manner. Typically, there is a physical change in the cross-sectional area of the flow path. The velocity of flow passing through this diminished cross-sectional area increases, which in turn causes a decrease in pressure on the downstream side. The pressure differential between the throat and downstream side of the restricted area is used to indicate flow rate.

There are three basic equations derived from Bernoulli's principle that apply to differential-pressure flowmeters. The first shows a square root relationship between the flow rate and differential pressure. In this equation, the velocity of flow (v) is equal to a flow path constant (k) times the square root of the differential pressure (h) divided by the density of the flowing fluid (d). Mathematically, this is expressed by the equation

$$v = \sqrt{(h/d)}$$

A second equation takes into account the density of the flowing fluid. In this expression, fluid density applies to both mass flow and volume flow measurements. The equation

$$W = kA \sqrt{(h/d)}$$

is used for volume (Q) flow measurements. The letter A in the expression refers to the cross-sectional area of the pipe in which flow must occur. The equation

$$W = kA \sqrt{(hd)}$$

is used to show mass flow rate (W). This expression takes into account the cross-sectional area (A) of the pipe, the flow path constant (k), and the square root of the differential pressure times the fluid density.

The square root relationship expressed by the equations for velocity, volume, and mass flow rates has a decided effect on flowmeters. The primary element, for example, produces a nonlinear output in differential pressure between the high and low ends of the scale. To improve the readability of this output, the secondary element should take the square root of the differential pressure values. Square root extractors generally tend to decrease the overall accuracy of the resulting measurement. In practice, flowmeters are selected so that they have a rather narrow range of flow-value indication. A ratio between the minimum and maximum range should not exceed 3 to 1, or at most 3.5 to 1, where maximum accuracy is important. Where repeatability and control-signal level values are of primary concern, a ratio of 10 to 1 is acceptable.

DIFFERENTIAL-PRESSURE FLOWMETERS

The head principle of operation is used extensively in industry for a number of flowmeter applications. These instruments, in general, are inexpensive and reliable plus have accuracies of ±0.25 to 2 percent of full scale, and can be selected from a number of different range values. Measurement and control applies to the flow of incompressible gasses, liquids, and slurries. Typical instruments in this classification use *orifice plates, venturi tubes, flow nozzles, target meters,* and *Pitot tubes.*

Orifice-plate Flowmeters

Measuring flow as a result of the differential pressure developed across an orifice dates back to the days of Galileo and Bernoulli in the early 1700s. Commercial use of the orifice principle, however, has been confined to recent times, when process applications created a need for this method of control. In spite of the wide range of instruments available, the *orifice-plate flowmeter* still ranks as one of the most widely used methods of measuring flow.

The orifice plate of a flowmeter serves as the primary measuring element or sensor of the instrument. This part of the instrument comes in a variety of different types and styles. Four representative types are shown in Figure 6-3. The paddle type of orifice plate in Figure 6-3(A) is available either with a *concentric* bore or with an *eccentric* or *segmental* bore. These plates are usually made of stainless steel, but they are available in other materials. Concentric orifice plates are used to develop differential pressure levels where liquid volumetric flow-rate metering is desired. The concentric universal-type of orifice plate in Figure 6-3(D) is similar in operation, but requires a different housing.

An eccentric orifice plate is shown in Figure 6-3(B). In this plate, the orifice is located at a tangent to the pipe wall diameter. This type of orifice is commonly used in horizontal pipe installations. Eccentric plates

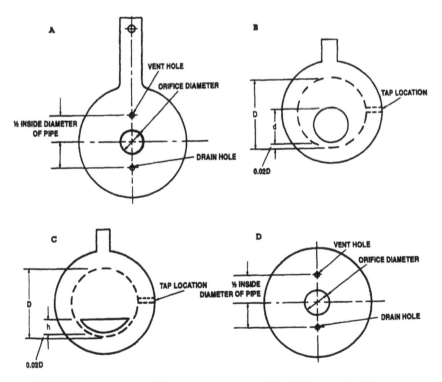

Figure 6-3. Four representative orifice plates: (A) Paddle type; (B) Eccentric type; (C) Segmental type; (D) Concentric universal type. (Courtesy Fischer & Porter Co.)

are used to meter gas, vapor containing some liquid, or liquids contaminated with solids. For the best accuracy, concentric orifices should be used whenever possible.

A segmental orifice plate is shown in Figure 6-3(C). This type of plate is used in applications similar to those fulfilled by the eccentric orifice. Specifically, this plate is used to prevent the damming up of material when used in horizontal lines. Segmental plates have reduced accuracy when compared with the concentric type of orifice.

A cross-sectional view of a typical orifice plate installation and the resulting flow diagram is shown in Figure 6-4(A). In this illustration, the orifice plate is of the concentric type. Fluid passing through the sharp, square-edge opening of the orifice forms a unique flow contour, as shown by the arrows. The diameter of the flow stream immediately downstream of the orifice is called the *vena contracta area*. The area of the stream at this point is equivalent to the diameter of the orifice.

Orifice plates are normally installed between a pair of *flanges,* as shown in Figure 6-4. The orifice plate is held in place by two hex-headed bolts that extend from each side of the flange. In some installations, it is possible to replace the orifice plate without interrupting the flow. A straight run of smooth pipe is installed before and after the orifice, the length of which is dependent on the *beta ratio* of the system. This ratio is determined by the flow velocity and differential pressure of the system.

In orifice-plate flowmeters, there is continuous contact between the flow material and the orifice plate. Any change in orifice diameter, surface, or edge shape will have a pronounced effect on the resulting differential pressure developed. The orifice must therefore be kept free of wear and corrosion in order to maintain an accurate indication of flow. Orifice-plate flowmeters are therefore best suited for clean measuring applications of liquid and gas.

Venturi-tube Flowmeters

The general theory of operation and performance of a *venturi tube* is similar to that of the orifice-plate flowmeter. This type of flowmeter is designed mainly for slurry measurements and for liquids that contain solids in suspension.

A venturi tube consists of a main barrel section that decreases in diameter and a discharge section that expands again to the original diameter. The restricted area increases flow velocity and decreases head pressure. Differential pressure measured at a distance of one-half of the pipe

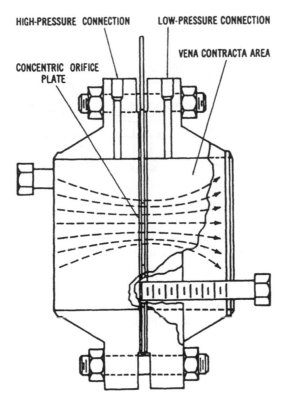

Figure 6-4. Orifice-plate flange cross-section view.

diameter upstream and at the center of the throat restriction are used to indicate flow rate.

Figure 6-5 shows a representative venturi tube for measuring clean fluid flow. The piezometer rings at the entrance and the throat section are used to measure pressure. This part of the venturi tube contains a set of holes distributed around a ring to indicate average pressure. When slurries or dirty liquids are to be measured, the piezometer ring is exchanged for a pressure ring tube.

A venturi tube is designed so that it has no sudden changes in contour, no projections, and no sharp corners placed in the flow path. Through this type of construction, dirty fluids can be readily measured without plugging up or clogging the flow path. Venturi tubes may be round (as in Figure 6-5), eccentric, or rectangular. They may be inserted into larger pipes or serve as single sections of pipe.

The major limitations of a venturi-tube type of flowmeter is its con-

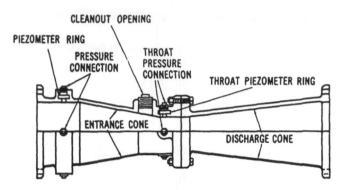

Figure 6-5. A Venturi tube.

struction cost. The accuracy of measurement is also less than a sharp-edged concentric orifice plate. It will, however, permit a rather accurate measurement of approximately two-thirds more fluid than a thin-plate sharp-edged orifice of the same size at the same differential head pressure. Flowmeters of the venturi type are primarily used to measure dirty flow streams and in high-velocity flow applications.

Flow Nozzles

Flow nozzles are the primary element of a type of flowmeter that also employs the *restriction principle* to develop differential head pressure in a flow path. A flow nozzle has a bell-shaped flare on the flow-approach section, which follows into a cylindrical throat section. Differential pressure is developed at taps located one pipe diameter in front of the approach and one-half diameter to the rear of the section. Figure 6-6 shows a cross-sectional view of a flow nozzle.

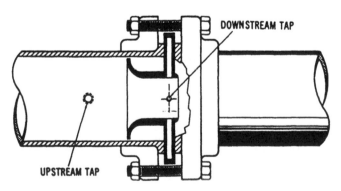

Figure 6-6. A flow nozzle.

Due to the streamlined contour of the flow nozzle inlet, high-velocity flows can be easily measured without causing erosion problems and clogging. Since the precision of the contour is not critical in measurement, flow nozzles can be expected to remain in calibration for long periods under adverse flow conditions. When solids are prevalent in the medium, flow nozzles are mounted vertically, with flow in the downward direction. Flow nozzles, in general, are more efficient than orifice plates. Approximately 60 percent more flow will pass through a nozzle than through an orifice plate, and the nozzle develops the same differential pressure as the orifice plate.

Applications of flow-nozzle instruments are found in high-velocity fluid flow and stream measurement, where erosion may cause a problem for other primary elements. The streamlined contour of the flow nozzle also tends to pass suspended solids through the restriction without a problem. Flow calculations for a flow nozzle are similar to those for a venturi tube or an orifice plate.

Target Flowmeters

The primary element of a *target flowmeter* is an annular orifice that is created in a section of pipe or tubing. This construction is shown in Figure 6-7. The annular orifice is formed by a circular disc that is placed in the center of the flow path. Flow through the open ring area between the disc and the inside tubing diameter develops a force on the disc that is proportional to the velocity head pressure. Mechanical motion of the disc is transferred through a connecting rod that passes through a flexible seal. Head pressure is measured by a secondary element that responds to the amount of mechanical motion produced by the connecting rod.

A cross-sectional view of an industrial flowmeter of the target type is shown in Figure 6-8. The target disc appears in the center of the orifice area. The disc is supported by the force arm. Mechanical motion of the target is transferred to the top of the unit, which serves as a force-balance transmitter.

Target flowmeters are, in general, used to measure the flow of liquids, vapors, or gases. This type of instrument is well suited for hard-to-measure applications, including condensates, hot materials, dirty liquids, or viscous flows. It usually gives consistent and dependable service for long periods. There are no differential-pressure taps to become clogged, and this type of instrument can be used in temperatures up to 700°F.

The target area of the primary element is available in sizes from

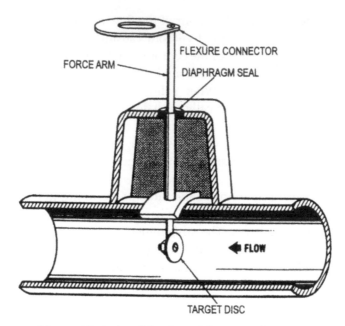

FLEXURE CONNECTOR

FORCE ARM

DIAPHRAGM SEAL

FLOW

TARGET DISC

Figure 6-7. A simplification of the target flowmeter.

0.6 to 0.8 times the diameter of the flow tube. Available sizes range from 0.5-in to 8-in pipe diameters. Flow around the annular orifice has the advantage of being less sensitive to changes in the Reynolds number and to upstream piping configurations. The greatest range of improvement in this area takes place when the target size is 0.8 times that of the flow-tube diameter.

Pitot-tube Flowmeters

A *Pitot tube* is the primary element of a laboratory type of flowmeter that compares the head of a flowing stream with a static head. This element of the instrument balances the velocity head of a flowing stream with a liquid head of mercury or water in a secondary measuring element, such as a manometer, as shown in the simplified diagram of a Pitot-tube flowmeter in Figure 6-9.

An elementary Pitot tube has two pressure passages in its construction. One faces into the direction of the flow stream and intercepts a small portion of the flow head. The second passage is perpendicular to the flow axis and reacts to static pressure. The difference in pressure between these two heads is proportional to the square of the stream velocity in the vicin-

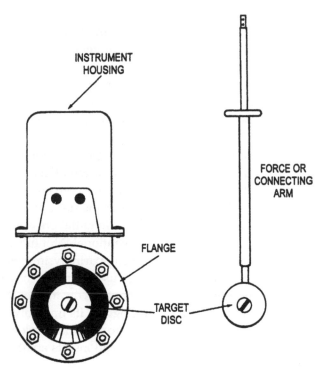

Figure 6-8. End view of a target flowmeter.

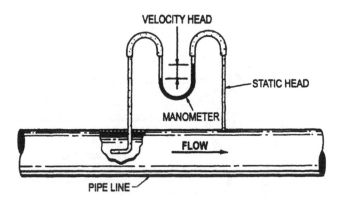

Figure 6-9. A simplification of a Pitot-tube flowmeter.

ity of the impact area of the Pitot tube.

A Pitot-tube flowmeter is often described as an *impact* type of instrument. The pressure passage facing into the flow stream is normally called the *impact area*. This type of flowmeter generally causes almost no pressure loss to the flowing stream because of the size of the impact tube. Pitot tubes are easily inserted into a flow stream by installation through a nipple in the side of the pipe or tube. In test work, the tube is installed through a gate valve with packing so that it can easily be moved across the stream to establish a flow velocity profile.

The *Venturi Pitot* in Figure 6-10 is a common variation of the Pitot tube that is often used 0 develop an increased differential pressure. Impact pressure is developed in the same manner as in a conventional Pitot tube. This pressure, which is the impact head minus the static head, is then compared with the venturi head pressure. The differential pressure in this type of instrument is greater than that of a conventional Pitot tube.

Pitot tube primary elements are used sparingly in most industrial flowmeter applications. They are primarily used in research work, laboratory testing operations, and to perform spot tests on an operating system. The tendency of a Pitot tube to become clogged, its limited velocity measuring range for fluids, and its sensitivity to abnormal velocity distribution effects tend to limit its industrial applications. Its ability to measure

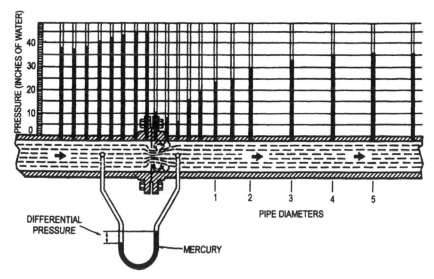

Figure 6-10. A Venturi Pitot-tube flowmeter.

high-speed air flow causes it to be a very important instrument in airplane operation. In general, a number of improvements in other flowmeters have reduced the industrial applications of this instrument.

VARIABLE-AREA FLOWMETERS

The *variable-area flowmeter* is one of the simplest and most elementary methods of determining flow rate. It consists of a metering float and a tapered tube that is larger in diameter at the top. Flow enters the tapered tube at the bottom and lifts the float. This action indicates flow rate on a calibrated scale. The density of the float is generally greater than that of the flowing substance being measured, so that the buoyant effect alone is not sufficient to lift the float.

Variable-area flowmeters operate on a principle that is closely related to that of a head-pressure type of instrument using an orifice. In an orifice type of instrument, there is a fixed aperture, and flow rate is determined as a function of the developed differential pressure. In a variable-area instrument, the tapered tube serves as a variable orifice that develops a constant pressure drop. Flow rate is indicated by the area of the tapered opening through which the flow must pass. This area is indicated by the position that the float takes in the tapered orifice.

There are three basic forms of variable-area flowmeters in use today. *Rotameters* are the most common form of flowmeter in this classification. This group employs a tapered upright glass or plastic calibrated tube with a dense float. *Orifice meters* are a variation of the rotameter that have a fixed orifice and a tapered float. *Piston meters* are the third division of this classification. They incorporate an accurately fitted piston that is lifted by flow pressure in a sleeve. All three forms of instruments are available as either direct-reading meters or can be used to generate an output for remote-indicator installations.

Rotameters

The *rotameter* is a variable-area type of flowmeter that consists of a tapered metering tube and a float that is free to move within the tube. This construction is shown in Figure 6-11. The entire unit is mounted vertically, with the small end of the tapered tube at the bottom. The flow material to be measured enters the bottom of the tube and exits at the top. In making its path through the tube, the flowing material causes the float to move in an upward direction. The position of the float is read on a calibrated scale that indicates gpm for liquids or standard cubic feet per minute (scfm) for gas flow.

Figure 6-11 shows a cross-sectional view of a rotameter. When no fluid or gas is flowing, the float comes to rest at the bottom of the tapered tube. The float is usually designed so that it blocks off the small end of the flow tube completely. When flow occurs, the buoyant effect of the applied gas or fluid and the flow force lift the float, causing the float to rise a pro-

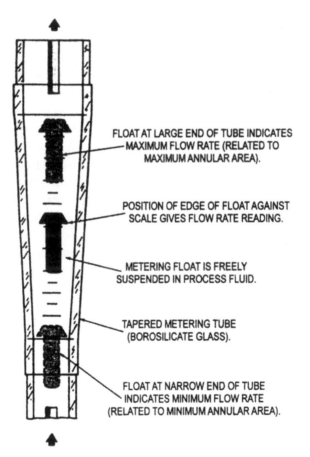

FLOAT AT LARGE END OF TUBE INDICATES
MAXIMUM FLOW RATE (RELATED TO
MAXIMUM ANNULAR AREA).

POSITION OF EDGE OF FLOAT AGAINST
SCALE GIVES FLOW RATE READING.

METERING FLOAT IS FREELY
SUSPENDED IN PROCESS FLUID.

TAPERED METERING TUBE
(BOROSILICATE GLASS).

FLOAT AT NARROW END OF TUBE
INDICATES MINIMUM FLOW RATE
(RELATED TO MINIMUM ANNULAR AREA).

FLUID PASSES THROUGH THIS ANNULAR OPENING BETWEEN PERIPHERY OF
FLOAT HEAD AND I.D. OF TAPERED TUBE. FLOW RATE VARIES
DIRECTLY AS AREA OF ANNULAR OPENING.

Figure 6-11. Cross-sectional view of a rotameter. (Courtesy Fischer & Porter Co.)

portional amount according to the flow rate. As the float moves toward the larger end of the tapered tube, the flow path cross-sectional area increases. The process continues until all of the flow material passes around the float. When the velocity head pressure of the flow plus its buoyant effect equals the float weight, dynamic equilibrium occurs. The float then maintains this position as an indication of the flow rate.

The *glass-tube rotameter* in Figure 6-12 fit is used to measure gas and liquid flow in line sizes of 2" in diameter and less. This type of instrument is used where the fluid is free of entrained solids. Accuracy ranges from 2 percent of the full-scale range for inexpensive instruments to 0.5 percent for precision instruments. Typical range spans of measurement are 12 to 1.

A *molded-plastic flowmeter* uses a *ribbed-bore taper tube* that guides and stabilizes a ball float. Through this type of construction, the float is held in the center of the flow area. It reduces turbulence and provides a smooth flow through the tapered area. The window area, which is attached to the flow surface, provides a quick and easy display of float position. Scale sizes include 10, 5, and 2 ins.

When rotameters are used to record flow rates on chart recorders or to transmit information to remote locations, they must employ a mechanism that will respond to the position of the float. One very common method is to attach the float of a rotameter to a permanent magnet. Magnetic coupling is then used to generate an output signal that is representative of flow rate. For electrical displays, magnetic output is often used to alter the impedance of a bridge circuit. For recording or pneumatic output, a magnetic following mechanism is often used.

Figure 6-13 shows a representative magnetic rotameter take-out mechanism that is used to drive the pen of a recorder. The extension of the float or *plummet* is attached to a permanent magnet. Changes in flow rate are transmitted as magnet positions inside of an extension tube. A second permanent magnet is located outside of the extension tube. This magnet is free to follow the movement of the internal magnet. The external magnet is connected to a lever and link that drives a pen mechanism. Through this unit, flow rate values may be graphically recorded on a chart for permanent records.

Orifice-Plug Flowmeters

Orifice-plug flowmeters employ a tapered float assembly and a fixed-diameter orifice. This type of instrument is often considered to be a compact, low-cost version of the larger glass-tube rotameter. Figure 6-14 shows

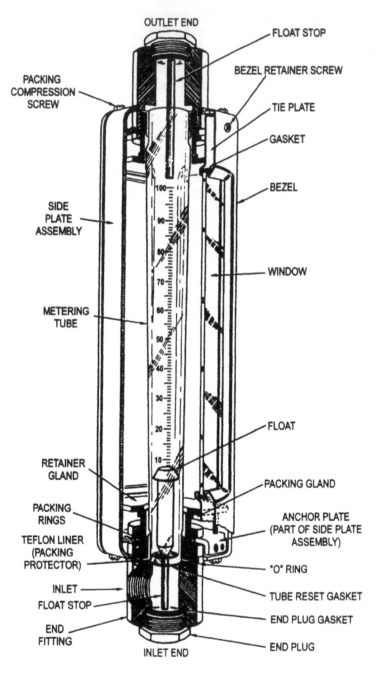

OUTLET END

FLOAT STOP

BEZEL RETAINER SCREW

PACKING
COMPRESSION
SCREW

TIE PLATE

GASKET

BEZEL

SIDE
PLATE
ASSEMBLY

WINDOW

METERING
TUBE

FLOAT

RETAINER
GLAND

PACKING GLAND

PACKING
RINGS

ANCHOR PLATE
(PART OF SIDE PLATE
ASSEMBLY)

TEFLON LINER
(PACKING
PROTECTOR)

"O" RING

INLET

TUBE RESET GASKET

FLOAT STOP

END PLUG GASKET

END
FITTING

END PLUG

INLET END

Figure 6-12. A glass-tube rotameter. (Courtesy Fischer & Porter Co.)

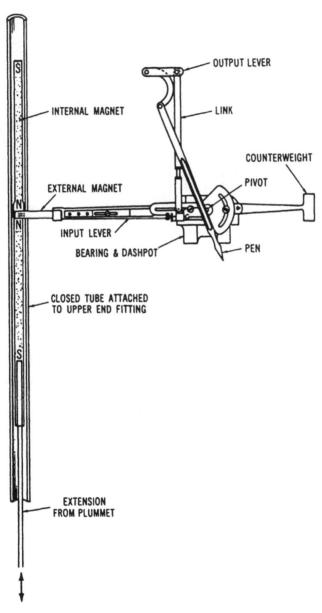

Figure 6-13. Magnetic rotameter take-out mechanism.

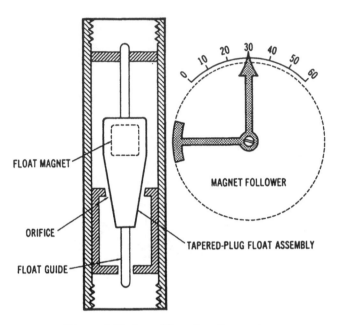

Figure 6-14. An orifice-plug flowmeter.

the construction of an orifice-plug flowmeter.

In operation, flow is admitted to the bottom of the instrument and exits at the top. The ratio of float weight to plug diameter is generally high for this meter. This results in flow rates larger than those of a glass-tube rotameter. An increase in flow lifts the tapered float assembly out of the orifice. The output of the meter is then developed by a permanent magnet located in the float. The position of the float is tracked by an external follower magnet attached to an indicator. The housing of the meter must be nonmetallic in order to allow a coupling of the magnetic fields. Flowmeters of this type are commonly used in applications where accuracy is not essential and compact size and cost are important.

Piston-type Flowmeters

Piston flowmeters employ a sleeve or cylinder, which is held in a cast body, and a fitted piston or metering plug. Orifices cut into the sleeve are uncovered by the piston until a sufficient area is opened to permit passage of the flow being measured. Metering of flow is achieved by location of the piston, which provides a direction indication of the orifice area and consequently the rate of flow. The position of the piston can be viewed through a window on a calibrated scale, by a magnetic follower, or by a mechanical hand-deflection mechanism.

Figure 6-15 shows a variable-area flowmeter of the piston type that is used to transmit flow-rate measurements to indicating, recording, or controlling equipment at remote locations. This particular instrument indicates flow rate on a uniformly graduated scale. The unit can be installed directly into horizontal or vertical lines or with an optional inverted flow scale. This flowmeter can also monitor flow in a downward gravity-feed line.

Fluid applied to a piston flowmeter enters the housing on the left cap (1) and exits from the right side. This creates a pressure differential across the piston orifice, moving the piston against the retaining spring (6). Because piston movement and orifice area are proportional to the rate of flow, the greater the rate of flow, the further the piston moves along the tapered metering cone (3). Externally, the flow indicator (5) encircling the flowmeter body (2) is magnetically coupled to the high-flux density magnet mounted in the piston assembly (4). A colored line on the indicator ring is read against the precalibrated scale (7) attached to the inner surface of the dust guard.

Piston flowmeter operation is influenced to some extent by the viscosity of the flow, and must be calibrated to accommodate each type of flow. One way of altering this characteristic is the sharp-edge orifice con-

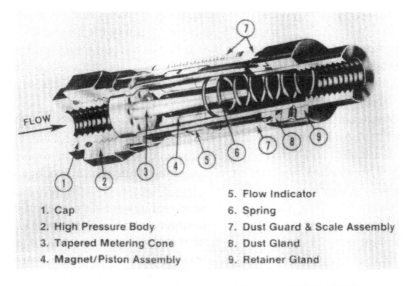

1. Cap	5. Flow Indicator
2. High Pressure Body	6. Spring
3. Tapered Metering Cone	7. Dust Guard & Scale Assembly
4. Magnet/Piston Assembly	8. Dust Gland
	9. Retainer Gland

Figure 6-15. A piston flowmeter. (Courtesy HEDLAND)

struction of the piston assembly. This design permits the instrument to have greater operating stability and accuracy over a wide range of viscosities. In a 3-in meter, a tripling of viscosity produces no measurable change in reading. The 1/4-in flowmeter, which is the most viscosity sensitive, produces a 1.5 percent change in the reading over the same range. The standard flow scale for fluids is calibrated in gpm or liters per minute (lpm) for a specific gravity of 1.0 for water and water-based fluids. For oil and petroleum-based fluids, calibration is at 0.876 specific gravity. A pneumatic instrument is also available, with a multipressure flow scale to indicate flow rate in scfm at pressures from 40 to 130 psi. Piston flowmeters are commonly used to measure specific flow materials and are selected according to a range of operating viscosities.

MAGNETIC FLOWMETERS

A *magnetic flowmeter* is an instrument designed to measure the volume flow rate of electrically conductive fluids passing through a pipe or tube. This instrument is particularly applicable for the measurement of fluids which are somewhat difficult to handle. Corrosive acids, sewage, detergents, tomato pulp, crude oil, and paper pulp are common applications.

A partial cutaway view of the magnetic flowmeter is shown in Figure 6-16. Notice the locations of the electrode assembly, the magnetic coils, and the nonmagnetic flow tube. This assembly is a volumetric fluid transducer that changes conductive fluid flow into an induced voltage when fluid flows through a magnetic field. The amplitude of the generated signal is directly proportional to the flow rate of the fluid.

Magnetic flowmeters are considered to be *obstructionless* metering instruments. An inherent advantage of this principle is that pressure losses are reduced to levels occurring in equivalent lengths of equal-diameter piping. This reduces and conserves pressure-source requirements compared with other metering methods.

Figure 6-17 shows a schematic representation of a magnetic flowmeter. Figure 6-18 is a pictorial representation of the basic operating principles involved. The flowmeter is constructed around a section of pipe that requires no orifices or obstructions within it. This means that the flowmeter itself has very little effect on the flow rate of the fluid through it. Around the metering section of pipe are wound two field coils: one

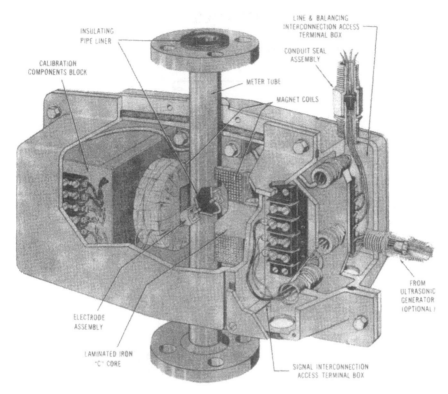

Figure 6-16. Partial cutaway view of a magnetic flowmeter. (Courtesy Fischer & Porter Co.)

above the piping and the other below it. When current passes through these coils, a magnetic field is produced in the direction shown in Figure 6-17. The flowmeter section of the pipe has two electrodes positioned so that the fluid passing between them is perpendicular to the magnetic field of the coils.

Consider the operation of the flowmeter when a constant magnetic field is applied. Under these conditions, operation is based on *Faraday's law of electromagnetic induction*. Simply stated, the voltage induced across a conductor moving at right angles through a magnetic field is proportional to the velocity of that conductor, or

$$v = 1/c \, Bdv$$

where

     c   = dimensional constant

B = flux density of the magnetic field
d = inside diameter of the pipe
v = velocity of the conductor

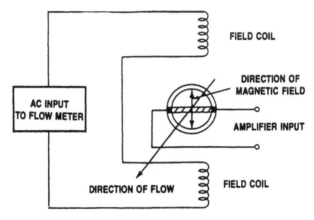

Figure 6-17. Schematic of a magnetic flowmeter.

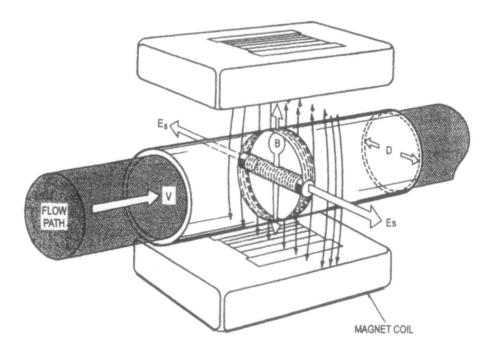

Figure 6-18. Pictorial representation of magnetic flowmeter operation. (Courtesy Fischer & Porter Co.)

The conductor is the element of fluid that lies directly between the two electrodes. As fluid passes through the pipes, it is moving perpendicularly through the magnetic field (B). The value of B depends on the amount of current running through the winding of the field coil. For the present, this will be considered as a constant. The inside diameter of the pipe is, of course, also a constant. For most liquids with a reasonable conductivity, the constant c is considered to be 1. As a result, V is dependent only on the velocity of the conductor and is directly proportional to it. The induced voltage V is the flowmeter output, which is used as the amplifier input.

The magnetic field is not constant. An ac voltage is applied to the field coils. This means that the output voltage will be an ac voltage rather than a dc voltage, but its magnitude will still be proportional to the velocity of the fluid. The desired signal output of the flowmeter will also be in phase with the magnetic-field current. When B is maximum, maximum voltage is induced. B is maximum when the current through the field coils is maximum.

Since an ac voltage produces the magnetic field, some unwanted voltage will be produced when there is no flow. When there is fluid in the flowmeter and no flow, a stationary conductor is placed between the two electrodes. A voltage will be produced between these two electrodes due to the changing field. In other words, this voltage is produced by transformer action, and is maximum when a maximum change in the magnetic field occurs. Maximum change in the field is produced by maximum change in current. This occurs in an ac circuit when the current is passing through zero. Therefore, this unwanted signal is 90° out of phase with the desired signal. The feedback circuit will provide for elimination of this signal.

Another unwanted signal is produced by transformer action. This voltage will be induced into the electrodes and their leads. These leads are formed into a twisted pair and shielded where possible to minimize these unwanted signals. They must, however, be exposed near the electrodes, and cannot be formed in a twisted pair around the pipe. This voltage must also be 90° out of phase with the desired signal. All signals 90° out of phase with the desired signal are called *quadrature voltages*. They will be canceled out in the feedback circuits, and only the amplified signals will be proportional to the flow rate.

The signal converter of a magnetic flowmeter can be an integrally mounted assembly or a remote assembly housed in an independent enclosure. There is some difference in the circuitry of these two assemblies.

Integrally mounted units are smaller and have optional features built on independent circuit boards. The remote units are more sophisticated and are somewhat larger. Remote assemblies have more optional features and can be used with a number of different flowmeter units. The remote unit is a typically microprocessor-based signal converter. A pulsed dc type supplies a pulsed, constant-current dc signal to the magnet coils of the flowmeter to establish the magnetic field. The frequency of the pulse signal is field selectable. A standard frequency is 7.5 Hz.

A 15-Hz pulse signal is also available as a noise-reduction feature. This unit can have a display, can be equipped to monitor flow in either direction, can serve as a data link, and can interface with a large, data-based computer system.

MAGNETIC-PISTON FLOWMETERS

Magnetic-piston flowmeters are a recent addition to the flowmeter family and have been used in industry for only a few years. This new class of obstruction-type instrument is primarily used to measure low levels of flow. Most flowmeters are designed for measurements in line sizes of 1 in or larger.

Figure 6-19 shows a simplification of a magnetic-piston flowmeter. This instrument consists of a Teflon housing, stainless steel, PVC (or a similar material), and an outlet port located 90° from the inlet.

Located in the flow path is a piston-shaped magnet that is free to travel up and down. Piston movement alters the amount of flow that passes through the outlet port. Heavy flow causes the piston to shift upward, which causes a greater opening in the outlet line. Reduced flow shifts the piston downward, which reduces the size of the outlet-line opening. The position of the magnetic piston is detected by a *Hall-effect sensor*. A stationary magnet of opposite polarity is located external to the flow stream and in line with the piston. Magnetic repulsion created by the external magnet opposes piston movement and provides resistance to the flow.

The Hall-effect sensor of a magnetic-piston flowmeter develops an analog voltage that is determined by the position of the magnetic piston. The generated voltage from the transducer is linearly proportional to the flux density of the applied magnetic field. The Hall-effect principle says that when a magnet is placed perpendicular to one face of a current-carrying conductor, a voltage will appear at the opposite sides of the conductor.

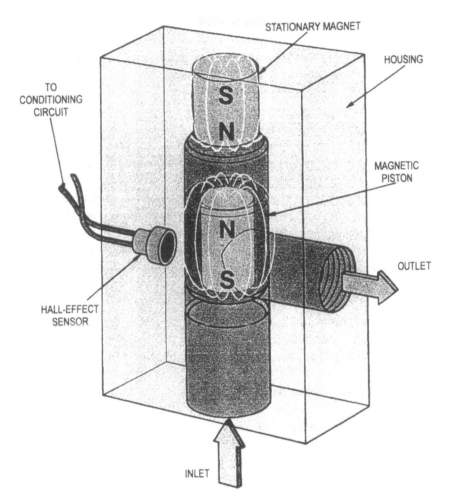

STATIONARY MAGNET

HOUSING

TO
CONDITIONING
CIRCUIT

S

N

MAGNETIC
PISTON

N

S

OUTLET

HALL-EFFECT
SENSOR

INLET

Figure 6-19. Simplification of a magnetic-piston flowmeter.

The generated voltage is proportional to the current flowing through the conductor and the flux density of the magnetic field.

In operation, flow lifts the piston off its seat, causing it to rise or fall as flow increases or decreases. Adjacent to the flow path, the Hall-effect transducer senses the resulting magnetic field and converts it to an analog signal. Since the magnetic field changes with piston position, the voltage produced by the transducer can be directly related to flow. The generated output of the sensor is a linear value that is measured in millivolts and allows a direct conversion to volumetric flow rate.

The vast majority of flowmeters are either inaccurate or simply not suitable for measuring low values of flow. Magnetic-piston flowmeters were designed to accurately measure fluids passing through small lines which have a flow value as small as 10 cc/min.

VELOCITY FLOWMETERS

Velocity flowmeters use primary measuring elements that respond to the speed at which a stream passes through a closed pipe as a measure of flow rate. Flowmeters of this type are capable of high-performance measurements that are useful in precision batching, blending, and billing applications. Accuracies are ±0.25 percent of flow rate over ranges from 20 to 1, to 70 to 1, depending on the size of the system. *Turbine, vortex-swirl,* and *vortex-strut* flowmeters are the primary types of instruments in this division.

Turbine Flowmeters

When a turbine rotor is placed in the path of a flowing stream, the imparting force of the flow produces rotary motion of the turbine rotor. The rotational speed of the turbine rotor can be used as an indication of flow rate. A simplification of a *turbine flowmeter* is shown in Figure 6-20.

Flow entering the simplified turbine starts at the left and exits at the right. The front support holds the rotor in place and conditions the flow pattern before it is applied to the rotor. Flow passing through the rotor causes it to turn with an angular velocity that is proportional to the volumetric flow rate, because a linear relationship exists between rotor speed and volumetric flow. The turbine flow then enters the outlet support, where it is reconditioned to a laminar flow before exiting the meter.

The resulting output signal of the simplified turbine flowmeter is developed by the signal pick-off coil. In the center of the pick-off coil is a small permanent magnet that influences the movement of the turbine rotor blades. When one of the blades passes through the magnetic field, it causes distortion by altering the path of the field. This change in reluctance causes the field to cut across the sensing coil, which generates a voltage by the induction process. The motion of the turbine blade generates an ac voltage into the pick-off coil. The frequency of the generated voltage is used as an indication of flow rate. The pickoff-coil assembly is commonly described as a *variable reluctance transducer.* This type of signal sensor is

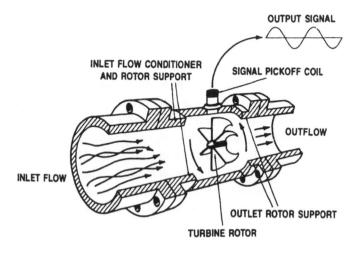

Figure 6-20. Simplification of a turbine flowmeter.

used in nearly all turbine flowmeters.

The resulting output or calibration factor (K) of a velocity flowmeter is expressed by the formula

$$K = T_K\, f/Q$$

where

T = time constant in gpm or lpm
Q = volumetric flow rate in gpm or lpm
f = generated frequency in hertz
K = number of representative pulses per volume unit, such as pulses per gallon or liter

In practice, the K factor for a specific instrument is established by the manufacturer prior to shipment. In this instrument, calibration remains accurate for long periods of operational time.

A typical turbine flowmeter instrument ranges in size from 15 to 300 mm (0.5 to 12 ins), with 0.1 percent repeatability and accuracies of ±0.5 percent over a 30 to 1 flow range. In this type of instrument, each output pulse corresponds to a discrete volume of material.

Vortex-swirl Flowmeters

Vortex-swirl flowmeters are velocity instruments that respond to the *vortex precession principle*. The primary element of this flowmeter does not obstruct the flow path, and measurement occurs when a rotating body of fluid enters an enlarged area within the instrument. A simplification of the vortex-swirl flowmeter is shown in Figure 6-21.

Fluid flow enters the swirl flowmeter on the left and is caused to rotate by passing through the stationary swirl-producing component. The center of the rotating fluid forms a twisting, high-velocity area, called a *vortex,* which tends to align itself in the center of the meter body area of this instrument. Upon entering the enlarged area, the vortex leaves the center and forms a helical path around the inside diameter of the meter body. This action is called the *precession function.* The frequency of helical precession has been proved to be proportional to the instrument's volumetric flow rate. A sensor placed in the enlarged area is then used to detect the frequency of the precession.

Just before exiting the flowmeter, all fluid is forced into the deswirl component. This part of the instrument restores the flow pattern to normal. The deswirl component is also used to isolate the measurement function from any downstream piping effects which could impair the precession function.

The sensor or secondary element of a swirl flowmeter is usually a thermistor that detects velocity changes in the processing vortex. The frequency with which this velocity change occurs is then transposed into a changing voltage. The thermistor in Figure 6-22 is placed in the precessional path so that it senses a change in temperature. Each helical precession causes a corresponding cooling in temperature due to the area of increased velocity. A constant-current source applied to the thermistor causes a change in voltage to be developed as a result of temperature vari-

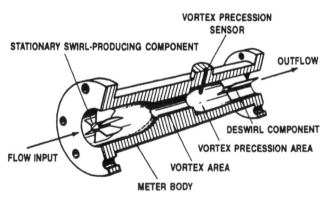

Figure 6-21. Simplification of a swirl flowmeter.

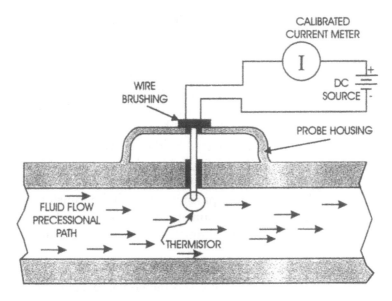

Figure 6-22. Thermistor flowmeter sensor.

ations. These voltage changes are subsequently amplified and filtered. A typical output signal is a 15-V peak-to-peak wave that varies in frequency with the volume of the flow rate.

The electrical output signal of a swirl flowmeter may be used to produce a direct reading on a hand-deflection meter or an electronic recording instrument, or it may employ a digital readout. This element of the flowmeter may be attached to the instrument for a local response or may be placed at a remote location. Electrical energy for operation must be supplied to both the primary and secondary elements of the instrument.

Swirl flowmeters are employed in a number of industrial process control applications. This type of instrument employs no moving parts, has good reliability, and has a minimum of maintenance problems. Each instrument is individually calibrated for a specific application and is linear over an operating range of 100 to 1. Repeatability on this instrument is ±0.25 percent, for accuracies of ±0.75 percent. In general, this method of flow measurement requires costly installation. Its advantage is low maintenance costs for long operational periods.

Strut-Vortex Flowmeters

The *strut-vortex flowmeter* is a common variation of the swirl flowmeter. This particular type of instrument responds to the *vortex-shedding principle.* An obstruction bar or strut placed in the flow path produces a number of predictable swirls or vortices downstream. These vortices separate or are shed from the strut and travel downstream in a predictable pattern. The number of vortices that appear downstream in a given period is directly proportional to the liquid flow rate. Flow striking this strut sheds small vortices that alternately change direction, as shown in Figure 6-23.

The sensing element of the strut flowmeter is normally a thermistor. Temperature variations detected by the thermistor are changed into a square-wave output voltage. The high-velocity center of each vortex causes a pronounced change in temperature compared with other flow areas. The number of vortices passing across the thermistor in a given unit of time is proportional to volumetric flow rate. The output frequency signal is amplified, filtered, and ultimately applied to a recorder, indicator, or digital readout.

A recent development in vortex flowmeters uses ultrasonic signals as a method of detecting flow rate. This method of sensing, shown in Figure 6-24, combines both the vortex-shedding principle and ultrasonic detection in its operation. An ultrasonic beam is transmitted across the vortex pattern as it travels downstream from the strut. As each vortex passes through the ultrasonic beam, it causes a change in signal amplitude. This produces a form of *amplitude modulation* (am) which is similar to that of a radio communications system. Each amplitude change in the signal represents a vortex passing through the beam. The total number of amplitude changes is then used to indicate liquid flow rate.

The modulated output signal of the primary element is detected and processed through electronic-signal conditioning equipment. A block diagram of a representative unit is shown in Figure 6-25. The entire assembly is attached to the flowtube body. To energize the instrument, 115/230-V, 60-Hz electrical energy is needed. Remote instruments located as far as 1,000 ft away may be actuated by the output of the instrument. An internal view of the instrument is shown in Figure 6-26. This type of instrument is commonly used as a high-performance flowmeter. It is specifically designed for long-term reliability and a wide range of operating characteristics.

POSITIVE-DISPLACEMENT FLOWMETERS

Positive-displacement flowmeters are instruments that divide specific and known amounts of a flow stream into parts and count them for an

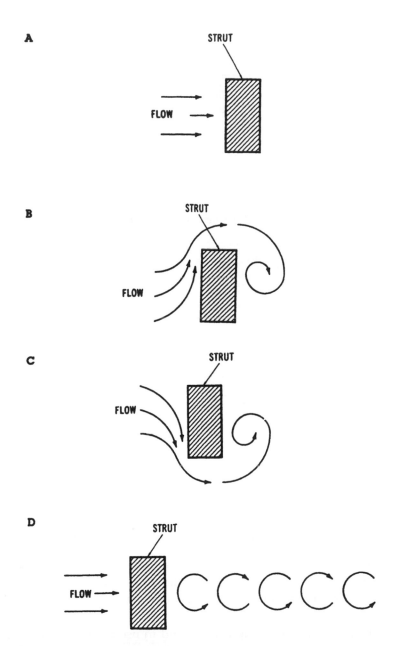

Figure 6-23. Vortex shedding principle: (A) Flow approaching strut; (B) Vortex generated clockwise; (C) Vortex generated counterclockwise; (D) Generated vortices.

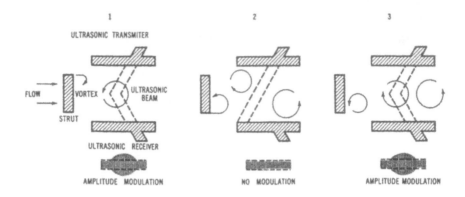

Figure 6-24. Ultrasonic vortex operation.

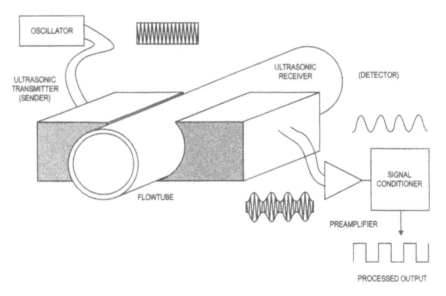

Figure 6-25. Block diagram of an ultrasonic flowmeter.

indication of totalized flow. Several ingenious methods have been devised to accomplish this operation. Totalized flow has applications in blending operations, automatic loading systems, distribution networks, and dispensing.

Of all flowmeters manufactured and in use today, the positive-displacement type of instrument is by far the most common. Typical examples include the millions of water, natural gas, and gasoline meters used to dispense gas and fluids for everyday consumption. Industrial uses for this

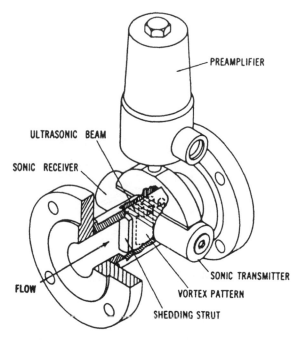

Figure 6-26. Internal view of an ultrasonic vortex flowmeter.

instrument have been restricted to metering applications in commodity-dispensing systems, but recent trends in automatic blending operations and continuous accounting systems tend to indicate an entirely different industrial role for totalizing flowmeters in the future.

A number of unusual positive-displacement flowmeters are available, as this type of metering action can be performed by pistons, rotary vanes, diaphragms, and bellows flowmeters. Meters of this type are available in sizes ranging from 0.5 to 16 in. They are simple to install and can have accuracies as high as ±0.1 percent. Range is 5 to 1 for liquids and 100 to 1 for gases. Most instruments of this type do not employ electrical, pneumatic, or hydraulic power in order to function. Energy for operation is extracted from the flow stream and appears as a pressure loss between the inlet and outlet of the meter.

Nutating Disc Flowmeters

The *nutating disc flowmeter* in Figure 6-27 is a popular instrument to measure liquid flow. The moving assembly is used to separate the applied fluid into measurable increments. It consists of a radially slotted disc, with an integral ball bearing at the bottom, and an axial pin, which divides the

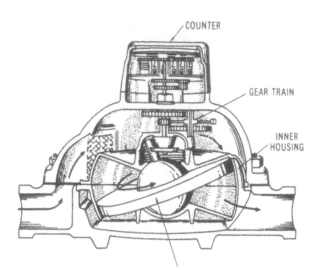

Figure 6-27. Nutating disk flowmeter (partition and slot in disk not shown).

metering chambers into four parts. Two of these appear above and below the disc on the input side, with two more located in transposed positions on the output side.

Liquid entering the metering chamber causes the disc to *nutate* or wobble as it rotates. With each nutation, a known volume of liquid is passed through the instrument. The free end of the axis of the disc moves in a circular path as the disc nutates, which drives a gear train and counter mechanism located at the top of the instrument. This type of flowmeter has an accuracy of ±1 percent. It is mainly used for small flow applications with a maximum capacity of 150 gpm.

Rotating-vane Flowmeters

Figure 6-28 shows a simplification of a spring-loaded *rotating-vane flowmeter*. This meter has the rotor offset toward the bottom of the housing. Through this type of construction, larger volume areas are present at the top. As a result, large volumes of flow can be made to move across the top with little or no return through the bottom.

Rotating-vane flowmeters are used in the petroleum industry to measure gasoline, crude oil, and low-viscosity fluids. Metering ranges are available from a few lpm or gpm to 25,000 barrels per hour for crude oil metering. Accuracies of ±0.1 percent are typical, and values as high as ±0.05 percent are possible with larger instruments. These instruments are

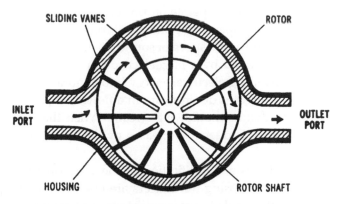

Figure 6-28. Simplification of a rotating-vane flowmeter.

available in a variety of different materials and can be used with fairly high-temperature flow streams and up to pressures of 1,000 psi.

An indication of the totalized flow passing through a rotating-vane flowmeter is often achieved by mechanical connection to the shaft of the rotating vane. Gear drive assemblies that drive a totalizing clock mechanism are common. In addition to rotary motion, the shaft may also be used to produce electrical signals. The frequency of the rotating shaft is used to indicate different values of liquid flow.

Bellows Flowmeters

The positive-displacement principle applied to the measurement of gas flow poses pronounced problems when using any must be continuous, without any value changes in pressure due to pumping action. In addition, gas measurement must not be influenced adversely by changes in temperature and pressure. To solve these problems, the *bellows* type of flowmeter was developed.

ULTRASONIC FLOWMETERS

One of the most unusual developments to occur in flowmeter technology in recent years is the application of ultrasonic measuring techniques. This technology is largely the result of several new materials used in transducers, and related developments in pipe design. This technique does not obstruct the flow path and can be achieved by portable instruments that are easily clamped onto a pipe at a moment's notice.

Ultrasonic flowmeters operate on the principle of sound propagation in liquid. This is the operating principle of sonar. A change in pressure travels through a liquid at the speed of sound. In an ultrasonic flowmeter, sonic pulses are generated by a piezoelectric transducer. This device is responsible for converting electrical energy into a vibrating signal. When the pulsed wave is directed downstream in a flowing liquid, its speed or frequency adds to that of the flowing stream. When the same signal is directed upstream, its frequency decreases by the speed of the flowing stream.

Figure 6-29 shows a simplification of the ultrasonic flowmetering principle. Here, two opposing transducers are inserted into a pipeline at a 45° angle. Using this angle is a convenient way of obtaining the average flow velocity along the path being measured. In practice, ultrasonic pulses of 1.25 MHz are radiated alternately between the two transducers. In the nontransmitting mode of operation, each transducer serves as a receiver or detector. An electrical signal is developed alternately by the two transducers.

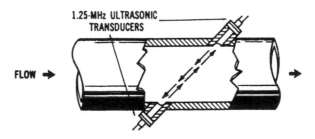

Figure 6-29. A simplification of the ultrasonic flowmetering principle.

Flow-rate measurement by an ultrasonic instrument is based on the time difference between upstream and downstream signal-pulse propagation. A *beat frequency* produced by mixing the output signals of the two transducers is proportional to the average fluid velocity along the measured path. In effect, the speed or frequency at which each transmitted pulse travels through the liquid is cancelled, which yields only the frequency difference. For example, if a pipe was full of liquid that was not flowing, the downstream frequency might be 1.25 MHz, with an equal upstream frequency. The frequency difference in this case would be 1.25 MHz minus 1.25 MHz, or zero. This would indicate no resulting flow.

When a fluid flow does occur, the downstream pulse frequency could increase in value to 1,250,500 Hz. The upstream frequency would likewise be influenced by the flow and could cause a corresponding decrease in value to 1,249,500 Hz. The resulting difference in frequency would be 1,250,500 Hz minus 1,249,500 Hz, or 1,000 Hz. This difference in frequency could then be equated to a flow rate of a specific value. Any further increase in flow rate would yield a greater difference in frequency and a higher flowrate indication.

The frequency difference signal of an ultrasonic flowmeter is typically in the range of 50 to 100 kHz. This signal is usually cleaned up and may be counted by a digital-display instrument or converted into appropriate voltage or current values that are displayed on an analog display.

The advantages of ultrasonic flowmeters depend a great deal on the specific application. In general, they are accurate to within ±0.5 percent of full scale, have linear ranges of 100 to 1, do not obstruct the flow path, and can be adapted to pipe sizes from 0.25 in to 30 ft. They are also independent of flow temperature, density, viscosity, and pressure. The major disadvantages are initial cost and sensitivity to fluid composition with a high percentage of particulates. As a general rule, the ultrasonic flowmetering principle provides an excellent solution to a wide range of flow measuring problems in industry.

MASS FLOWMETERS

During the last decade, *mass flowmeters* have become an important evaluation tool. This type of instrument measures mass-related flow and indicates values in lb/hr or kg/m. It is identified as one of the most accurate of all liquid measurement systems and responds well to high flow rates. In general, this type of flowmeter is somewhat expensive. Its operation is based on a natural phenomenon called the *Coriolis force*.

Coriolis Mass Flowmeters

The *Coriolis effect* or force occurs because of atmospheric pressure acting on flow, causing a vibration in the flow tube. The fluid then moves with a velocity in a circular motion outward. The force developed by this effect can be harnessed to accomplish flow measurement of material mass. This linear and angular motion can be further demonstrated by using a garden hose. Take several feet of hose in your hands and hold it in front of you in a U-shape. With the hose filled with water that is not flowing, swing it back and forth and you can see the even movement. Now do the same thing with water flowing through the hose. The twisting or angular action will be readily apparent.

Instruments that employ this effect are true mass meters, and measure the mass rate of flow directly, as opposed to volumetric flow. Since mass does not really change, the instrument itself does not have to be adjusted for variations in liquid properties. Similarly, there is no need to adjust for changing temperature or pressure conditions. This instrument has become very useful for measuring liquids whose viscosity changes with increases in flow.

Figure 6-30 shows one of the designs of the Coriolis flowmeter. The internal workings of this instrument consist of a tube that is shaped in the form of the Greek letter *Omega* (Ω). The flow tube is enclosed in a sensor housing connected to an electronics unit. The sensing unit can be installed directly into any process and the electronics unit can be located up to 500 feet away from the sensor itself.

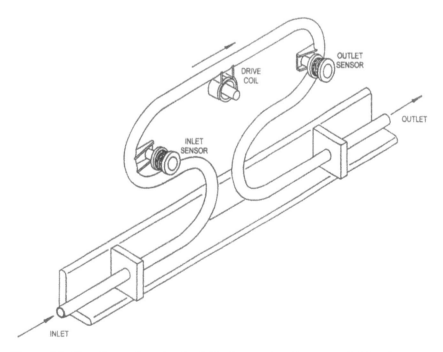

Figure 30. One flow tube of a Coriolis flowmeter. (Courtesy Schlumberger Industries)

Inside the sensor housing, the Omega-shaped flow tube is vibrated at its natural frequency by a magnetic device located at the bottom of the tube. The vibration is similar to that of a tuning fork, covering less than 0.1 in and completing a full cycle about eighty times per second. As liquid flows through the tube, it is forced to take on the vertical movement of the tube. When the tube is moving in an upward direction (half of its cycle), the liquid that is flowing in the meter resists being forced up by pushing down on the tube. This process is shown in Figure 6-31.

Having already been forced upward, the liquid flowing from the meter resists vertical motion and pushes up on the flow tube. This action causes the tube to twist and move downward. During the second half of its vibration cycle, it twists in the opposite direction.

The amount of twist in the tube is directly proportional to the mass flow rate of the liquid flowing through the tube. Magnetic sensors located on each side of the flow tube measure the tube velocities, which change with each twist of the tube. These sensors feed this information to the electronics unit and it is then converted to a voltage that is proportional to mass flow rate.

The back and forth twisting of the flow tube as it moves through the vibrational cycle causes the inlet motion to lag behind that of the outlet. Measuring the lag produces a highly accurate flow measurement.

Dual-tube Mass Flowmeters

Dual-tube mass flowmeters have two identical parallel tubes. The

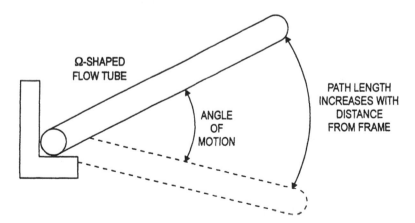

Figure 6-31. Vibration of a Coriolis flow tube. Motion stops and reverses at each end of the vibration cycle.

sensors that measure the movement in a single-tube mass flowmeters are mounted on the meter frame (the point of reference for the instrument). External vibrations, such as from a nearby instrument, can shake the frame, which in turn shakes the tube. The result is flow-measurement inaccuracies.

In the two-tube design, the sensor is mounted on one of the tubes instead of the frame. Since the tubes are identical, they have the same response to external vibrations, so there is no difference in movement between its measuring tube and its point of reference. The second tube cancels out the effects of external vibrations.

Thermal Mass Flowmeters

Thermal mass flowmeters have traditionally been used for gas measurements. Some liquid flow-measurement designs are now available. This particular type of mass flowmeter operates independent of density, pressure, and viscosity.

Thermal meters used a heated sensing element isolated from the fluid flow path. The flow stream conducts heat from the sensing element and the flow temperature is proportional to the mass flow rate. In these units, all or part of the flow passes through a sensing tube. A constant heat rate enters the flow through the sensor tube from two externally wound resistance detectors, as shown in Figure 6-32. These detectors both heat the tube and sense the tube temperature. When heat is lost, the first detector that meets the flow senses the drop. The flow then passes under a heater which heats it to a predetermined temperature. The second detector mainly reads the temperature following the heating process, but can also heat the flow if the temperature is not at the prescribed level.

The only requirement for the thermal mass flowmeter is that the gases should be very clean (e.g., bottled gases), since the construction design makes these units very sensitive to any dirt particles. For flows that do contain any particles, the immersed style of flowmeter should be used.

Immersion-probe Flowmeter

An *immersion-probe flowmeter* consists of a thermal mass-flow probe and electronics. The probe contains two resistance temperature detectors that are inserted into the sensing tube. These sensors can be easily cleaned when dealing with various particles. The one drawback of this particular type of flow measuring system is that if the flow is not uniform or flat, as shown in Figure 6-33, point measurement will vary

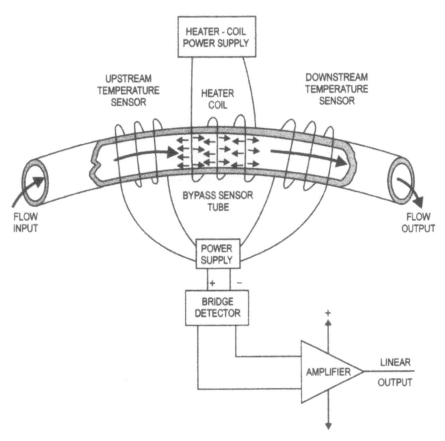

Figure 6-32. Operating principle of a heated-tube thermal mass flowmeter.

throughout the flow system.

The first sensor constantly measures the ambient temperature of the flow and maintains the line temperature differential for the second sensor. If the temperature of the flow varies, the first sensor detects this and applies a reference to it, creating a temperature-compensated basis for measurement.

The second sensor detects the flow and heats it to a predetermined difference in temperature above the first. As the gas flows past, a certain amount of heat is transferred from the heated sensor to the gas. The heat transfer rate is proportional to the mass velocity of the flow, making the second sensor the key to an immersion-probe flowmeter.

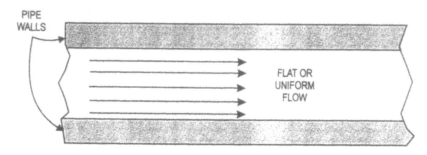

PIPE
WALLS

FLAT OR
UNIFORM
FLOW

Figure 6-33. Flow profile for an immersion-probe flowmeter.

CHOOSING A FLOWMETER

The output of both sensors operate as part of a *Wheatstone bridge*. The current of the bridge is then converted to a standard analog output signal by the integrally mounted electronics. The schematic in Figure 6-34 shows the sensors and the Wheatstone bridge.

To obtain accurate flowmeter data, selection of the device is crucial. Sometimes the choice of a particular flow-measuring instrument is based only on the type of flow that is used. In other instances, the decision might rest on the size of the line, piping before and after the meter, accuracy in measurement, range, feasibility, and a half-dozen other factors.

As we have said before, each flow sensor has specific limitations regarding the fluid to be metered. Increasingly important is the ability to interface meters with a computer for instantaneous flow readout, or remotely controlled flow which allows for unattended operation. To meet these demands and others, new types of flowmeters are being introduced and the older designs are being improved and updated.

The following is a list of the major types of flowmeters and some things to consider when selecting an instrument for a specific application.

Differential Pressure or Head. This particular type of meter is susceptible to a buildup of solid particles or particulates in the flow. In some cases, the orifice must be replaced periodically to assure accurate readings. Venturi-tube instruments can be costly and may cause permanent pressure loss in the flow path. This type of instrument obstructs the flow to some extent. In general, the output is a mechanical indication of flow.

Variable Area. These flowmeters typically provide only a visual indication of flow. They can be accomplished inexpensively, and they ob-

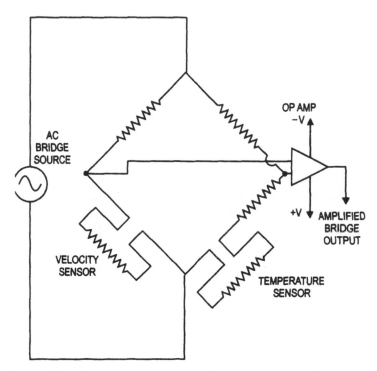

Figure 6-34. Wheatstone bridge circuit of an immersion flowmeter.

struct the flow path to some extent. These meters primarily produce a mechanical indication of flow.

Magnetic. Fluids being measured by these flowmeters must be electrically conductive. Instruments that respond to this principle are somewhat costly. This type of instrument does not obstruct the flow path, and is very reliable for long operational periods.

Velocity. These meters only work with clean liquids or gases, and can be severely damaged by high flow (which changes the calibration). The meters are usually expensive. There is some obstruction of the flow path.

Positive Displacement. Fluid temperature and viscosity affect the accuracy of this type of meter. It contains numerous moving parts and responds well to totalizing measurements.

Ultrasonic. These meters are somewhat expensive. Accuracy depends on the concentration of particulates in the flow, but the instrument has no obstruction in the flow path. This type of meter produces an electronic output signal and can be easily installed.

Mass. This segment of meters can operate only if the flow is uniform or clean.

SUMMARY

Nearly all of the products manufactured by industry are influenced in some way by the flow of materials. Flow measurement is achieved either by inference or by direct displacement of quantities.

The systems concept can be applied to flow measurement. The source of the system is generally a bulk storage tank or a reservoir that provides potential energy. A path is provided by pipes or hoses and control alters the flow. Work done as a result of the volume of flow is based upon the demands of the system.

Flow equipment is classified according to the type of measurement it performs. Rate flow and totalizing flow are the two general classes of measuring instruments.

Head flowmeters measure a change in pressure that occurs when liquids or gases are forced through a restriction in the flow path. In general, head flowmeters express a square-root relationship between the velocity, volume, and mass. The output is nonlinear, and is corrected with square-root extractors or through the use of narrow ranges of indication. Orifice plates, venturi tubes, flow nozzles, target meters, and Pitot tubes are some of the commonly used primary elements in head flowmeters.

In variable-area flowmeters, a tapered tube or float provides an orifice type of function. Flow rate is determined by the position that the float takes in the tapered orifice. The variable-area principle is used in rotameters, orifice plugs, and piston meters. Flow indications may be registered by float position, piston-position hand deflection, or chart recording.

Magnetic flowmeters measure the volume flow of electrically conductive fluids passing through a pipe or tube. Corrosive acids, sewage, detergents, and liquefied-pulp flow streams are typical applications. Voltage is induced into a conducting flow stream when it moves at right angles to a magnetic field. The instantaneous voltage generated at the electrodes represents the average fluid velocity at a given instant. The conductivity of the measured fluid must be at least 5 to 20 ps/cm. The voltage-sensing element may be mounted on the side of a pipe or may be submerged in the flow stream in large-piping applications. The advantage of this method of measurement is its obstructionless-flow characteristic.

Velocity flowmeters respond to the speed at which a stream passes through a closed pipe. Turbine flowmeters produce rotary motion as a result of the velocity of a flowing stream. Output is developed by turbine-blade distortion of a magnetic field. Vortex-swirl meters respond to velocity flow by employing the vortex-precession principle. A swirl component causes the flow to produce a high-velocity twisting vortex. When the vortex enters an enlarged area, it forms a helical path around the meter body. Flow is determined by measurement of the helical precession frequency. Strut-vortex flowmeters respond to velocity flow by plating an obstruction in the flow path. A predictable number of vortices appearing on the downstream side of the strut are counted and are used to indicate flow rate. Ultrasonic sensing is also used to count vortices in strut instruments.

Positive-displacement flowmeters divide the flow stream into parts and count them as an indication of totalized flow. Nutating disc flowmeters divide a flowing fluid into four measurable increments that are counted as an indication of totalized flow produced by rotation. Rotating-vane flowmeters count the number of predictable volumes of fluid that pass through rotating areas. Shaft motion is used to actuate mechanical counters or generate electrical signals. Positive displacement of a bellows within a closed chamber is also used to measure totalized gas flow. Two or more chamber units are needed to produce a smooth continuous flow of gas through the instrument.

Ultrasonic flowmeters respond to the propagation of pulsed signals in a flowing liquid. Flow-rate measurement is based on the time differences that occur in upstream and downstream frequencies. These instruments are accurate within ±0.5 percent and do not obstruct the flow path.

Mass flowmeters can measure very high rates of solids, liquids, or gases in lbs/hr or kg/min. Operation of this type of meter, based on the Coriolis force, measures flow directly, instead of by volume. A tube shaped like the Greek letter Omega vibrates at its natural frequency by a magnetic device. This action causes the tube to twist and move up and down. The amount of twist in the tube is directly proportional to the mass flow rate of liquid flowing through the tube. A highly accurate mass flow rate is then produced.

Thermal mass flowmeters are used mainly for measuring gas. A heated sensing element conducts heat from the flowstream. The heat of the flow is proportional to the mass flow rate. Immersion-probe flowmeters contain two sensors measuring ambient and line temperature differential. The outputs of the sensors operate as part of a Wheatstone bridge. This current is converted to a standard analog output signal.

Chapter 7

Analytical Process Systems

OBJECTIVES

Upon completion of this chapter, you will be able to
1. Explain how analytical instruments are classified according to energy-source interaction with the substance under test.
2. Identify some instruments in the electric- or magnetic-field interaction group.
3. Explain the function of mass spectrometry.
4. Show how conductivity is used to analyze materials.
5. Identify some instruments in the thermal- or mechanical-energy interaction group.
6. Show how chromatography is used to analyze gas and liquid samples.
7. Explain how viscosity, density, and specific gravity are determined.
8. Identify some instruments in the electromagnetic-radiation interaction group.
9. Explain how turbidity, photonephelometry, and photometry are used to analyze gas and liquid samples.
10. Identify some instruments in the chemical-energy interaction group.
11. Explain how colorimetry, spectrometry, and combustion are used to analyze chemical samples.
12. Define pH and show how it is analyzed.

KEY TERMS

Acidity—A measure of the hydrogen ion content of a solution.
Adsorption—A separation process involving the interaction of functional groups of a material being analyzed with specific structural features

on the surface of a stationary material.

Alkalinity—A measure of the hydroxyl ion content of a solution.

Alpha rays—Helium ions, or one of the particles emitted by radium.

Aspirator—An apparatus that produces suction to move or collect materials.

Base—A water-soluble, alkaline substance that is capable of reacting with an acid to form a neutral salt and contains a number of free hydroxyl ions. Typical solutions are household lye, bleach, ammonia, borax, and baking soda.

Beta rays—Electron emissions by a radioactive material. **Chromatography**—A method of analyzing gas or liquid by separating its constituents and causing them to seep through an adsorbent material.

Combustion—A process that produces burning as a result of adding chemicals such as oxygen to certain materials; causes heat and light to be produced.

Conductance—A measure of the ability of a material or solution to pass electrical current; the reciprocal of resistance. Density The mass per unit volume of a substance. Detector A device that is used to change or pick out a signal of one type from another signal of a different type.

Dyne—A metric unit of force that causes a mass of one gram to accelerate one cm/s/s.

Fluorescence—The simultaneous emission of light waves from a material that absorbs shorter X rays or ultraviolet waves.

Gamma rays—Electromagnetic radiation that comes from changes in the nucleus of an atom due to nuclear collisions or radioactive changes.

Half-cell—Part of a pH electrode; develops a voltage that is proportional to the hydrogen ion content of the solution in which it is placed.

Infrared—Wavelengths of the electromagnetic spectrum from 0.3 to 0.00075 pm.

Ion—Any electrically charged particle of molecular, atomic, or nuclear size.

Ionization—The process by which a neutral atom or molecule is split into positive and negative ions.

Linear variable differential transformer (LVDT)—A transformer with a moving core that is used to detect motion or position changes.

Mass spectrometer—An instrument for analyzing gases, liquids, and solids by introducing a sample into an ionizing source and forcing it to pass through a tube that is surrounded by a magnetic field.

Nephelometry—A photometric measuring process that determines the amount of suspended solids in water.

pH—A measure of the acidity or alkalinity of solutions based on a measurement of the concentration of hydrogen ions; values less than 7 are considered to be acidic, while values greater than 7 are considered to be basic.

Photoelectric cell—A device whose electrical properties undergo a change when exposed to light.

Photon—A unit of electromagnetic radiation or a packet of light energy.

Poise—The unit of dynamic viscosity; a material requiring a shear stress of force on one square centimeter to produce a shear rate of one reciprocal second has a viscosity of 1 poise.

Reagent—A substance that takes part in one or more chemical reactions or biological processes and is used to detect other substances.

Resistivity—The opposition that a material has to the flow of current; expressed in m/cm.

Scintillation—The production of visible light or electromagnetic radiation when some materials are struck by high-energy particles.

Siemens—A unit of electrical conductance, formally called the mho, that represents the ability of a substance to pass 1 A of current when 1 V is applied.

Solvation—A chemical reaction in which ions in a solution are normally combined with at least one molecule of solvent.

Specific gravity—A ratio between the density of a liquid to that of water, or the density of a gas to that of air.

Spectrometer—An instrument used to determine the index of refraction.

Spectrophotometer—An analytical instrument used to measure the relative intensities of light in different parts of the spectrum.

Stoichiometry—A branch of science that deals with the laws of proportions, or the conservation of matter and energy, as applied to chemical activity.

Stoke—An expression of kinematic viscosity; 1 stroke equals viscosity, in poise, divided by density, in g/cc.

Synchronous motor—An ac motor that operates at a constant speed regardless of the applied load.

Thermal conductivity—The transfer of heat from one part of a solid to the remainder of the solid.

Thermistor—A solid-state device that uses changes in its internal resistance to measure temperature.

Titration—A method of determining the strength of a solution or the con-

centration of a substance by adding a small amount of a known reagent to the test sample.

Toroid—A doughnut-shaped core piece of magnetic material that has a wire
coil threaded through the center hole.

Turbidity—A disturbance or disorder that occurs in liquid when a solid
material is present; causes the solution to have a cloudy appearance
when viewed by the human eye.

Ultraviolet—Electromagnetic radiation of a shorter wavelength than visible violet light.

Viscosity—The property of a fluid that offers internal resistance to its flow.

X rays—Exceedingly short electromagnetic waves; produced by special
tubes that interact with radioactive substances.

INTRODUCTION

Industrial manufacturing is a general term used to describe either the
entire process of converting raw materials into something of more value or
the combining of ingredients into a final product. Numerous processes are
utilized by industry to get a product into its final form. In addition to the
regular manufacturing processes, there is also a critical need to analyze the
physical and chemical properties of a product at various steps during its
production. *Quality control, specification testing,* and *final inspection* are usually the result of analytical process systems.

Some of the most significant developments in process equipment in
the past decade have taken place in the area of analytical testing instrumentation. The expansion of process equipment in this area is primarily attributed to two significant considerations. One is the demand for a more accurate,
thorough, and rapid means of testing "in-process" product quality. Second,
there is the pressing need for instrumentation that will detect materials and
variable conditions that contribute to air, water, noise, thermal, and other
forms of environmental pollution.

Analytical evaluation was once one of the slowest production operations in the entire manufacturing process. The availability of new analytical
equipment has changed processing that was formerly achieved by manual
testing procedures into continuous analytical operations that can be accomplished automatically during production. In this chapter, some of the common analytical process instruments that are used to evaluate product quality will be discussed.

ANALYTICAL PROCESS CLASSIFICATIONS

The number of different analytical processes used by industry is almost endless when compared with the processes discussed in preceding chapters. Density, specific gravity, viscosity, electrical and thermal conductivity, combustibility, acidity level, and alkalinity are only a few of the processes on this list. The type of equipment employed to evaluate these different processes also makes a formidable list. In general, this means that a more workable grouping of analytical process instrumentation is needed for an analysis to be meaningful. One such classification divides analytical processes into *matter-* and *energy-interaction* groups. Nearly all of the physical and chemical analysis that is of major concern to industry can be placed into these groups.

In theory, a unique energy state exists in each combination of atoms and molecules. Each atomic configuration that exists in the molecules of crystals, solids, liquids, or gases can be readily defined through different electron energy states. This physical state can be easily determined by observing an interaction that takes place between the substance under test and an external source of energy. The external source of energy may be an *electric* or *magnetic field, thermal* or *mechanical energy, electromagnetic radiation,* or *chemical energy.* These four energy-source groups will serve as the primary divisions of our analytical process classifications.

The classification of analytical process systems by energy-source grouping is valuable. Many analytical properties can, for example, be measured or inferred by more than one type of interaction. Through proper selection of equipment, it may be possible to analyze different processes through the use of a single instrument. This would reduce costly duplications and simplify many application problems. A general understanding of each energy-source group and some representative instruments that are employed by analytical process systems will serve as the basis of this presentation.

ELECTRIC- OR MAGNETIC-FIELD INSTRUMENTS

Electric- or *magnetic-field* instruments provide a powerful method of analytical evaluation. These instruments generally are used to analyze a resulting current, voltage, or magnetic flux that occurs when a test sample is exposed to an electric or magnetic field. Through this type of processing, a number of distinguishing chemical and physical features of the sample material can be sorted out and measured. A *mass spectrometer*, which detects constituent *ions* in a sample, according to mass and change, is one instrument in this group. Vapors, gases, liquids, and solid samples can be analyzed through this type of instrument. Moisture analyzers, gas detectors, oxygen analyzers, and nuclear magnetic spectra analyzers are some of the other representative instruments found in this classification.

Electric- and Magnetic-Field Reaction Theory

The production of net electric charges on atoms or molecules by ion bombardment, radiation, electrolysis, or magnetic induction establishes a measurable condition between the test sample and the energy source. Ionized gases can, for example, be accelerated by applying an electric or magnetic field. The resulting output can be collected and measured as an electric current by mass spectroscopy. Ions in a test solution can be transported or deposited under the influences of various applied potentials. Electric conduction and moisture analysis is achieved through this type of processing. Induced magnetic properties give rise to a number of rather specialized techniques which are precise and quite selective for the determination of elements and compounds. Oxygen analyzers and nuclear magnetic instruments operate by means of this principle

Mass Spectrometers

A *mass spectrometer* is a general type of analyzer that is used for qualitative evaluation of gases, liquids, and solids. Evaluation is made by introducing a sample into an ionizing source, accelerating it by use of an electrostatic or magnetic field, and separating it according to mass, detection, and recording. These instruments employ a number of electronic components in each element of the system. Figure 7-1 shows a simplified diagram of a mass spectrometer.

In practice, mass spectrometers are called on to determine the physical makeup of compounds that contain several different materials in varying amounts. Analysis of this type requires the use of principles or behaviors that are characteristic of many materials. Mass spectrometer operation is based on the following basic responses:

First, acceleration (a) of a particle of material is directly proportional to the applied force (F) and inversely proportional to its mass (m) in an instrument constant (k). The formula

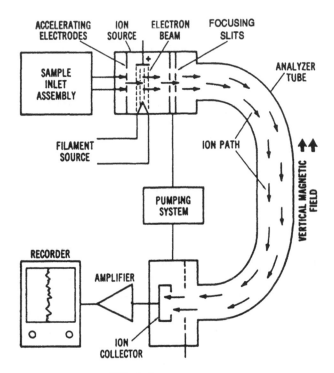

Figure 7-1. A simplified diagram of a mass spectrometer.

$$a = k \, F/m$$

is used to determine the acceleration of a particle.

Second, ions subjected to a uniform magnetic field will produce a curved path, with the radius (r) depending upon their mass (m), accelerating potential in volts (V), magnetic field strength (B), and electronic charge (e). The formula

$$r = 144 \sqrt{mVe} \, /B$$

expresses this relationship.

Third, when the kinetic energies of two particles are equal but their masses are different, it will take a different amount of time for them to travel an equal distance.

Fourth, the angular velocity by which a charged particle moves through a magnetic field is based on frequency (f), electronic charge (e), mass (m), magnetic-field strength (B), and the speed of light (c) in cm/s.

The formula

$$f = 1/2\pi \, eB/mc$$

is used to determine angular velocity, in Hz.

Last, when radio frequency (rf) is applied to particles traveling at different speeds, varying amounts of energy are imparted. This produces a form of selective acceleration.

Spectrometer Operation

In mass spectrometer (MS) operation, molecules are both ionized and fragmented. These particles are then sorted into groups according to their *mass-to-charge ratio*. A plot of the count of ions against the mass of different types of ions is called a *mass spectrum*. This record is a valuable tool in the analysis of chemical compounds.

A mass spectrometer is divided into several different components. The first is the *sample inlet assembly*. The system must also have an *ion source*. A *spectrometer tube* is used to sort out ions that pass through the system, and a detector is then needed to determine the ion content of the sample solution that passes through the spectrometer tube. Detector output is amplified and used to drive a recorder or indicating device. The entire system is generally controlled by a dedicated computer that handles data and manipulates the operation of the entire system.

The sample inlet system of a mass spectrometer must be versatile enough to handle gas, liquids, and solids. This part of the system is usually operated at 200°C and at a vacuum pressure of 0.02 *torr*. A *torr* is 1/760 of atmospheric pressure, and is equivalent to 1 mm of mercury. A sample can be introduced into a spectrometer in a variety of different ways. *Heated batch inlets* cause the sample to expand under pressure to approximately 50 μm of mercury. The sample then bleeds into the ion source through a molecular or viscous *leak plate*. *Direct sample introduction* is achieved by forcing the sample into a vacuum lock attached to a heated probe. Direct insertion of the sample is achieved by venting the sample through a vacuum lock. Capillary inlets are used to insert continuous gas samples.

The ion source of a spectrometer varies a great deal between different instruments. Its primary responsibility is to produce ions and to give them kinetic energy for introduction into the spectrometer tube. The ion source in Figure 7-1 employs a heated filament similar to that of a vacuum tube. Ions are produced by *electron bombardment*. The *electron-impact source* in Figure 7-2 is the most widely used ion source. In this source, gas molecules from the inlet are introduced into a stream of accelerated electrons. The electrons and molecules collide, causing ionization and fragmentation. The resulting ions are then injected into the spectrometer by the accelerating slits. Another ion source responds to *chemical ionization*. In this source, the molecules being analyzed are ionized by an *ion-molecule reaction*. A *reagent gas* is first ionized by electrons. These ions then react with the source molecules in a number of different ways. This type of ionizing source transfers less energy to analyte molecules, causing less fragmentation. Fragmentation is also controlled by the choice of reagent gas. Other types of ion sources are *field ionization, spark source, atom bombardment,* and *thermal ionization.*

The spectrometer or analyzing tube of a mass spectrometer is used to separate ions emanating from the ion source as efficiently as possible. Quantitatively, this function determines the resolving power of the instrument. This is defined as the ratio of M to ΔM. In this ratio, M and M plus ΔM are the mass numbers of two neighboring peaks of equal intensity in the spectrum. A key factor in this operational procedure is the ability of the instrument to distinguish between M and M plus ΔM. Generally, this is

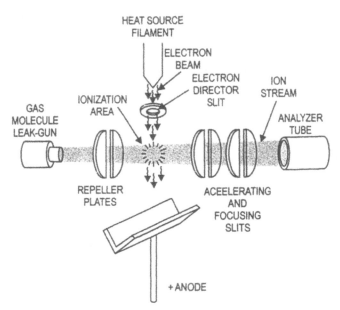

Figure 7-2. Electron-impact source for a mass spectrometer.

achieved when the valley between peaks is no more than 10 percent of the maximum intensity of M or M plus ΔM. A *magnetic-deflection analyzing tube* is used to achieve this operation. In the spectrometer tube, a controllable magnetic field causes ions to deflect along curved paths according to their mass-to-charge ratio. Tube design is such that only those ions that follow the path coinciding with the arc of the magnetic field are brought to focus on the detector. Each ion path is therefore separated into different mass/electron charge streams. In an actual instrument, the radius or curvature of ions in a magnetic field is 4 to 6 in (10 to 15 cm). The analyzer tube in Figure 7-3 shows a representation of this operation.

The detector of a mass spectrometer is primarily an electronic function. Most instruments employ an *electron multiplier* or a *channel electron multiplier array* as a detector. Both of these devices respond to the release of electrons from a piece of solid material by ion impact. The resulting emitted electrons are then collected and amplified, and the signal voltage developed by the amplifier causes current to flow through the detector load resistor. This output represents a spike or *pulse* of energy that is ultimately displayed or recorded.

Mass spectrometer operation must be kept under high vacuum, with diffusion pumps that are backed by rotary oil pumps, If the spectrometer is connected to a *gas chromatograph,* an interface is placed between the inlet system and the chromatograph to allow sample transport while removing the carrier gas. An effusion membrane with jet separators is used for this purpose. Finally, most high-resolution spectrometers employ dedicated computers for data handling and component control. The computer may be used to compare the developed system data with a library of recorded data from thousands of known compounds. Through this procedure the instrument can identify chemical compounds quickly and with a high degree of accuracy.

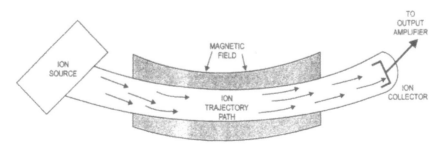

Figure 7-3. Analyzer tube of a mass spectrometer.

Industrial applications of the mass spectrometer are quite numerous. Common uses include such things as compound identification, quantitative analysis, leak detection. molecular-structure determination, and isotope-ratio determination.

Electrical Conductivity

All materials conduct electricity to some degree. Even the best insulating materials break down and go into conduction when a suitable voltage is applied. As a rule, most insulators necessitate more voltage to break down than is available from an energy source. Industrial applications of electrical conductivity are primarily concerned with the movement or flow of charged particles in an aqueous solution. The measure of a solution's ability to conduct electricity is called *conductivity*. Conductivity is often described as the reciprocal of *resistivity*.

Traditionally, the unit of conductivity has been the *mho/cm*. A mho is the reciprocal of an ohm of resistance. A resistivity of 100/cm has a conductivity of 0.01 mhos/cm. The mhos/cm unit of conductivity is now being replaced by an international term called the *siemens/cm* (S/cm). A common expression of conductivity in industrial solutions is the *microsiemens/cm* (μS/cm). The resistivity of this solution would be expressed as megaohms/cm (MΩ/cm). A cm^3 of high-purity water has a resistivity of 18.3 MΩ/cm at 25°C. This could be expressed as a conductivity of 0.054645 πS/cm at 25°C. A solution such as salt water may have resistivity as low as 30Ω/cm. Expressed as conductivity, this would be 0.033333 S/cm. Highly conductive solutions are best expressed as conductivity values while solutions with low levels of conductance are more commonly expressed as resistivity values.

Conductivity Analyzers

Electrical-conductivity analyzers are designed to determine the concentration of dissolved chemicals in various liquid solutions. These instruments are of particular interest to industry in the analysis of liquids and compounds dissolved in water. Operation is based on the measurement of ion concentrations by a rather simple and inexpensive process. Instruments of this type employ an energy source, a conductivity cell or probe, a measuring circuit, and a readout or display device. Accuracies of ±0.05 percent of full scale or ±1 digit or less can be achieved with laboratory and portable instruments.

The fundamental principle of electrical conductivity is based on the movement of electrons in metallic conductors and ions in an *electrolytic solution*. Electrical-conductivity analyzers primarily respond to both types of conduction. For electrolytic solutions, electrons are supplied from the source to the solution under test. From this point on, conduction is by positive and negative ions through the sample solution. Positive ions travel toward the cathode electrode, where they are reduced, and negative ions move toward the anode, where oxidation occurs. The conductivity of a solution depends on the concentration and mobility of its ions.

If dc were used as an energy source for a conductivity analyzer, the solution near each electrode would progressively change. In addition, electrode reaction would ultimately set up a voltaic reaction that would produce a *back* or *counter voltage*. The term *polarization* is commonly used to describe this action. To solve this problem, ac electricity is commonly used as the energy supply source in this type of analyzer.

Figure 7-4 shows an ac Wheatstone bridge of an electrical-conductivity analyzer. In this unit, 1 to 10 V of 60-Hz ac is applied as the source. The measuring element of this unit is a *null detector*, such as a zero center-reading galvanometer, a millivolt meter, an oscilloscope, or a differential amplifier that drives a microprocessor. Potentiometer R_s is called the *standard arm* of the bridge. It is primarily used to change the sensitivity range of the instrument. Resistor R_2 is a calibrated slide-wire potentiometer. Resistance R_x is determined by the conductivity of the sample solution into which the probe is placed.

Figure 7-4. An ac Wheatstone bridge electrical-conductivity analyzer.

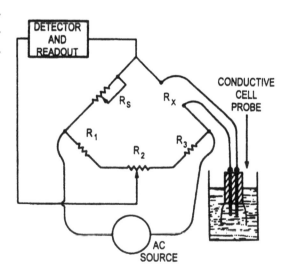

Standard conductivity analyzers have a measuring range from a fraction of a µS to more than 1 S. A siemens represents the ability of a substance to pass 1 A of current when 1 V is applied. In practice, this type of instrument is used to measure very low-level chemical contamination. Distilled or demineralized water, for example, has a conductance of 0.2 $µS/cm^3$ at 25°C. The addition of a few grains of salt will cause the conduction of the solution to change as much as 500 percent.

A cross-sectional view of the conductivity cell or probe of a conductivity analyzer is shown in Figure 7-5. This particular probe contains six electrodes. Two voltage electrodes supply energy to the sample, and conductivity is measured between the two current electrodes. The remaining two electrodes are linked into the measuring electrodes in such a way as to limit the stray current of the cell via the sampling solution. Should fouling of the probe occur, only a negligible amount of current will be drawn from the voltage electrodes because of their connection to a high-impedance source. The voltage will therefore remain unaffected, which ensures that the current output is correct for a particular value of conductivity. Correct operation is possible as long as the fouling impedance of the electrode

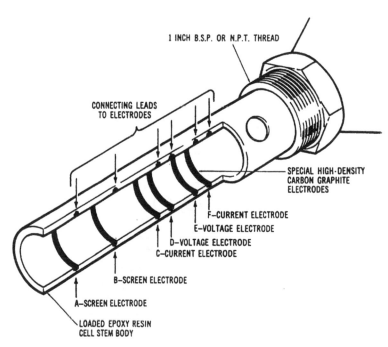

Figure 7-5. A cross-sectional view of a conductivity probe.

does not increase beyond the dynamic amplitude range of the source.

Oxygen Analyzers

Oxygen analyzers are conductivity instruments that are used in a variety of in-line process installations, for portable or field-based measurements, and in laboratory analysis equipment. Measurements achieved by this type of instrument are useful for flue-gas monitoring, food packaging, hazardous-environment monitoring, heat treating, boiler feedwater evaluation, and pollution engineering. These instruments respond to the electrical conductivity of an applied oxygen sample. The percentage of oxygen in a sample can be monitored on a continuous basis, and oxygen levels can be displayed on a meter, identified on a digital display, recorded on a chart, used to actuate an alarm, displayed on a computer, or used to alter a mixing process.

The sensor of an oxygen analyzer produces an output signal as the result of conduction produced by oxygen applied to a special cell. The sensor in Figure 7-6 has been specially designed for oxygen detection. The silver anode and gold cathode of the cell are protected from the test sample by a thin layer of Teflon. An electrolytic solution of potassium chlorine is held in the sensor by a saturated membrane. When a test sample is applied to the cell, oxygen will diffuse from the sample to the cathode and

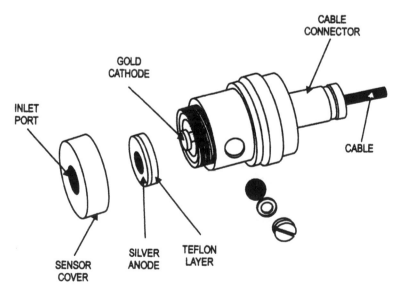

Figure 7-6. An oxygen sensor.

be reduced by chemical reaction from $O_2 + 2H_2O + 4e$ to $4OH$. At the same time, anode reaction will cause $4Ag + 4Cl$ to change to $4AgCl + 4e$. As a result, a conduction current will flow between the dc-energized anode and cathode. The magnitude of this current will be proportional to the amount of oxygen appearing in the test sample. The electrolyte solution of the sensor will remain charged for up to six months before recharging is needed. Recharging is achieved by simply adding a new supply of the liquid electrolyte. These cells may be used in flow chambers, for submersion applications, and in fast-recovery gas-flow assemblies. This sensor and instrument are found in a wide variety of process analysis applications.

Oxygen analyzers can also be used to determine the *dissolved oxygen* (DO) content of an aqueous solution. An example of such a sensor and its electronics is shown in Figure 7-7. When a potential of 0.8 V is applied across the anode and cathode, any oxygen that passes through the membrane of the cell will be consumed or reduced at the cathode. This action causes current conduction between the anode and cathode. The magnitude of the current produced is proportional to the amount of oxygen in the sample and is therefore a measure of the amount of oxygen present.

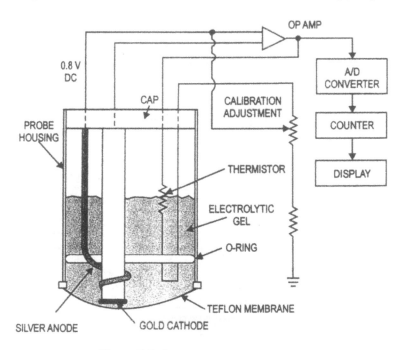

Figure 7-7. An oxygen sensor system.

The DO of a sample solution is measured in parts per million (ppm) or a percentage of the saturated value. A measure of DO in natural water is taken to assure its presence. Oxygen is needed to improve appearance, reduce odors, and maintain aquatic life. In boiler feedwater, DO is measured to assure that it is maintained at a minimum level, because minute traces of oxygen in feed water causes increased corrosion in steam-generating equipment.

Electrodeless Conductivity Analyzers

When high-conductivity or corrosive solutions are to be analyzed, *electrodeless* instruments can be utilized. Figure 7-8 shows a simplification of an *electrodeless conductivity cell.* This type of analyzer measures conductivity through a closed loop provided by the sample solution. The sensor assembly consists of two toroidal windings placed on each side of a nonconductive pipe. One winding is excited by an audio-frequency signal supplied by an oscillator in the control assembly. The second winding is connected to a detector circuit. The flow of a conductive material through the probe determines the amount of magnetic field transferred into the detector coil.

Since the electrodeless sensor does not make direct contact with the sample solution under test, it does not have any of the common problems normally associated with contacting-type conductivity sensors. Polariza-

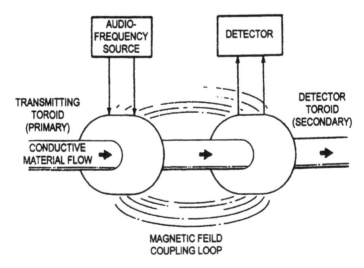

Figure 7-8. A simplification of an electrodeless conductivity analyzer probe.

tion, oil fouling, process coating, or electroplating do not affect the performance of electrodeless sensors. The probe is normally coated with a vinyl ester material that is nonconductive for most process fluids. This type of insulation eliminates ground loops, which can affect the accuracy of the conductivity process. Each sensor generally has an integral temperature compensator to automatically adjust the conductivity reading to a 25°C reference temperature. The probe is used to develop an output signal that is applied to a current amplifier. A standard 40- to 20-mA output signal is developed by the control unit of this system. The indicator or display of this instrument is normally calibrated to read percentage of electrolytic concentration instead of μS of conductivity. This type of probe is generally more costly than its contacting-electrode counterpart. The advantages of electrodeless sensors generally outweigh the additional cost of this method of measurement when compared with the cost of direct-contact electrodes. Noncontact conductivity sensors are designed for use in difficult environments and for measurements that are not easily obtained by direct-contact sensors.

THERMAL- OR MECHANICAL-ENERGY INSTRUMENTS

Thermal energy or *mechanical energy* as an external source to produce an interaction with a substance is used in the second group of analytical instruments. This technique is quite unique when compared with the other three instrument divisions. For example, it may involve the use of massive amounts of a sample or a minute portion of some material in its analysis procedure. Measurement of this type is made in terms of energy transmission, work done, or changes in the physical state of the sample. Thermal conductivity, dew point, thermal expansion, density, specific gravity, and viscosity are some of the measurements achieved through this type of analysis.

Thermal-conductivity Analyzers

The ability of some gas molecules to become highly excited, which causes molecular vibration, twisting, and rotation, permits them to conduct large amounts of heat away from an external heat source with which they collide. The resulting cooling effect on a heated body can be used to determine the quantity of particular molecules present in a sample. In practice, thermal-conduction analysis is achieved by comparing the change in heat dissipation that occurs between a reference value and the sample being analyzed.

The principle components of a thermal-conductivity analyzer are the measuring cell, regulated power supply, Wheatstone bridge, and case temperature control unit. A cross-sectional view of a representative measuring cell is shown in Figure 7-9. As a rule, this part of the analyzer is made of a rather large mass of metal that serves as a *stabilizing heat sink*. Passages formed in the cell provide a flow path for the sample and reference materials. Cavities extending from the top of the cell into each passage are used to house the filament heat source. The construction of the sample and reference passages is primarily the same.

The heat source of a thermal-conductivity analyzer may be either a hot-wire filament or a glass-bead thermistor, depending on the sample being measured. The source may be placed directly in the flow path or recessed in the cavity. A recessed source provides reduced sample noise, but is somewhat slower to respond to changes in flow. In practice, a cell employs one or more heat sources in each passage. The sensitivity of the cell is improved with each additional source element. Some cells are built to accommodate as many as eight pairs of sources, for a high level of sensitivity.

The electrical part of a thermal-conductivity analyzer is shown in Figure 7-10. A Wheatstone bridge, a regulated power supply, a bridge bal-

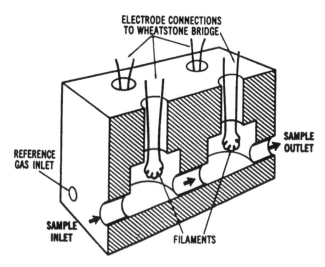

Figure 7-9. Cross-sectional view of a four-element thermal conductivity cell.

ance ammeter, and an unbalance ammeter are included in the circuit. The power supply must be capable of supplying between 100 and 300 IDA to the bridge circuit. Measurement stability is directly dependent on power-supply regulation.

Analyzer operation is based on a comparison of the thermal conductivity of a sample gas and a reference gas. When a metered sample of 50 to 200 cm^3 passes through the measuring cell, it flows across the heated filaments or thermistors. This action causes heat energy to be conducted away from the source. The amount of heat conduction that takes place is based on the thermal conductivity of the flowing gas. If the sample has a lower thermal conductivity than the reference gas, less heat is conducted

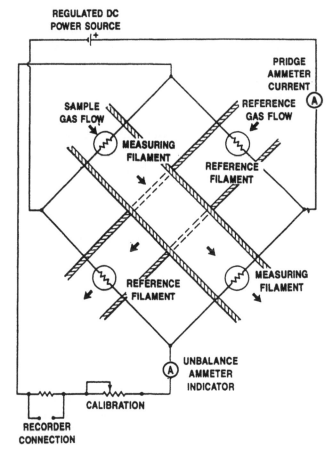

Figure 7-10. Electrical components of a thermal gas analyzer.

away and the resistance of each heat source increases, causing the bridge to be unbalanced. A sample with higher thermal conductivity than the reference gas will produce an unbalance due to a decrease in resistance. The degree of bridge unbalance is then calibrated in terms of gas composition.

Gas Leak Detector

The *gas leak detector* typically utilizes the thermal-conductivity principle in its operation. This instrument is more often used to detect gas leaks than in analysis applications. It is energized by 120- V, 60-Hz ac and is equipped with rechargeable batteries and an audio sound system. It detects all gas mixtures that have a thermal-conductivity value greater than air.

When the probe of the instrument is passed into a suspected leak area, it draws in a sample by pump action. This sample is then applied to a special low-volume thermal-conductivity resistance cell. At the same time, a reference sample, such as air, is drawn into the second thermal cell. The Wheatstone bridge then compares the two resistance values and applies the difference signal to an operational amplifier. The resulting output is then measured and displayed on the indicating meter. The audio system is energized by the same output opAmp, and produces an audio tone when a certain level of gas is sensed by the probe. Instruments of this type are a necessity for maintaining leak-free gas systems.

Chromatography

Chromatography is an instrumental procedure, based on the *adsorption* principle, that is used to separate different components of a chemical or gaseous substance. Liquid chromatography deals with the separation of liquid constituents of a sample solution. Gas chromatography is responsible for the separation of the individual constituents of a gaseous solution. In general, the process includes component separation, identification, and quantitative measurement. This procedure is particularly useful when test samples possess similar physical and chemical properties that make other analytical techniques difficult or impractical.

Since the early 1950s, chromatography has found widespread usage in process analysis applications. These instruments are used to analyze different mixtures and to determine the components of gas, vapor, or liquid streams. The sensitivity, speed, accuracy, and simplification of this method of analysis has resulted in a phenomenal number of new applications. This technique is now considered to be the most widely used analytical method in industry.

The functional elements of a modern chromatographic instrument are shown in the block diagram in Figure 7-11. The basic components of this system apply to both gas and liquid chromatography. The method of analysis in this case is a discrete-sample laboratory procedure. Analysis begins when a representative sample is injected into the sampling port of the instrument. The sample also enters the mobile phase of the operation at this time. Pressure reduction and temperature stabilization are usually achieved in the sampling port. 'The sample then goes into a constant-volume loop, where it is held and eventually injected into the chromatographic column. The column *effluent* or output is then monitored by a system detector that changes chemical information into an electrical signal. The signal is amplified and applied to a recorder, computer memory, or graphic display.

Chromatography is a separation technique based on the differential affinities of a dissolved substance and its movement through some stationary material packed in a cylindrical column. The stationary phase of this operation is determined by the content makeup of the solid or liquid material of the column. This is determined by the physical structure of the material through which the dissolved substance must pass. The mobile phase of chromatography usually deals with the flow of a sample through a cylindrical tube packed with an adsorbent particulate material. The term *adsorbent* refers to the holding power or attracting characteristics of each particle of the packing material. When a discrete sample is applied to the column inlet, it first encounters the stationary phase of operation. When

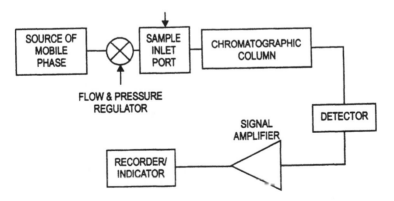

Figure 7-11. Block diagram of a chromatographic instrument.

the sample starts to move through the stationary material, it enters the mobile phase of operation. The mobile phase is responsible for individual sample constituents separating from one another according to their relative affinities for the adsorbent material. Substances with a low affinity for the adsorbent material pass through the column more rapidly and emerge first. Substances with a greater affinity for the adsorbent material remain in the column longer and emerge later.

The chromatographic separation process is illustrated in Figure 7-12. Operation begins when the sample solution is placed in the inlet at the top of the column. The solution molecules are equally distributed when first introduced. As the solute begins to move (goes into the mobile phase), the compounds become separated. Typically, some type of detector that responds to the effluent is placed at the bottom or outlet of the column. This detector monitors the quantity of column effluent. The electrical output of

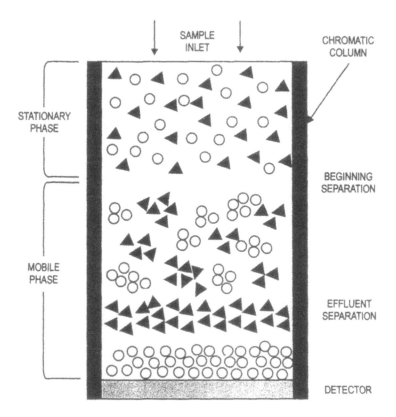

Figure 7-12. Chromatographic separation.

the detector is used to produce a display that indicates the flow rate of the mobile phase of the operation. The display is generally called a *chromatogram*. Chromatograms are used to identify different constituents of the sample solution. A representative chromatogram is shown in Figure 7-13.

Gas chromatography (GC) is an important industrial analytical process. In this type of chromatography, the mobile phase of operation uses an inert gas, such as helium, nitrogen, hydrogen, or argon, to carry the sample through the column. The carrier gas is supplied from a gas cylinder and its flow rate is regulated. Separation of the test sample is accomplished in the mobile phase of the operation. This is commonly called the *vapor phase* in gas chromatography. The vapor phase must be temperature controlled in a specific temperature range in order to separate the composition of a sample. Packed columns, capillary columns, and porous-layer columns are used in this operation. The analyzed mixture is injected through a rubber membrane into a stream of preheated carrier gas by means of a small-volume syringe.

Gas samples of less than 1 µl can be measured in this instrument. After the sample is volatilized and carried into the column, the separation operation begins. Column effluents are then sensed by a detector. *Flame*

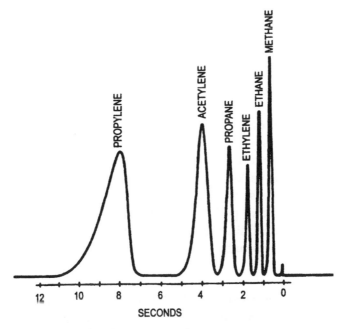

Figure 7-13. A gas chromatogram.

ionization detection (FID) is commonly used in GC. This type of detector responds to thermal ionization. An electrode situated above a hydrogen flame, which is ordinarily nonconductive, monitors thermal conductivity. Any gas emerging from the column will cause the hydrogen flame to become conductive. The flame-electrode assembly will produce an output signal that identifies the eluted gas components when conduction occurs. This type of detector is responsive to all organic compounds.

Liquid chromatography (LC) is generally used for the separation of unstable samples that have large nonvolatile molecules. The resolving power of LC is usually lower than that of GC, but its simplicity is quite advantageous. In LC, the mobile phase of operation is percolated through a chromatographic column by gravity, a pump, or centrifugal force. Mobile-phase composition is an important variable in liquid chromatography. Sample solubility, the competition between mobile phase and sample for available adsorbent surfaces, and the solvation of some chromatographic materials all influence the separation process, and the construction of an LC column must take all of this into account. The effluent of a column is selectively collected and prepared for further analysis. The extent of separation is monitored either in collected fractions or by a continuous detector. Collected fractions are detected by a physical process, while continuous detection is largely accomplished by optical properties such as spectrophotometry, fluorometry, reflective index recording, and light scattering. In addition, electrochemical cell detectors that respond to conductivity, coulometry, and polarography are used. In general, there is a lack of universal detection for liquid chromatography comparable to that in gas chromatography.

High-performance liquid chromatography (HPLC) has become especially important for the separation of complex mixtures of nonvolatile materials. Essentially, this form of chromatography employs the use of very small particles in the packing of its chromatographic columns. It also uses high pressure to increase the rate of flow and shorten the separation time. HPLC is widely used in pollution engineering, water chemistry, chemical/petrochemical analysis, food and cosmetics production, drug screening, and biomedical research.

Viscosity

Viscosity is the property of a fluid that offers resistance to flow. Nearly everyone has had an opportunity to work with liquids that are thick and sticky. This type of liquid is considered to have a high *viscosity rating.*

It obviously takes more force to move a highly viscous fluid than one of a lower viscosity rating.

The viscosity rating of a fluid is primarily attributed to the attraction of molecules within its structure. Fluids with a high viscosity normally have a strong attraction between individual molecules. As a result, fluids tend to form layers or films within their structure. When flow occurs, each layer is forced to move against or over another layer, and the resulting opposition to flow is called *fluid resistance*. The viscosity of a fluid determines its fluid resistance.

There are several different things to take into account in an investigation of viscosity. *Absolute viscosity* is a resistance measure of internal deformation, and refers to the force required to move one parallel surface at a speed of 1 cm/s past a second parallel surface separated by a fluid film of 1 cm. This is further shown in Figure 7-14.

The fundamental unit of absolute viscosity is the *poise*. It is expressed mathematically by the formula

$$\text{Absolute viscosity (poise)} = \text{dyne-second}/\text{cm}^2$$

A smaller unit of absolute viscosity is the *centipoise*. One centipoise equals 0.01 poise.

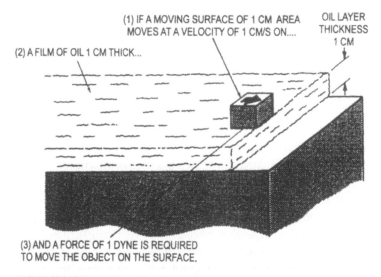

(1) IF A MOVING SURFACE OF 1 CM AREA MOVES AT A VELOCITY OF 1 CM/S ON....

OIL LAYER THICKNESS 1 CM

(2) A FILM OF OIL 1 CM THICK...

(3) AND A FORCE OF 1 DYNE IS REQUIRED TO MOVE THE OBJECT ON THE SURFACE.

(4) THE VISCOSITY IS EQUAL TO ONE POISE.

Figure 7-14. Principle of absolute viscosity.

Kinematic viscosity is a ratio of the absolute viscosity to the mass density of a liquid. When absolute viscosity is divided by the density of the liquid, it is expressed in *stokes*. A *centistoke* is 0.01 stoke. Conversions between absolute and kinematic viscosity are expressed by

$$centipoise = centistoke \times density$$

or

$$centistoke = centipoise/density$$

In both formulas, density is in g/cm^3.

Viscosity Instruments

Viscosity measurements are achieved by an interaction between the material under test and mechanical energy. A falling piston or ball moving through the sample, rotating spindles, or displacement of a float are some of the methods of measuring viscosity. Instruments that employ these principles may be used in laboratories to make periodic measurements, or they may be attached to a system for continuous evaluation operations. Viscosity measurements are used to determine values such as the flowability of fluids or the concentration size of solids in a slurry, in mixing operations, and in food processing.

The falling-ball principle, as shown in Figure 7-15(A), is one of the simplest methods of measuring viscosity. A sample of fluid is placed in a graduated container and a sized metal ball is then dropped into the top of the container. The time it takes the ball to travel a certain distance is measured. The viscosity of the fluid is directly proportional to the fall time. The falling piston and cylinder in Figure 7-15(B) is a variation of the falling ball. These methods of measurement are used in laboratories to evaluate the viscosity of test samples.

Continuous viscosity measurements are often achieved by the *rotameter principle*. This instrument uses a positive-displacement pump to maintain a constant flow of fluid through the instrument. By maintaining a constant flow rate, the instrument is sensitive only to changes in viscosity. The float is positioned on a scale calibrated in centistokes. This type of instrument is called a *single-float viscorator* by its manufacturer.

A variation of the rotameter principle is employed in the *concentric viscorator*. This instrument is designed for continuous in-line viscosity measurements. It contains a viscosity-sensitive float mounted inside a differential-pressure regulator float. The regulator float maintains constant

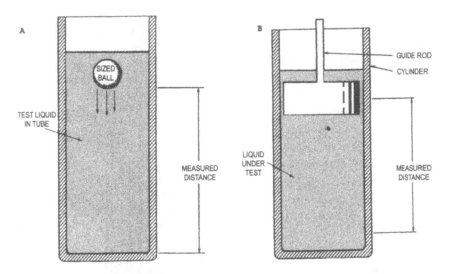

Figure 7-15. Viscosity measurements: (A) Falling-ball principle; (B) Falling-piston principle.

flow, through an orifice, to the inner viscosity float. Movement of the inner float is therefore directly proportional to the viscosity of the fluid. An extension rod attached to the viscosity float is used to actuate a recording instrument.

A cross-sectional view of a concentric viscorator is shown in Figure 7-16. Fluid entering the inlet port divides into two paths. The portion of fluid flowing upward produces a pressure drop based on its weight across the differential-pressure float. Changes in flow rate cause the float to rise, thus creating a larger annular area around the float to compensate for increased flow.

Fluid flowing downward enters the differential-pressure float through an orifice at the bottom of the instrument. The pressure across the viscosity float is of a constant value based on the difference between P_2 and P_3. Since the differential float compensates for changes in flow rate, the viscosity float only responds to fluid-viscosity values. The meter float extension is used to actuate a recording instrument or to initiate a control function.

Viscometers

Viscosity can be measured by inserting a vibrating member or rotating spindle into a test sample. Upon entering the sample, the mechanical member experiences a damping force that is a function of the viscosity of the sample. The power required to maintain vibration or rotation at a constant-amplitude oscillation is directly proportional to the viscosity of the material under test. The advantages of oscillation viscosity testing include precision measurement and high sensitivity to small viscosity changes at low levels.

A laboratory rotational viscometer with digital readout is often used. This instrument measures viscosity by sensing the *torque* required to rotate a spindle at a constant speed while it is immersed in the test sample. The developed torque is proportional to the viscous drag on the immersed spindle, and thus the viscosity of the sample. This type of viscometer is rugged, mechanically sound, designed for long use in extreme environmental conditions, and retains its calibration for extremely long periods of time. Its easy-to-read digital display is a percentage value

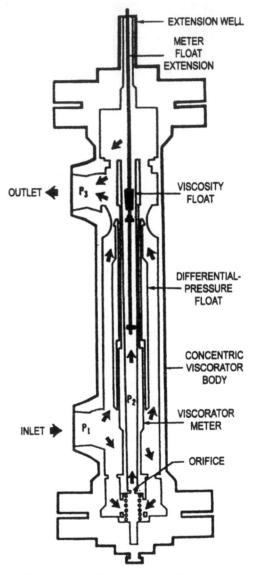

Figure 7-16. Cross-section of a concentric viscorator. (Courtesy Fischer & Porter Co.)

that readily converts into centipoise, More sophisticated versions of this instrument have improved accuracy, continuous sensing, spindle-speed monitoring, computer compatibility, and can be programmed for automatic operations.

Rotational viscometers are also available for in-line measurement of process viscosity. An in-line viscometer permits automatic control of a manufacturing process. This instrument may be designed to operate in a fully flooded system under pressure. When placed in-line, it generates a linear 4- to 20-mA signal that is proportional to viscosity. This output signal is compatible with standard industrial control equipment, recorders, data loggers, and computers.

A diagram of the internal workings of an in-line rotational viscometer is shown in Figure 7-17. In operation, the sample stream flows into the sample chamber at the bottom of the diagram. After passing through the inlet, it is forced into a rotor assembly that is driven by a synchronous motor. The turning force of the rotor causes the process sample to flow into a measuring annulus between the rotor and stator. The viscous drag of the sample on the stator is resisted by the torque tube assembly, causing angular motion of the synchronous rotor assembly. The linear ac voltage developed by this action is amplified to produce an output signal that is proportional to the viscosity of the applied sample. Measuring is continuous, and the instrument instantly responds to a change in sample viscosity. The response time of this instrument is directly related to the flow rate of the test sample. The assembly can either be directly installed in-line or placed in a bypass line for intermittent operation. Applications include viscosity measurements for alcohol-based materials, dairy products, fuel oil, latex-based adhesives, polymers, tomato puree, and other fluids.

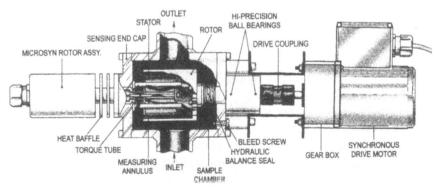

Figure 7-17. Internal workings of an in-line rotational viscometer (Courtesy Brookfield Engineering Laboratories Inc.)

Density and Specific Gravity

In industrial process analysis it is often necessary to know the *density* or *specific gravity* of a material being tested. The density of a material refers to its weight per unit of volume. Typical measures of density weight are g/cm^3 and lbs/ft^3. Specific gravity is a measure that compares the density of a sample with the density of water at a particular temperature. Specific gravity is a dimensionless value.

Specific gravity and density are often used interchangeably when evaluating liquids and solids. The weight of a ft^3 of aluminum, for example, is 167 lbs at room temperature. An equal volume of water at the same temperature is 62.3 lbs. The specific gravity of aluminum is therefore 167 divided by 62.3, or 2.68. This means that aluminum is 2.68 times as heavy as water.

Specific-gravity values may be greater or smaller than 1. A value indication of less than 1 indicates that the sample is lighter than water. When the value is greater than 1, the material is heavier than its water counterpart.

The density of a material can also be determined by the product of its specific gravity value and the density of water. Mathematically, this is expressed by the formula

$$\text{density [weight]} = 62.3 \; lbs/ft^3 \times SG$$

or

$$\text{density [weight]} = 1 \; g/cm^3 \times SG$$

The density of a material with a specific gravity of 0.5 would be 62.3 lbs/ft′ times 0.5, or 31.5 lbs/ft^3. This means that if any two values of density, weight, or specific gravity are known, the third value can be determined by calculation.

Specific Gravity and Density Instrumentation

Specific gravity and density measurements are primarily achieved by some type of interaction that takes place between a test sample and a form of mechanical energy. Such things as a float, displacement, purged air, and weight are in common use today.

One of the simplest ways of measuring liquid density or specific gravity is with a *float hydrometer*. This instrument has a weighted float that displaces a volume of liquid equal to its own weight. The float mechanism is usually made of hollow glass or a metal tube and is weighted at one end to make it float in an upright position. The position of the hydrometer float depends on the density of the liquid. A less-dense liquid causes the float to position itself lower in the liquid because a greater volume of liquid is displaced. A density or specific-gravity scale appears on the upper portion of the float. A reading is taken by noting the point on the scale to which the liquid rises. A hydrometer generally has a thermometer housed in the float mechanism. The temperature of the test sample is taken so that any density changes due to ambient temperature can be corrected.

An electrical hydrometer is shown in Figure 7-18. This particular instrument employs a LVDT as a transducer. The float of this instrument will displace its own weight in liquid. When submerged in a liquid, the float will rise or fall according to the specific gravity of the liquid in which it is submerged. The float positions the LVDT, which in turn produces an electrical signal in proportion to the displacement of its metallic core above or below its center. The output of the LVDT is connected to a calibrated meter which displays the specific gravity of the fluid being tested.

The displacement instrument in Figure 7-19 is used to determine the specific gravity or density of a liquid. With the displacement element completely immersed in the sample solution, the resulting buoyant force is directly dependent on the weight of the displaced liquid. Mechanical energy is therefore a function of specific gravity or liquid density.

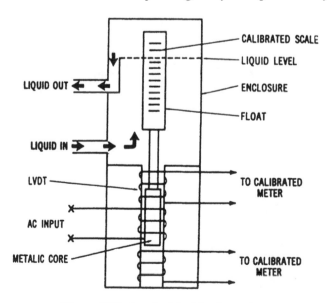

Figure 7-18. An electrical hydrometer.

In an operating process system, a test solution is first admitted to the displacer chamber. When the sample level is of a constant value, the resulting buoyant force raises the displacer accordingly. The torque lever in Figure 7-20, attached to the displacer, monitors position changes by turning a proportional amount. The opposite end of the torque lever acts as an actuating mechanism for the indicator. In direct-reading instruments, the lever simply moves an indicating hand on a calibrated specific-gravity scale. The same type of mechanism may also be used to actuate a pneumatic or electrical measuring instrument.

In an electrical specific-gravity instrument, displacer action is used to change the inductance of a coil or the core of a LVDT. The resulting output is amplified and used to drive a digital display, chart recorder mechanism, or computer. The output signal voltage is calibrated in density units or specific gravity.

The density of liquid or fluid flowing through an operating system can also be measured by using a *vibrating densitometer*. This instrument

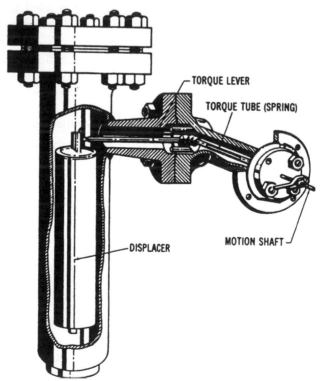

Figure 7-19. Displacer level instrument.

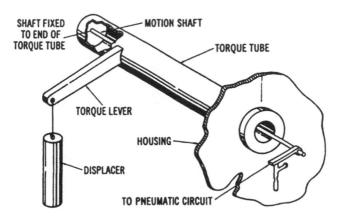

Figure 7-20. Operation of the torque lever of a displacer instrument.

was developed for extremely accurate fluid metering, pipeline interface detection, blending operations, and automatic process control applications. Figure 7-21 shows the installation of a vibrating densitometer in a process line.

The vibrating densitometer in Figure 7-22 shows two vertical tubes housed in a metal enclosure. The two tubes and end pieces form a mechanical resonant vibrating system. Fluid entering at the bottom of the left tube travels to the top, crosses to the right tube, flows down, and exits at the bottom. The two tubes are mechanically vibrated at a resonant frequency by an electronic oscillator, and the fluid flowing through the two tubes is a function of the resulting vibration. The resonant frequency of the vibrating tubes will vary with the density of the fluid. The driver piezoelectric element attached to the left tube is used to vibrate the tube at the resonant frequency. The pickup piezoelectric element attached to the right tube responds to vibration of the entire assembly, which depends on the density of the fluid passing through the instrument. A high-density fluid tends to slow down the vibrations, while a lower density causes it to return to the natural resonant frequency. The resulting output frequency is amplified and converted into a signal that is proportional to the density of the fluid passing through the instrument. The output can be converted into a 4 to 20 mA analog signal or a digital signal with a changing frequency that can be counted.

Fluid density or specific gravity can be determined by instruments that respond to the pressure of liquid in a container with a fixed height. Instruments of this type are commonly referred to as *hydrostatic-head devices* or *bubbler instruments.* This approach to density measurement is very simi-

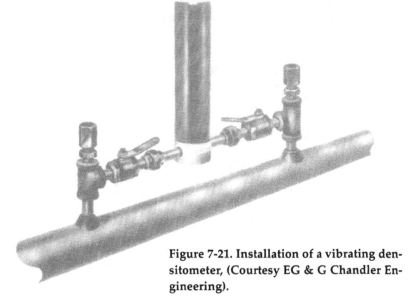

Figure 7-21. Installation of a vibrating densitometer, (Courtesy EG & G Chandler Engineering).

Figure 7-22. Vibrating densitometer diagram, (Courtesy EG & G Chandler Engineering).

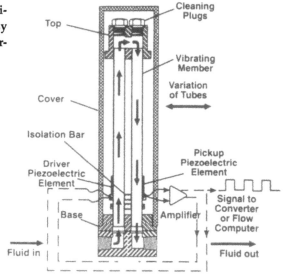

lar to that of the head-level measuring technique. In level measurement, the specific gravity of a liquid was known and level was determined by a difference in pressure. In density measurements, the head or tank level is maintained at a constant value, with different specific-gravity values producing changes in pressure.

In principle, the pressure of liquid at a given position is equal to the height of the liquid (H) times the density (p). In its simplest form, the pressure of a tank of liquid at a constant height varies directly with its density. A pressure transmitter placed at the bottom of a constant-level tank can therefore be used to determine liquid density, as shown in Figure 7-23(A).

A *liquid purge installation* is shown in Figure 7-23(B). Two taps connected to a vertical line are purged with a liquid reference fluid, such as water. In effect, a differential pressure is produced by the two water columns because of the position location. The purge rate of water is quite small, so only a minimum of dilution occurs. Density measurement of slurries is commonly achieved by this method.

A common specific-gravity test is achieved by the air-bubbler installation shown in Figure 7-23(C). The difference in head pressure developed between the reference liquid (water) and the process liquid is an indication of specific gravity. The reading or display of this instrument is normally calibrated directly in specific gravity.

One of the most widely used methods of density measurement is shown in Figure 7-23(D). In this unit, two bubbler tubes are installed in the

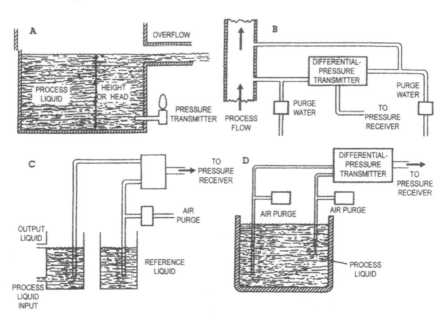

Figure 7-23. Liquid density measuring methods: (A) Measurement by pressure; (8) Liquid purge of an in-line process; (C) Differential bubbler and reference/ sampler; (D) Differential bubbler for single-vessel measurement.

sample solution at different positions. With one tube lower than the other, the difference in pressure will be the same as the weight of a constant height of the liquid. The resulting differential pressure is therefore equal to the weight of a constant volume of the liquid, and can be represented directly as specific gravity. This method of measurement is usually accurate to within 0.3 to 1 percent of the specific gravity.

ELECTROMAGNETIC-RADIATION INSTRUMENTS

Instruments that respond to an interaction between electromagnetic radiation and matter represent an important means of analytical evaluation. Information provided through this technique is based on the fact that *photons* of energy are emitted or absorbed whenever changes occur in the quantitized energy states occupied by electrons of associated atoms and molecules.

Electromagnetic-radiation instrumentation is unique because of its dependence on the frequency of the energy excitation source. Energy of the highest frequency (or shortest wavelength), for example, tends to produce high energy levels that are suitable for penetration into matter. Gamma rays, the shortest utilizable wavelengths, are commonly used to cause an interaction with atomic nuclei. X rays, by comparison, interact with inner-shell electrons. Visible light and ultraviolet energy are less penetrating because of their increased wavelengths. This causes an interaction between valence electrons and some weak interatomic bonds. Infrared radiation and microwave frequencies, having even longer wavelengths, cause an interaction between weak interatomic bonds. This alters the molecular structure by producing vibration and rotational spin.

The interaction principle of electromagnetic-radiation instrumentation is primarily the same for all instruments of this classification. Structurally, each type of instrument is unique because of its energy excitation source. Instruments of this classification are grouped according to the type of energy source used. X rays, ultraviolet radiation, infrared light, and visible light are often used in analytical instruments. These instruments are commonly used to determine the elemental and molecular composition of gases, liquids, and solids.

Nuclear Radiation Instruments

Nuclear radiation is used by industry in two general ways. One method makes use of radiation to effect specific chemical and biological changes. This is used in the irradiation of plastics to alter molecular bonds, and in food and drug preparation. Second, industry uses radiation for inspection purposes and a large number of measurement techniques. Radiation detection is a continuous operation when radioactive materials are in use.

Radiation-detector Principles

Radiation detectors are devices that sense electrical disturbances created by the emission of radioactive particles. These particles are invisible. Neither their numbers nor their energies can be measured directly. A radiation detector must work with the effects produced by the passage of such rays through matter. Electrical disturbances occur when *alpha* and *beta particles* or gamma *photons* invade certain materials. The energy transformation created by each particle is commonly used as an indication of total radiation power.

There are two ways in which radiation detectors measure the effect of particle collisions. One type of detector responds to the amount of ionization that takes place in a certain volume of matter when it is exposed to radiation. The second method responds to brief flashes of light energy that are produced when some materials are struck by high-energy particles. This principle is called scintillation.

Ionization Detectors

Ionization detectors use a small chamber that houses an active form of a solid material, liquid, or gas. When high-speed alpha or beta particles are applied to the chamber, collisions with the active material cause it to release outer-shell electrons. As a result, the stable atoms of the chamber material become ionized. The number of ionizing collisions that take place is then used as a measure of the amount of energy entering the chamber.

Figure 7-24 shows a drawing of a gaseous-diode type of ionization chamber. When ionization occurs by radiation exposure, gas molecules are changed into positive and negative ions. Ions and free electrons are collected by the respective electrodes of the diode tube and cause a resulting current in the external circuit. At low voltage levels the resulting current is extremely small. It tends to arrive in small bursts or pulses that occur at irregular intervals. The resistor-capacitor circuit placed in series with the chamber helps to maintain the output at a fairly constant level. The time constant of this circuit is normally several seconds, in order to show a representative current. This type of detector is often called a *current chamber*. Voltage measured across the resistor with a high-impedance *voltmeter*, oscilloscope, or strip-chart recorder indicates the radiation intensity in *roentgens per unit of time*.

Scintillation Counters

Scintillation counters are an interesting form of radiation detector. A simplified drawing of a scintillation counter is shown in Figure 7-25. A specific form of *phosphor* is optically coupled to a *photomultiplier tube* with both ends enclosed in a light-free shield. Radiation applied to the unit releases energy by exciting atoms of the phosphor crystal. This energy, which appears as bursts of light, is then released from the phosphor. The intensity of the light flash is directly proportional to the amount of radiation applied to the phosphor crystal. A major portion of this emitted light is directed toward the cathode of the photomultiplier tube, and photoelectrons released from the cathode are then multiplied to produce a negative-going pulse of voltage in the output circuit. This type of detector translates light energy into an electrical value which can then be read on a meter, oscilloscope, or strip-chart display unit.

In Figure 7-25, notice the location of the phosphor emitter with respect to the photosensitive cathode. Electrons released from the photo-cathode are immediately attracted by the positive potential of the first dynode. Upon reaching the first dynode, each electron ejects several other electrons by secondary emission. This multiplication process is then repeated by each succeeding dynode. A total multiplication on the order of 1 million may be produced by the photomultiplier tube. The current through the cathode load resistor develops a voltage which serves as the

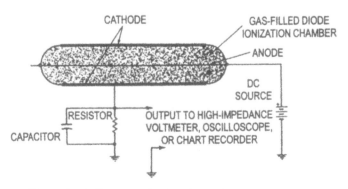

Figure 7-24. Gaseous-diode tube ionization detector.

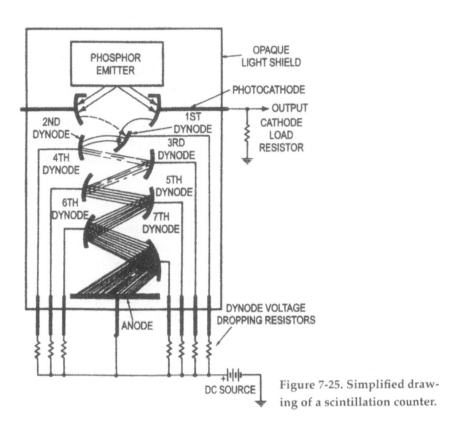

Figure 7-25. Simplified drawing of a scintillation counter.

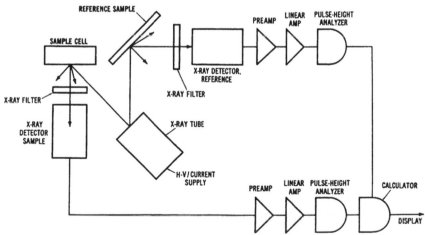

Figure 7-26. Sulfur content analyzer, block diagram.

output of the counter. The unit is calibrated and the output is read on a scale as roentgens per unit of time.

Scintillation counters are unusual because they have a high detection-recovery time. Units are available that can respond to events occurring only 1 μs apart. The response time of the unit is primarily determined by the characteristic response of the phosphor employed. Units are available that are well suited for a number of industrial detection applications in liquid, solid, and gaseous form. The phosphor crystal of each unit must be transparent to the wavelength of light it produces so that all flashes can be detected by the photomultiplier.

X-ray Radiation Instruments

X-ray radiation is used in some analytical instruments to measure the elemental content of liquids and solids. As a rule, this type of instrument is very specialized and is designed to measure only one type of element. A sulfur content analyzer that uses the X-ray radiation principle is shown in Figure 7-26. Content evaluation is achieved by loading the sample in the front drawer, closing the drawer, pressing a button, and reading the percentage of sulfur content by weight on a digital display. Precision measurements take approximately five minutes, while high-speed evaluations can be achieved in a few seconds.

When primary X rays emitted from a tube are applied to the test sample, they can cause the emission of secondary X rays. The secondary X rays can be distinguished from others because of their *fluorescent energy* content. This energy is dependent on the atom content of the elements in the sample and is proportional to the square of the atomic number. Since the intensity of fluorescent X rays depends on the number of atoms per unit volume, element content measurement is based on the intensity of the fluorescent X-ray content produced.

When the applied primary X rays have an energy level greater than the minimum excitation energy of a particular atom being analyzed, fluorescent X rays are produced. When the primary X-ray energy level is less than the excitation energy of an atom, secondary fluorescent X rays are not produced. As a result, selective excitation can be used to choose only the particular element being analyzed.

A block diagram of the sulfur content analyzer is shown in Figure 7-26. When high voltage and current are applied to the X-ray tube, emission is produced and directed toward both the sample cell and reference sample: Only the fluorescent X rays of sulfur are permitted to pass through the

X-ray filters. The respective detectors then respond to fluorescent energy. They produce voltage pulses of a short duration that are proportional to the energy of the X rays. High-gain operational amplifiers are used to boost the amplitude of each detected signal, and respective pulse height analyzers in each line respond only to the sulfur pulses which will be of the greatest amplitude. The two signals are then combined and counted and the resulting output signal drives a digital display.

The basic X-ray principle, with selective excitation, can be used to analyze the content of numerous elements. Generally, each type of analyzer is designed to evaluate only one specific element. Analysis is done by fixed-sample testing and usually does not lend itself to continuous process evaluation applications.

Ultraviolet Analyzers

Ultraviolet analyzers are either of the extractive-sample type or are of the in-line process type, but the basic theory of operation is the same for both. In the extractive-sample method of measurement, a small sample is withdrawn from the process line and applied to an analyzer mounted externally at a remote location. In-line analyzers are attached to a process line and measurements are observed at the operating site.

Figure 7-27 shows a block diagram of an extractive type of ultraviolet analyzer used to monitor gas and particulate samples of an exhaust stack. The stack sample is drawn into the detection chamber by a circulating pump. Ultraviolet light passing through the sample is then directed onto a beam-splitting mirror. One beam is applied to the measuring cell while the second beam is directed to the reference cell. Filters in front of the measuring cell are used to isolate a specific wavelength of light for measurement, and the measuring photocell produces an equal amount of current according to the intensity of that wavelength of light. The reference light beam is also filtered, which isolates it from the wavelength of the measured beam. The reference beam then strikes the photocell, which causes a corresponding current. The current from each photocell is then amplified, compared, and applied to the display unit. Sulfur-dioxide levels in stacks are commonly measured by this method.

Infrared Analyzers

The infrared region of the electromagnetic spectrum lies between the visible-light region and the microwave region. Wavelengths of 0.78 to 100 um are included in this part of the spectrum. This region is ex-

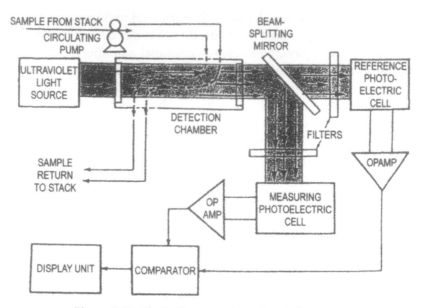

Figure 7-27. Block diagram of an ultraviolet analyzer.

tremely important, because molecular vibration and rotation frequencies occur within this range of the spectrum. Each molecule can be characterized by a unique set of "fingerprints" based on its absorption of vibrations. The depth of absorption may also be used to measure the number of molecules present in a structure.

A common infrared instrument used in industry to determine pollution levels is the *non dispersive infrared analyzer*. This unit determines levels by selective infrared absorption of the component in question. Representative samples are forced to flow through an analysis chamber and are subjected to infrared radiation. Comparisons between absorption levels of a known gas and the sample are observed and analyzed by this unit.

The *non dispersive infrared* (ndir) unit is classified as a *dual-beam optical analyzer*. This type of unit must produce a suitable source of infrared radiation, the emission spectrum of which covers the absorption of the vapors or gases being measured. The absorption of radiation at certain wavelengths is then used to determine the concentration of a gas.

Figure 7-28 shows a diagram of a positive filter type of nondispersive infrared analyzer. The motor near the light source is used to interrupt the beam that is applied to each chamber simultaneously, with equal on and off periods. Radiation from the beam passes through the chambers, and one optical path passes through a reference gas. The two output signals are then applied to photon light detectors. The output signals are amplified, compared, detected, and applied to a digital display, computer, or strip-chart recorder.

The detector of an ndir analyzer represents the major characteristic difference in these units. In addition to photon light detectors, capacitor microphones, gas cells, and solid-state detectors are all in common use today. The basic operating principle is the same for practically all units, but the detection techniques vary a great deal. Units of this type are presently being used to analyze automotive gas emissions. As a general rule, ndir analyzers are used in fixed installations and laboratories because of their cost and delicate sensitivity.

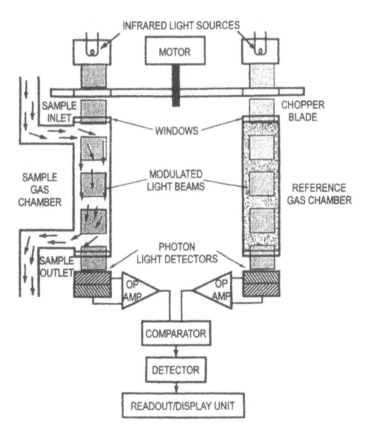

Figure 7-28. Simplification of a photonephelometer.

Photometric Analyzers

A number of analyzers are commercially available that measure and record process variables through the emission of visible light. This type of instrument employs a light source or beam that passes through the test sample and is applied to a *photodetector*. The resulting output of the detector is amplified and used to actuate a display device, a recorder, an alarm, or a computer.

Applications of the *photometric analyzer* are quite numerous. Gas and liquid pollution-analysis instrumentation is very common. This type of equipment is often portable and capable of operating unattended for long periods of time. Laboratory instruments are also available and are used to analyze small test samples in a controlled environment. A number of in-process instruments are available for industrial sample analysis. This type of instrument may be called on to monitor mixing operations or to evaluate sample material as it flows through a system.

Photonephelometers

A photometric measuring process called *nephelometry* is commonly used to determine the amount of suspended solids in water. This type of analyzer utilizes the *Tyndall effect* to determine particle concentrations. Particles that are invisible in a direct light path can be easily observed at right angles to the light source. In essence, the particles reflect certain levels of light, and the process of measuring the amount of reflected light, which is directly proportional to the number of particles in suspension, is nepheometry. Nephelometry is a measure of the *turbidity* of materials suspended in a sample solution. Turbidity refers to something that is confused, disordered, crowded, or obscure, such as the cloudiness that the human eye might see when viewing a liquid sample solution.

Figure 7-29 is a simplified diagram of a photonephelometer that is used to view the turbidity of a sample solution. The liquid sample cylinder is completely masked with an opaque material except for small windows in the center and on the sides. The center window serves as a direct entrance for the light source. Photocells mounted on each side of the cylinder detect the level of reflected light from the suspended particles, and the intensity of the reflected light is used to determine the turbidity of the solution.

Turbidity is an expression of the optical property of a sample solution that causes light to be scattered and absorbed rather than transmitted in straight lines through the solution. In a water sample, the presence of suspended matter, such as clay, mud, organic matter, algae, rust, or cal-

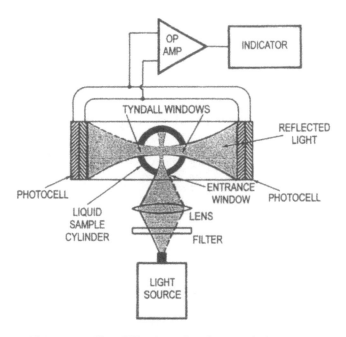

Figure 7-29. Simplification of a photonephelometer.

cium carbonate, causes turbidity. The *nephelometric turbidity unit* (NTU) is an expression of turbidity in terms of the amount of light scattered at right angles to the path of light passing through a test sample.

A simplification of a ratio turbidimeter is shown in Figure 7-30. This instrument has a tungsten-filament lamp that directs light through a lens into a sample cell. Light energy projected through the fluid is either scattered or absorbed by the particulate matter. Three silicon photodiode detectors are used to measure the resulting output light. Light is deflected to the 90° detector, the forward scatter detector, or passes through the sample to the transmitted-light detector. This type of design compensates for color in the sample, light fluctuations, lens haze, and dust *on* the optics. Electronic signal conditioning of the three detected signals results in excellent linearity, color rejection, and elimination of stray light.

CHEMICAL-ENERGY INSTRUMENTATION

Instruments that respond to chemical-energy reactions represent a unique part of the analytical instrumentation field. Measurement through

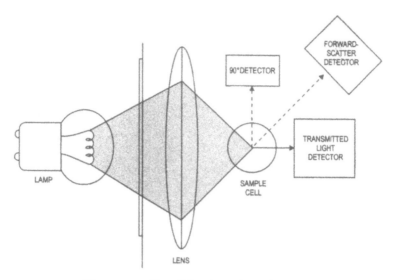

Figure 7-30. Simplification of a turbidimeter.

this procedure is achieved by such things as *reactant* or *sample consumption,* measurement of *reaction product, thermal-energy liberation,* and *solution equilibrium.* The inherent relationship of one chemical element or compound to another, plus the *stoichiometric* and *thermodynamic* behavior of certain materials, permits a positive method of identification and analysis. In addition, there is the separation of some substances, which produces measurable potentials that are indicative of substance concentration or composition.

Chemical-energy instrumentation has a high degree of specificity. It permits composition determinations of liquids, slurries, and the like when analysis by other techniques is less suitable. *Colorimetry, combustion-analysis reaction product analyzers,* and *pH measurements* are included in this group of instruments. Applications include automatic, in-line process evaluation and laboratory analysis equipment. In general, instruments in this classification are tailored to the sample being tested and to the type of information desired. This opens the way for a wide range of specialized instrumentation. Only a few of the more representative industrial analysis instruments will be presented here.

Colorimetry Instrumentation

Quantitative analysis of liquid samples is often achieved photometrically by measurement of the color wavelength of a test sample. In this method of analysis, a known quantity of a specific reagent is mixed with the *test* sample. The term *titration* is used to describe this operation. Color changes in the titrated sample are then evaluated by comparing it with a color disc or color cube, or checking it with a *colorimeter*.

Testing by the colorimetric method is inexpensive, fairly accurate, and adaptable to a wide range of test applications. This test procedure is commonly used in industry to determine water hardness, alkalinity, dissolved oxygen, and chlorides. Long-path viewing makes this analysis procedure more meaningful and convenient. When an adaptor is in place, viewing tubes are inserted through the top of the color comparator and color intensity is reflected to the viewing port through an angled mirror. Since the viewing tubes remain in a vertical position, sampling, handling, and viewing problems are reduced.

In colorimetry measurement, comparisons performed in the analysis of a test sample are made visually by the operator, Human vision is somewhat imprecise, however, when trying to distinguish between similar colors. An electronic colorimeter can be used to improve accuracy in this operation of the analysis procedure.

An electronic colorimeter has a light source, a sample cell, a filter module, and a silicon photodiode to distinguish between different colors. Interchangeable filter modules are loaded into the instrument to automatically configure it to provide a direct readout for a group of tests measured at a specific wavelength of light energy. Filter modules are available in wavelengths ranging from 420 to 810 nanometers (nm). A test sample is loaded into the top of the instrument, the sample compartment is closed, and the *READ* key is pressed. The results are displayed directly on the liquid-crystal display. Absorbance or transmittance percentage units are selected for display by pressing the desired front-panel key. This type of instrument can be readily adapted to a wide range of analysis tests for numerous applications.

Spectrophotometer analyzers are designed to measure the color of titrated samples through variations in light wavelengths. This type of instrument uses a circular variable interference filter with a wavelength range of 400 to 700 nm. In this area of the light spectrum there is a definite relationship between wavelength and color.

Many spectrophotometers employ the dual-detector principle shown in Figure 7-31. In practice, a titrated sample is placed in one light-beam path and a reference sample is placed in the alternate light beam. The output voltage developed by each photo diode is applied to a differential

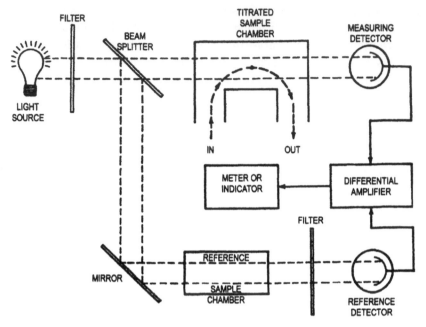

Figure 7-31. Dual-beam dual-detector spectrophotometer.

amplifier. The amplified output signal voltage is read on a calibrated indicator, digital display, or used to drive a chart recorder.

In spectrophotometer operation, the color of a reference sample is compared with the color of the test sample. When the color wavelength of the sample and reference are equal, the difference signal is at zero or in balance. Differences in color produce a scale reading above or below the zero indication level. Special meter scales are often supplied with the instrument for each reference sample. A variation of this instrument uses alternate light filters of selected wavelengths instead of the color reference sample. Operation is achieved by rotating a light filter disc for color comparison with the test sample.

A variation of the dual-beam spectrophotometer is the single-beam instrument, shown in Figure 7-32. This instrument uses improved optics and microprocessor technology to enhance its operating features. The light source is a hot nichrome filament that produces high energy in the range of 400 to 600 nm. Light from the source is mechanically chopped and applied to mirrors M_1 and M_2. Reflected light passes through the sample compartment and the entrance slit. A set of four long-pass filters are sequentially passed through the light source to eliminate wavelengths other than those of the first order. The collimating mirror (M_3) directs light to the grating prism and then to the second collimating mirror (M_4). From M_4, light passes through the exit slit and is focused on the photo detector element by the ellipsoid mirror (M_5). The electrical signal of the photo detector is amplified and applied to a microprocessor. *A/D* conversion, chopper-motor control, timing-signal generation, memory, and light-beam scanning are all controlled by the microprocessor. An output signal of approximately 1 mV is used to drive the display device.

Combustion Analyzers

Combustion analyzers are frequently Used in industrial applications to determine the presence and measure the concentration of combustible gases and vapors on a continuous basis. In this method of analysis, *catalytic combustion* is often used to detect the concentration of combustibles in a process stream. Instrumentation includes a combustion chamber, a thermal sensing element, bridge circuitry, and amplification that responds to resistance changes. The output is indicated as a percentage of the combustible material present in the sample.

Figure 7-33(A) shows a block diagram of a combustion gas analyzer. In analyzer operation, a test sample is extracted from the main flow and applied to the detector analyzer furnace. As the sample passes through the combustibles detector, a dc voltage signal is generated. This voltage is

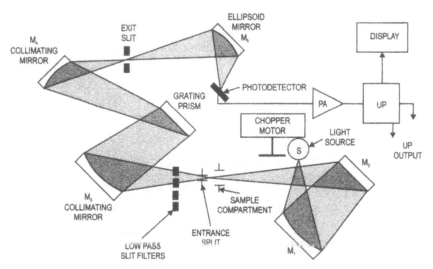

Figure 7-32. Single-beam spectrophotometer optics.

applied to a resistance bridge that contains a detector filament resistor and a reference filament resistor. These resistors are heated to a temperature well below the ignition temperature of the gases present.

The detector filament resistor is coated with a catalyst that causes the combustible gases to burn on its surface. The reference filament resistor is identical with the detector filament resistor except for the catalytic coating. Reaction is therefore the same for both resistors with respect to gas conductivity, specific heat, and velocity. This circuit action is used to cancel these variables.

When the temperature of the detector filament resistor increases due to the burning action on its surface, a corresponding increase in resistance takes place. This particular unit is designed to respond to any combustible gas that contains oxygen. Since the generated heat varies with different gases, the detector filament resistor must be calibrated to respond to a specific gas.

Figure 7-33(B) shows a simplification of the working parts of a combustion oxygen analyzer. The inlet probe of this unit draws the flow sample into the analyzer by an air-operated aspirator that creates a vacuum by forcing air out the other end of the flow loop. Flue gas is drawn into the inlet pipe to fill the vacuum created by the outgoing air flow. Approximately 5 percent of the flue gas is lifted into the furnace cylinder by convection.

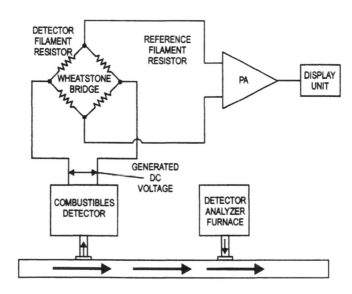

Figure 7-33. Combustion analyzer: (A) Block diagram.

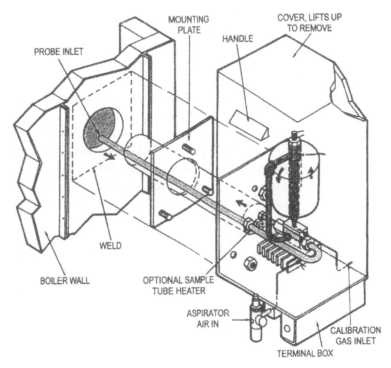

Figure 7-33 (*Cont'd*). Combustion analyzer: (B) Analyzer parts.

The oxygen sensor is housed in a tube located in the center of the furnace cylinder. The sample gas flows up through the sensor by thermal convection and exits at the top of the cell to be returned to the main flow path. As the sample gas passes through the sensing element, an electrical voltage signal is generated. The signal is then applied to a bridge circuit where its output is amplified and used to actuate an analog or digital instrument.

pH Analyzers

The measurement of *pH* is perhaps one of the most universal applications of analysis in all of industrial instrumentation. Any industry that uses even the simplest chemical reaction in a manufacturing process will more than likely improve its performance by monitoring pH levels. One of the key factors in keeping consistent, uniform products is the ability to measure and maintain pH at the proper level.

It is extremely important in many chemical process applications to know if a particular chemical solution has an *acid* content or a *base* or *al-*

kaline content. Typical acid *solutions* are vinegar (which is acetic acid), the citric acid of fruit juice, and the dilute sulfuric acid of a battery. Ammonia water, by comparison, is a rather weak base solution; concentrated lye mixtures form a very strong base solution. Acid and base solutions have entirely different chemical reactions when they exist in solutions. Successful control of chemical processing therefore necessitates that acid and base levels be carefully controlled to ensure a desired outcome.

A more specific definition of pH refers to the number of ionized or free hydrogen ions (H+) and hydroxyl ions (OH–) in a solution. Acid has an abundance of H+ ions, while base has large numbers of free OH– ions. Therefore, pH is a measurement of the ratio of hydrogen and hydroxyl ions in a solution. If equal amounts of base and acid are present, the solution is considered to be *neutral* or at the center of the pH scale.

The term pH is derived from p, the mathematical symbol for a *negative logarithm*, and H, the chemical symbol for hydrogen. A formal definition of pH shows that it represents the negative logarithm of the hydrogen ion activity of a solution. This is expressed mathematically by the formula

$$pH = - \log (H+)$$

In this expression, pH represents the quantitative information needed to show the degree of ion activity in a representative sample solution. Since the relationship between hydrogen ions and hydroxyl ions in a specific solution is constant for a given set of conditions, either value can be determined by knowing the other. This means that pH is a measure of both acid and base, even though, by definition, it is a selective measure of hydrogen ion activity. Since pH is a logarithmic function, a change of 1 pH unit represents a tenfold change in hydrogen ion content.

The numbering system of a pH scale ranges from 0 to 14. The number 7, located at the center of this span, is considered to be an indication of a neutral solution. Acid levels occupy the positions from 7 down to 0, with the smaller numbers indicating the highest acid levels. Numbers 7 to 14 represent the base scale, with the largest numbers indicating the highest base levels. Figure 7-34 shows a pH scale, its mV equivalents, some representative industrial processes, and the pH level of some common household products.

The probe or electrode of a pH instrument is often thought of as a battery whose voltage varies with the pH level of the solution in which it is placed. It essentially contains either two separate probes in a single housing or two distinct probes. In either case, one probe or part of the common probe is sensitive to hydrogen. A special glass bulb or membrane material is used that has the ability to pass H+ ions inside of the sensitive bulb. When the electrode is placed in the solution, a voltage that is proportional to the hydrogen ion concentration is developed between an inner electrode and the outer electrode or glass bulb material. This pH-sensitive probe is often called a *half-cell*.

A second discrete electrode or the alternate part of the common electrode is used to develop a reference voltage. The reference electrode is primarily responsible for producing a stable voltage that is independent

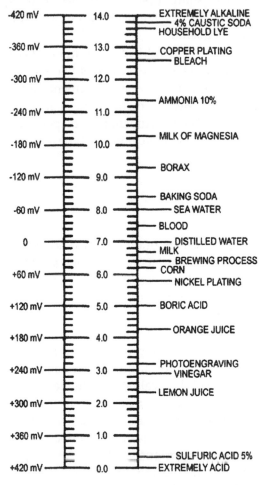

Figure 7-34. pH scale.

of solution properties. This probe or half-cell develops a fixed voltage value when placed in the solution. When the reference half-cell and the pH glass-bulb half-cell are combined, they form a complete probe.

The voltage value produced by a common probe assembly is a linear function of the pH being measured. A pH value of 7.00 produces a v, while at pH 6.00 it produces +0.06 V or 60 mV. Notice that this voltage value has a positive polarity. If the voltage were of a negative polarity, it would indicate a pH reading of 8.00.

The probe of a pH instrument generally produces 0.06 V or 60 mV for each change of 1 pH unit. A probe voltage of +0.3 V or 300 mV indicates a pH value of 2.00. This shows that +300 divided by 60 equals 5 units, or 7 minus 5 equals 2. A pH of 2.00 indicates a strong acid level of something similar to the juice of a lemon.

Figure 7-35 shows a simplified diagram of a pH measuring instrument. In this diagram, pH is observed on either a hand-deflected scale or a strip-chart recorder. Quantitative measurements of pH are produced by this type of instrument. Measurement can be of either discrete samples or in-line process values. The instrument may be permanently attached to a process system, or it may be of the portable type that can be readily moved to different locations.

SUMMARY

Analytical processes are often classified according to an interaction that occurs between the material under test and an external source of energy. Electric or magnetic fields, thermal or mechanical energy, electromagnetic radiation, and chemical energy are the four divisions of analytical processes used in this presentation.

Electric- or magnetic-field instruments respond to a resulting current, voltage, or magnetic flux to produce an output indication. Mass spectrometers are commonly used for quantitative and qualitative evaluation of gases, liquids, and solids. Evaluation is made by introducing a sample into an ionizing source, accelerating it by an electrostatic or magnetic field, and separating it according to mass, detection, and recording.

Electrical conductivity is designed to determine the concentration of dissolved chemicals in various solutions. An energy source, a conductivity cell or probe, a measuring circuit, and a readout display device are included in this type of instrument. The instrument responds to the

Figure 7-35. Parts of a pH measuring instrument.

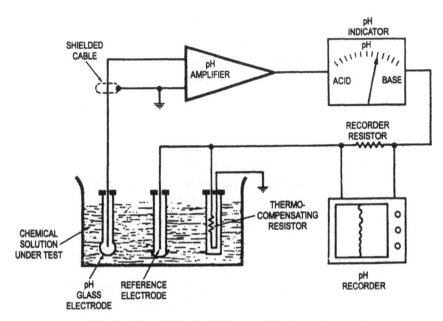

Figure 7-35. Parts of a pH measuring instrument.

movement of electrons in metal and ions in liquid solutions. Trace-moisture analyzers monitor the content of a gaseous stream through a special type of conduction cell. Oxygen analyzers are used to monitor flue gas, residual oxygen in food packaging, and the presence or absence of oxygen in hazardous environments; in heat treating; and in automobile exhaust evaluation. Electrodeless conductivity analyzers respond to conductivity through a closed loop created by the sample solution.

Thermal- or mechanical-energy instruments respond to test samples in terms of energy transmission, work done, or changes in the physical state of a sample. The ability of some gas molecules to become highly excited, which causes vibration, twisting, and rotation, permits them to conduct large amounts of heat away from an external thermal source. Thermal conduction is based on the comparison of heat dissipation that occurs between a reference value and a sample value. The principle components of a thermal analyzer are the measuring cell, regulated power supply, Wheatstone bridge, and case temperature control. Gas-leak detectors utilize this principle by drawing a sample into a test cell and air into a reference cell. Resistance values of the cells are compared and develop an output signal across a bridge circuit.

Chromatographs are thermal instruments that respond to the adsorption principle to separate different components of a test sample. Essentially, this type of instrument has component separation, identification, and quantitative measurement. The separation function is achieved by a chromatograph column that contains granulated silica gel, activated alumina, or carbon. The eluted output of the column is applied to a thermal conductivity cell that causes an imbalance in a bridge circuit.

Viscosity measurement is achieved by an interaction between the test sample and some form of mechanical energy. A falling piston or ball, rotating spindles, and float displacement are some of the common methods of measurement. Vibration viscometry is a variation of the principle that responds to damping as a function of viscosity.

Specific gravity and density measurements rely on an interaction between mechanical energy and a test sample. Float hydrometers, displacements, purged air, and hydrostatic head instruments are employed to achieve this type of measurement.

Analytical instruments that respond to an interaction between electromagnetic radiation and matter represent an important classification of instrumentation. This procedure is based on the fact that photons of energy are emitted or absorbed whenever changes occur in the quantitized states occupied by the electrons of associated atoms and molecules. This form of instrumentation is based on the frequency of the energy excitation source. Nuclear radiation; X rays; ionization detectors; scintillation counters; and ultraviolet, infrared, and photometric analyzers are included in this classification.

Chemical-energy instrumentation responds to such things as reactant consumption, reaction products, thermal-energy liberation, and solution equilibrium. Colorimetry instrumentation responds to the wavelength of a test sample after it has been mixed with a reagent. Spectrophotometers measure the wavelength of titrated test samples. Combustion analyzers are used to measure the concentration of combustible gases and vapors on a continuous basis. Catalytic combustion is often used to achieve this type of evaluation.

A pH analyzer responds to the acid. base, or alkaline content of a test solution through chemical-energy interaction. The term pH specifically refers to the number of H+ or OH– ions in a solution through electrochemical voltage generation. A pH half-cell and a reference half-cell are placed into the test sample to produce a voltage that is indicative of the pH content.

Chapter 8

Microprocessor Systems

OBJECTIVES

Upon completion of this chapter, you will be able to
1. Define the terms *bit, word,* and *byte.*
2. Differentiate between digital, large-scale, and microcomputer systems.
3. Profile how microprocessors fit into industrial process control systems.
4. Explain the purpose of the arithmetic logic unit.
5. Describe the relationship between sequence controllers and the microprocessor unit.
6. Compare and contrast read/write and read-only memories.
7. Describe the role of read-only memory in a computer system.
8. List some of the functions that are basic to a microprocessor.
9. Discuss programming procedures.
10. Explain the role of computer integrated manufacturing.

KEY TERMS

Accumulator—A temporary register that is designed to store data that are to be processed.

Address register—A microprocessor function that temporarily stores the address of a memory location that is to be accessed for data.

Architecture—The physical component layout of a microprocessor or computer assembly.

Arithmetic/logic unit (ALU)—A circuit capable of performing a variety of arithmetic or logic functions.

Assembly—language A machine-oriented language in which mnemonics are used to represent each machine language instruction.

Base—The number of symbols in a number system. A decimal system has a base of 10, a binary system has a base of 2, an octal system has a base of 8, and a hexadecimal system has a base of 16.

Basic—A high-level programming language; acronym for Beginner's All-purpose Symbolic Instruction Code.

Baud—A unit of data transmission speed equal to the number 01 bits that can be manipulated per second.

Binary—A system of numerical representation that uses only two symbols, 1 and 0.

Bit—A contraction of the term *binary digit.*

Bus—A channel or path used to transfer signals between devices or components.

Byte—A special term used for an 8-bit word; a byte usually consists 01 eight bits.

CIM—Computer integrated manufacturing.

Clock—A timing device that generates the basic periodic signal used to control the timing operations 01 a synchronous computer.

Data register—A group of electronic devices and circuits that store data.
    Digital—A value or quantity related to numbers or discrete values.
    Enable—To make a computer operation possible, such as permitting data to be transferred from one location to another by applying a signal to the enable input.

Encoding—The process of converting an input signal into binary data or changing one digital code to another.

Erasable programmable read-only memory (EPROM)—A read-only memory in which information has been entered by applying electrical pulses to selected cells. This information can be erased by applying ultraviolet light to the device through a quartz window.

Execute—One of the operational cycles of a central processing unit. **Facsimile (FAX)**—The transmission of images over a telephone line. **Fetch**—An operational cycle of the central processor.

Firmware—Permanently installed instructions in computer hardware that control operation.

Flowchart—A symbol diagram used to aid in program development.

Gate—A circuit that performs special logic operations; examples include AND, OR, NOT, NAND, and NOR.

Instruction—Information that tells a microprocessor or computer what to do.

Interface—A process or piece of equipment in a computer that brings things together.

Large-scale integration (LSI)—The combining of several thousand circuits or components on a single IC chip.

Logic—A decision-making capability of computer circuitry.

Machine language—Instructions written in binary form that a computer can execute directly.

Memory—A process that can store logic 1 and logic 0 bits in such a manner that they can be accessed or retrieved.

Microprocessor—An IC or set of ICs that can be programmed to process data.

Mnemonic—An abbreviation that resembles a word or reminds one of a word.

Network—A group of computers that are connected to each other by communication lines to share information and resources.

Nonvolatile memory—Memory that is capable of retaining data after the power-supply voltage has been removed.

Operand—A quantity being operated on or performed in a processing system.

Operational code (opcode)—The part of a computer instruction word that designates the task to be performed.

PLC—Programmable logic controller.

Program—A list of ordered tasks that is used to control the operation of a processor or computer system.

Programmable read-only memory (PROM)—A memory device that can be programmed by the user.

Radix—The base of a numbering system.

Random access memory (RAM)—Memory that can be both read and changed during computer operation. Unlike other semiconductor memories, RAM is volatile. If power is lost, all data stored in RAM is lost.

Read-only memory (ROM)—A form of stored data that can be sensed or read, but is not altered by the sensing process.

Read/Write (R/W)—A memory operation that can be read from or written into with equal ease.

Register—A digital circuit that is capable of storing or moving binary data.

Software—Instructions such as a program that controls the operation of a computer.

Substrate—A piece of N or P semiconductor material that serves as a foundation for other parts to be formed on in the construction of an integrated circuit.

Volatile memory—A type of memory that necessitates electrical power in order to store information. If the power is removed, all information will be lost.

Write—A memory operation where a new word or bit of data is placed in a particular memory location.

INTRODUCTION

Computer technology has brought a significant number of changes to industrial process instrumentation. Centralized computers have been used for years to perform a range of control operations in automatic process applications. Elaborate calculations can be performed quickly, data can be stored and retrieved at a moment's notice, and deductions can be made from data that will influence a functioning process. All of this has caused industry to become more dependent on the computer and has placed a higher premium on computer operational time.

Around 1963, minicomputers appeared on the scene to provide computer technology for industrial applications not requiring the capacity of a large-scale computer. These units were significantly less costly than big computers, and instant success demonstrated the usefulness of this computer to present an immediate information display for the operator. Information is generally placed into this type of unit through a typewriter keyboard.

Microcomputers were introduced by a number of different manufacturers in the early 1970s. These computers were an extension of the technology employed by the minicomputer. Large-scale integration (L8I) technology was used to place thousands of discrete solid-state components on a single integrated circuit (IC) chip. The entire central processing unit (CPU) of a microcomputer is achieved today by a single IC chip called a *microprocessor unit* (MPU). A microprocessor, memory, input/output, interface chips, and a power source form a microcomputer system. This combination of components can be built on a single printed-circuit board and easily housed in the cabinet of the instrument being controlled. Microcomputer technology has revolutionized industrial processing by providing inexpensive, small-capacity computers that can be built into instruments.

COMPUTER BASICS

The term *computer* is a broad general term which describes a number of operating systems. It primarily refers to a device that will perform automatic arithmetic computations when provided with the appropriate information. It is, however, much more than a fast calculator. A computer can choose, copy, move, compare, and perform other nonarithmetic operations on alphabetic, numeric, and other types of symbols used to represent different things. The computer manipulates these symbols through a routing map called a *program*. The program is a detailed set of instructions that directs the computer to function in a way that will produce a desired result. Computers are fast and accurate symbol-manipulating systems that are designed to automatically accept and store input data, process them, and produce a resulting output.

Computers use numbers represented by the presence or absence of voltage at a particular level. A high voltage level or pulse is taken to represent a "1" state, while no voltage indicates the "0" state. A voltage value or signal pulse is a *bit* of information. The term bit is a contraction of the letters *bi* from the word binary and the letter t from digit. A *binary system* has two states or conditions of operation and uses the *base* 2 counting system. A group of pulses or voltage-level changes produces a *word*. A *byte* consists of eight bits.

A computer system, regardless of its complexity, has a number of fundamental parts that are the basis of its operation. These parts may be arranged in a variety of different ways and still achieve the same primary function. The internal organization and design of each block of a computer differ between manufacturers. Figure 8-1 shows a block diagram of a computer.

Input

Computers use a variety of devices for their *input*. The input is primarily responsible for direct human to machine communication or *interfacing*. Input data supplied to the computer are translated into some type of number code or machine language before operation progresses. These data may appear as magnetized material on a floppy disk or on a hard disk-drive unit. The keyboard or workstation connected directly to a computer is an example of a direct input device. Other inputs can be a mouse, a marking pen, a touch screen, or a microphone/sound-generating station.

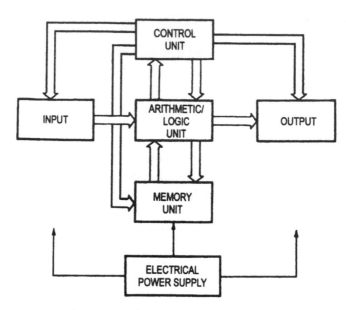

Figure 8-1. Block diagram of a computer.

Processors

The *processor* of a computer is considered to be the heart of the system. Processors house sections for storage, arithmetic and logic, and control. The *storage* or memory function of a processor is responsible for *input data storage, working storage space, output storage*, and *program storage.* Data fed into the input storage area are held in memory until ready to be processed. The working storage space is like a sheet of scratch paper. It is used to hold data being processed and any intermediate result of processing. The output storage area is designed to hold the finished results of the processing operation until they can be released. The program storage area is responsible for holding all of the internal instructions of the processor. These functions of the processor are not fixed by built-in boundaries, but can vary from one application to another. A specific storage area may be called on to store input data in one application, output results in another, and processing instructions in a third. The program of the unit determines how storage space is utilized for a specific application.

The *arithmetic-logic unit* (ALU) and control section of a processor are combined to form the CPU. Essentially, all calculations are performed and all comparisons or decisions are made in the ALU of the CPU. Once data are fed into input storage, they are held and transferred as needed to the ALU where processing takes place. No processing occurs in the input storage area. Any intermediate results generated in the ALU are temporarily placed in the working storage area. These data may move from primary storage to the ALU and back again to the working storage area many times before processing is finished. Once the operation is complete, the final results are supplied to the output storage section and ultimately transferred to the output device. The type and number of ALU operations performed by the computer are determined by the design of the CPU.

The control section of a computer is responsible for selecting, interpreting, executing program instructions, maintaining order, and directing system operations. The control section does not process data; it acts like a central nervous system for other data-manipulating components of the computer. When processing begins, the first program instruction is selected and fed into the control section from the program storage area. It is then interpreted and sent to other components to execute the necessary action. Further program instructions are then selected and executed on a continuous basis until the processing is completed.

Memory

To supplement the limited capacity of the primary storage section, most computers have a *secondary* or *auxiliary memory unit*. Memory devices of this type are connected directly to the processor. This is considered to be an *on-line operation*. The memory unit accepts and retains data and/or program instructions from the processor. It then writes data back to the processor as needed to complete the processing task. *On-board* or *chip* memory is usually built into the assembly board of a personal computer. In addition, a computer can have soft-disk memory and hard-disk memory that extend the memory of the system to an unlimited capacity. This memory remains intact after the disk is removed from the system. Since the processor no longer has direct and unassisted access to this memory, its storage capability is considered to be *off line*.

System Output

The output of a computer is an *interfacing* component. This component is responsible for communication between the operator and the computer system. The output takes machine-coded data from the processor and converts them into a form that can be used by the operator, or serves as an input for another processing cycle. In some cases, this may be a series of magnetized areas on a disk or an electrical-voltage level change that drives a device. In personal computer systems, the output drives a display screen or a desktop printer. In an industrial machine, the output may control a manufacturing operation in the development of a product, or place data on a graphic display. Output devices are generally located near the processor unit, but sometimes the output may be an auxiliary device independent of the computer. These devices are called *peripherals.* Printers, plotters, workstations, terminals, and magnetic devices are common peripherals. Output that is not attached to the CPU is called *off-board* output. Personal computers may have auxiliary circuit boards that house special output devices. The device being controlled by the output usually dictates the location of its circuitry.

MICROCOMPUTER SYSTEMS

Microcomputer systems are one of the most significant developments in industrial processing in the last decade. The potential capabilities of the microcomputer have not been fully realized at this time. Benefits such as faster development times, smaller equipment size, increased reliability, easy serviceability, and lower product cost are only a few of the major considerations to keep in mind.

As a general rule, a microcomputer is quite complex when looked at in its entirety. It has a microprocessor, memory, an interface adapter, and several distribution paths called *buses.* To simplify this system, it is better to first look at a stripped-down version of its physical makeup. This is shown in Figure 8-2.

The simplified version of the microcomputer is similar to the basic computer diagram in Figure 8-1. The microprocessor, in this case, has replaced the ALU and control unit. Information within this system is primarily of two types: instructions and data. In a simple addition problem, such as $7 + 3 = 10$, the numbers 7, 3, and 10 are data and the plus sign is an instruction. Data are distributed by the data bus to all parts of the system. Instructions are distributed by the control bus in the same manner, but through a separate path. The address bus of the unit forms an alternate distribution path for the distribution of address data. It is normally used to place information into memory at an appropriate address. By an appropriate command from the microprocessor, data may be removed from memory and distributed to the output.

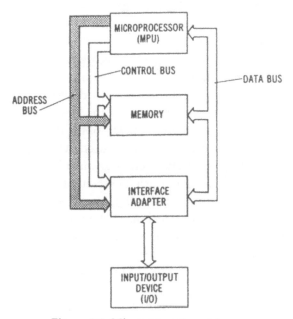

Figure 8-2. Microcomputer system.

In this chapter, *numbering*, which deals with the data and information utilized by the system, will be presented first. This area of investigation deals with the numbers that are actually manipulated by the computer. Second, *hardware*, the physical makeup of a microprocessor, will be discussed. *Memory* is the third major area of concern. Read/write memory and read-only memory will be discussed in this section. The fourth area will be directed toward *input/output* techniques. This deals with the placing of information into the computer from an outside source and converting its output into a usable signal. The fifth and last major area deals with microcomputer *software*. Program development, execution, and MPU languages are also presented in this section. Some understanding of how the microcomputer can be practically implemented to perform a number of useful control functions will also be set forth.

NUMBERING SYSTEMS

The most common numbering system in use today is the *decimal* system. Ten digits are used in this numbering system to achieve counting:

0, 1, 2, 3, 4, 5, 6, 7, 8, and 9. The number of discrete digits of the system is commonly called its *base* or *radix*. The decimal system, therefore, has a base or radix of 10.

Nearly all modern numbering systems are described as having *place value*. This term refers to the placement of a particular digit with respect to others in the counting process. The largest digit that can be used in a specific place or location is determined by the base of the system. In the decimal system, the first position to the left of the decimal point is called the *units* place. Any digit from 0 to 9 can be used in this place. When number values greater than 9 are to be used, they must be expressed in two or more places. The next position to the left of the units place is the *tens* place in a decimal system. The number 99 is the largest digital value that can be expressed by two places in the decimal system. Each place added to the left extends the capability of this system by a power of 10.

A specific number value of any base can be expressed by the addition of weighted place values. The decimal number 1,346, for example, would be expressed as

$$(1,000 \times 1) + (3 \times 100) + (4 \times 10) + (6 \times 1)$$

Note that the values increase progressively for each place extending to the left of the starting position or decimal point. These place or position factors can also be expressed as *powers* of the base number. In the decimal system, this would be 10^3, 10^2, 10^1, and 10^0, with each succeeding place being expressed as the digit number times a power of the radix of the number system base. The decimal number 3,421 is expressed this way in Figure 8-3.

The decimal numbering system is commonly used today and is convenient in our daily lives. Electronically, however, it is difficult to employ. Each digit of a base 10 system, for example, requires a specific value associated with it. Electronically, a system using this numbering method would necessitate a complex detection process to distinguish between different numbering values. The problems associated with defining and maintaining these ten levels are difficult to solve.

Binary Numbering System

Practically all electronic digital systems in operation are of the binary type. This type of system has 2 as its base or radix. The largest digital value that can be expressed in a specific place by this system is the number

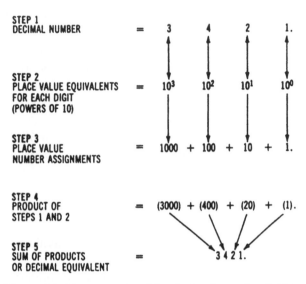

STEP 1
DECIMAL NUMBER   =   3   4   2   1.

STEP 2
PLACE VALUE EQUIVALENTS   = 10^3 10^2 10^1 10^0
FOR EACH DIGIT
(POWERS OF 10)

STEP 3
PLACE VALUE   =   1000 + 100 + 10 + 1.
NUMBER ASSIGNMENTS

STEP 4
PRODUCT OF   =   (3000) + (400) + (20) + (1).
STEPS 1 AND 2

STEP 5
SUM OF PRODUCTS   =   3 4 2 1.
OR DECIMAL EQUIVALENT

Figure 8-3. Components of the decimal number 3,421.

1. This means that only the numbers 0 and 1 are used in the binary system. Electronically, the value of 0 can be expressed as a very low voltage or no voltage. The number 1 can then be indicated by some voltage-value assignment larger than or more significant than 0. Binary systems that use this voltage-value assignment are described as having *positive logic*. *Negative logic,* by comparison, has voltage assigned to 0 and no voltage assigned to the number 1. In the discussion that follows, only positive logic will be used.

The two operational states of a binary system, 1 and 0, can be considered as natural circuit conditions. When a circuit has no voltage applied, it is considered in the off or 0 state. An electrical circuit that has voltage applied or is operational is considered to be on or in the 1 state. By using transistors or ICs, it is possible to change states in less than a microsecond. Components of this type make it possible to manipulate millions of 0s and 1s of information in a second.

The basic principles of numbering that are used by the decimal system also apply to binary numbers. The radix of the binary system, for example, is 2. This means that only the digits 0 and 1 can be used to express a specific place value. The first place to the left of the starting point (in this case the *binary point*) represents the units or 1s location. Places that follow to the left of the binary point refer to the powers of 2. The digital values of numbers to the left of the binary point are 2^0 (1), 2^1 (2), 2^2 (4), 2^3 (8), 2^4 (16), 2^5 (32), 2^6 (64), 2^7 (128), and so on.

As a general rule, when different numbering systems are used in a discussion, they must incorporate a *subscript* number to identify the base of the numbering system being used. The number $110._2$ is a typical expression of this pattern. This would be described as one-one-zero, instead of the decimal equivalent, one hundred and ten.

The number $110._2$ is equal to 6 to the base ten, or 6.10_{10}. Starting at the first digit to the left of the binary point, this number would have a place value of

$$(0 \times 2^0) + (1 \times 2^1) + (1 \times 2^2) \text{ or } 0 + 2._{10} + 4._{10} = 6._{10}$$

The steps in conversion of a binary number to an equivalent decimal number are shown in Figure 8-4.

A simplified version of the binary to decimal conversion process is shown in Figure 8-5. In this method of conversion, write down the binary number first. Starting at the binary point, indicate the decimal equivalent powers-of-two numbers of each binary place location where a 1 is indicated. For each 0 in the binary number, leave a blank space or indicate a 0. Add the place-value assignments and record the decimal equivalent.

The conversion of a decimal number to a binary equivalent is achieved by repetitive steps of division by the number 2. When the *quo-*

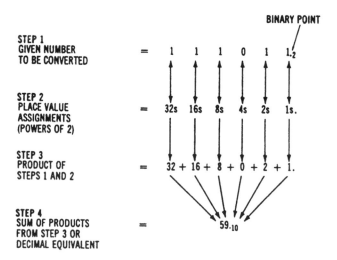

Figure 8-4. Conversion of a binary number to a decimal number.

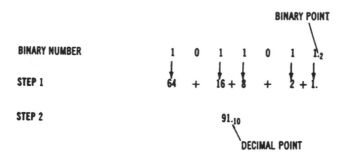

Figure 8-5. Binary conversion process.

tient (answer) is even with no remainder, a 0 is recorded. When the quotient has a remainder, a 1 is recorded. The steps needed to convert a decimal number to a binary number are shown in Figure 8-6.

The conversion process, in this case, is achieved by writing down the decimal number $35._{10}$. Divide this number by the base of the system (2). Move the quotient of Step 1 to Step 2 and repeat the process. The division process continues until the quotient becomes 0. The binary equivalent is the remainder values in their last-to-first placement order.

Binary Coded Decimal Numbers

When large numbers are to be indicated by binary numbers, they become awkward and difficult to use. For this reason, the *binary coded decimal (bcd) method* of counting was devised. In this type of system, four binary digits are used to represent each decimal digit. To illustrate this procedure,

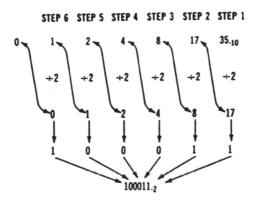

Figure 8-6. Conversion of a decimal number to a binary number.

the number $329._{10}$ will be converted to a bed number. In straight binary numbers, $329._{10}$ equals $101001001._2$.

To apply the bed conversion process, the base 10 number is first divided into discrete digits according to place values. This is shown in Figure 8-7, Step 1. The number $329._{10}$ yields the digits 3-2-9. Converting each digit to binary would allow the displaying of this number as $0011\text{-}0010\text{-}1001._{bcd}$. Decimal numbers up to $999._{10}$ can be displayed and quickly interpreted by this process with only twelve binary numbers. The dash line between each group of digits is important when displaying bed numbers.

The largest digit to be displayed by any group of bed numbers is 9. This means six digits of a number coding group are not being used at all in this system. Because of this, the *octal,* or base 8, and the *hexadecimal,* or base 16, systems were devised. Digital systems still process numbers in binary form but usually display them in bcd, octal, or hexadecimal values.

Octal Numbering Systems

Octal, or base 8, numbering systems are used to process large numbers through digital systems. The octal system of numbers uses the same basic principles outlined for the decimal and binary counting methods.

The octal numbering system has a radix of 8. The largest number displayed by the system before it changes the place values is 7, meaning that the digits 0, 1, 2, 3, 4, 5, 6, and 7 are used in the place position. The place values of digits starting at the left of the octal point are the powers of eight. The number 265'8 is changed to an equivalent decimal number, as shown in the procedure outlined in Figure 8-8.

Converting an octal number to an equivalent binary number is similar to the bed conversion process discussed previously. The octal number

GIVEN DECIMAL NUMBER		$3\ 2\ 9._{10}$	
STEP 1 GROUPING OF DIGITS	(3)	(2)	(9)
STEP 2 CONVERSION OF EACH DIGIT TO BINARY GROUP	(0011)	(0010)	(1001)
STEP 3 COMBINE GROUP VALUES		$0011\text{-}0010\text{-}001._{BCD}$	

Figure 8-7. Conversion of a decimal number to a bed number.

is first separated into discrete digits according to place value. Each octal digit is then converted into an equivalent binary number using only three digits. The steps of this procedure are shown in Figure 8-9.

Converting a decimal number to an octal number is a process of repetitive division by the number 8, as shown in Figure 8-10.

After the quotient has been determined, the remainder is brought down as the place value. When the quotient is even with no remainder, a 0 is transferred to the place position. The number $4,098._{10}$ is converted to an octal equivalent in Figure 8-11.

Converting a binary number to an octal number is important in the conversion process found in digital systems. Binary numbers are first pro-

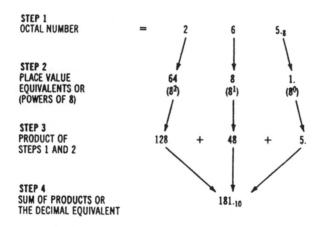

STEP 1
OCTAL NUMBER   =   2   6 $5._8$

STEP 2
PLACE VALUE   64   8   1.
EQUIVALENTS OR (8^2) (8^1) (8^0)
(POWERS OF 8)

STEP 3
PRODUCT OF   128   +   48   +   5.
STEPS 1 AND 2

STEP 4
SUM OF PRODUCTS OR $181._{10}$
THE DECIMAL EQUIVALENT

Figure 8-8. Conversion of an octal number to a decimal number.

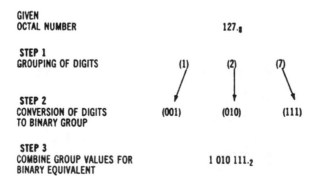

GIVEN
OCTAL NUMBER $127._8$

STEP 1
GROUPING OF DIGITS   (1)   (2)   (7)

STEP 2
CONVERSION OF DIGITS   (001)   (010)   (111)
TO BINARY GROUP

STEP 3
COMBINE GROUP VALUES FOR $1\ 010\ 111._2$
BINARY EQUIVALENT

Figure 8-9. Conversion of an octal number to a binary number.

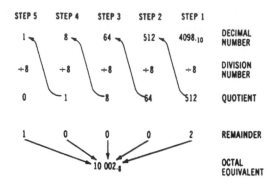

Figure 8-10. Conversion of a decimal number to an octal number.

cessed through the equipment at a very high speed. An output circuit then accepts this signal and may convert it to an octal signal which can be displayed on a readout device.

Assume that the number $10110101._2$ is to be changed into an equivalent octal number. The digits must first be divided into groups of three, starting at the binary point. Each binary group is then converted into an equivalent octal number, and these numbers are combined, while remaining in their same respective place, to represent the equivalent octal number. These conversion steps are outlined in Figure 8-11.

Hexadecimal Numbering System

The hexadecimal numbering system is used in digital systems to process large number values. The radix of this system is 16, which means that the largest number used in a place is equal to 15. Digits used to dis-

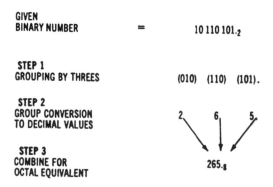

Figure 8-11. Conversion of a binary number to an octal number.

play this system are the numbers 0 through 9, and the letters A, B, C, D, E, and F. The letters A through F are used to denote digits 10 through 15, respectively. The place values of digits to the left of the hexadecimal point are the powers of 16: 16^0 (1), 16^1 (16), $(16)^2$ 16^2 (256), 16^3 (4,096), 16^4 (65,536), 16^5 (1,048,576), and so on.

The process of changing a hexadecimal number to a decimal number is primarily achieved by the same procedure outlined for other conversions. Initially, a hexadecimal number is recorded in proper digital order as shown in Figure 8-12. The place values or powers of the base are then positioned under each respective digit in Step 2. The values of Steps 1 and 2 can then be multiplied together to indicate discrete place-value assignments. In a hexadecimal conversion, Step 3 is usually added to simplify letter-digit assignments. Step 4 is the addition of the products. Adding these values together in Step 5 gives the decimal equivalent of a hexadecimal number.

The process of changing a hexadecimal number to a binary equivalent is a simple grouping operation. Figure 8-13 shows the operational steps for making this conversion. In Step 1, the hexadecimal number is separated into discrete digits. Each digit is then converted to an equivalent binary number using only four digits per group. Step 3 shows the binary groups combined to form the equivalent binary number.

The conversion of a decimal number to a hexadecimal number is achieved by the repetitive division process used with the other number systems. In this procedure, the divisor is 16 and remainders can be as large as 15. Figure 8-14 shows the necessary procedural steps for achieving this conversion.

STEP 1:	HEXADECIMAL NUMBER =	1	2	A	$F._{16}$
STEP 2:	PLACE VALUE EQUIVALENTS (POWERS OF 16) =	4096s	256s	16s	1s.
STEP 3:	PLACE VALUE DIGITS =	1	2	10	15.
STEP 4:	PRODUCT OF STEPS 2 AND 3 =	4096 +	512 +	160 +	15.
STEP 5:	SUM OF PRODUCTS OR DECIMAL EQUIVALENT =	4783_{10}			

Figure 8-12. Conversion of a hexadecimal number to a decimal number.

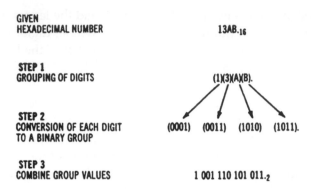

Figure 8-13. Conversion of a hexadecimal number to a binary number.

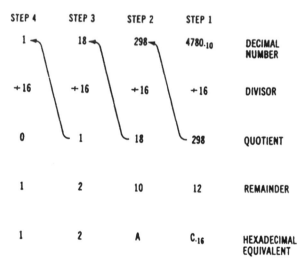

Figure 8-14. Conversion of a decimal number to a hexadecimal number.

Converting a binary number to a hexadecimal equivalent is a reverse of the hexadecimal-to-binary process. Figure 8-16 shows the fundamental steps of this procedure. Initially, the binary number is divided into groups of four digits, starting at the binary point. Each group number is then converted to a hexadecimal value and combined to form the hexadecimal equivalent.

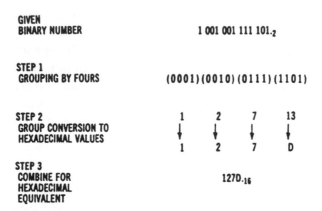

Figure 8-15. Conversion of a binary number to a hexadecimal equivalent.

THE MICROPROCESSOR UNIT

A microprocessor is a scaled-down computer that fits on a single IC chip. Typical chip sizes measure approximately 0.5 cm^2. In this small space the chip may contain over a million components. Several companies are now manufacturing these chips in a variety of different designs. The performance of these chips makes them compatible with the most demanding applications in business and industry.

A microprocessor is essentially a digital device that is designed to receive data in the form of 1s and 0s. It may then store these data for further processing or perform arithmetic and logic operations in accordance with stored instructions. The results will then be delivered to an output device. The microprocessor was primarily designed for use in low-level personal computers and is largely responsible for the term *microcomputer*. A microcomputer has several chips, including the microprocessor, mounted on its main circuit board. This board permits the computer to perform its arithmetic, logic control, and memory functions.

New developments in microprocessor technology have changed the internal construction of this device. In 1975, the first personal computers using a microprocessor began to appear on the market. Upgraded versions of these chips were soon introduced. Chips like Zilog's Z80, MOS Technology's 6502, Intel's 8080, and Motorola's 6809 all had built-in eight-line data paths or data buses. This type of chip is designed to retrieve from storage, manipulate, and process a single eight-bit byte of data at a time. A sixteen-line address bus is built into these chips for memory locations and instructions. This makes it possible for the chip to have 64,000 or 64K of memory.

A block diagram of a microprocessor shows that it contains a number of basic components connected together in an unusual manner (see Figure 8-16). Included in its construction are the ALU, an accumulator, a data register, address registers, a program counter, an instruction decoder, and a sequence controller. The following section presents a detailed look at the response of these primary functions of an MPU.

Arithmetic-logic Unit

All microprocessors contain an arithmetic-logic unit. In a sense, the ALU is a calculator chip that performs mathematical and logic operations on the data words supplied to it. Its keys are made to work automatically

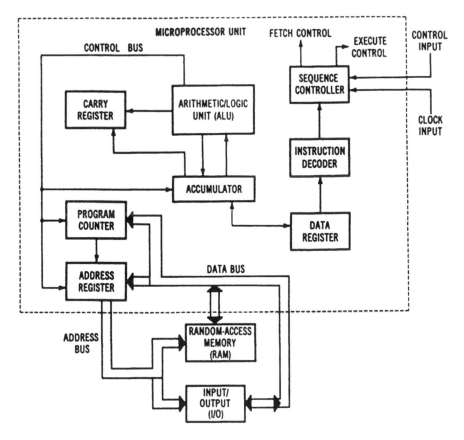

Figure 8-16. Simplified diagram of a microprocessor.

by control signals developed in the instruction decoder.

The ALU simply combines the contents of its two inputs, which are called the *data register* and the *accumulator*. As a rule, addition, subtraction, and logic comparisons are the primary operations to be performed by the ALU. The specific operation to be performed is determined by a control signal supplied by the instruction decoder.

The data supplied to the inputs of an ALU are normally in the form of 8-bit binary numbers. Upon receiving this data at the input, it is combined by the ALU in accordance with the logic of binary arithmetic. Since a mathematical operation is ultimately performed on the two data inputs, the latter are often called *operands*.

To demonstrate the operation of the ALU, assume that two binary numbers are to be added. In this case, consider the addition of the number $6._{10}$ and $8._{10}$. Initially, the binary number 00000110_2 is placed in the accumulator. The second operand, 00001000_2 representing the number $8._{10}$, is then placed into the data register. When a proper control line to the ALU is activated, binary addition is performed, producing an output 00001110_2 ($14._{10}$) which is the sum of the two operands. This value is then stored in the accumulator, where it replaces the operand that appeared there originally. It is important to note that the ALU only responds to binary numbers.

Accumulators

The *accumulators* of a microprocessor are temporary registers which are designed to store operands that are to be processed by the ALU. Before the ALU can perform, it must first receive data from an accumulator. After the data register input and accumulator input are combined, the logical answer or output of the ALU appears in the accumulator. This function is basically the same for all microprocessors.

In microprocessor operation, a typical instruction would be to "load the accumulator." This instruction enables the contents of a particular memory location to be placed into the accumulator. A similar instruction might be "store accumulator." In this operation, the instruction causes the contents of the accumulator to be placed in a selected memory location. Essentially, the accumulator serves as both an input source for the ALU and a destination area for its output.

Data Register

The *data register* of a microprocessor serves as a temporary storage location for information applied to the data bus. Typically, this register will accommodate an 8-bit data word. An example of a function of this register is operand storage for the ALU input. 10 addition, it may be called on to hold an instruction while the instruction is being decoded. It may also temporarily hold data prior to the data being placed in memory.

Address Registers

Address registers are used in microprocessors to temporarily store the address of a memory location that is to be accessed for data. In some units, this register may be *programmable*. This means that it permits instructions to alter its contents. The program can also be used to build an address in the register prior to executing a memory reference instruction.

Program Counter

The *program counter* of a microprocessor is a memory device that holds the address of the next instruction to be executed in a program. As a general rule, this unit counts the instructions of a program in sequential order. When the MPU has fetched instructions addressed by the program counter, the count advances to the next location. At any given point during the sequence, the counter indicates the location in memory from which the next piece of information will be derived.

The numbering sequence of the program counter may be modified so that the next count may not follow a numerical order. Through this procedure, the counter may be programmed to jump from one point to another in a routine. This permits the MPU to have branching capabilities should the need arise.

Instruction Decoders

Each specific operation that the MPU can perform is identified by an exclusive binary number known as an *instruction code*. Eight-bit words are commonly used for this code. Exactly 2^8 or 256_{10} separate or alternative operations can be represented by this code. After a typical instruction code is pulled from memory and placed in the data register, it must be decoded. The *instruction decoder* examines the coded word and decides which operation is to be performed by the ALU.

Sequence Controllers

The *sequence controller* performs a number of vital functions in the operation of the microprocessor. Using clock inputs, this circuitry maintains the proper sequence of events required to perform a processing task.

After instructions are received and decoded, the sequence controller *issues* a control sign that initiates the proper processing action. In most units, this controller has the capability to respond to external control signals.

Buses

The registers and components of most microprocessors are connected together by a bus-organized type of network. The term *bus,* in this case, is a group of conductor paths that are used to connect data words to various registers. A simplification of registers connected by a common bus line is shown in Figure 8-17.

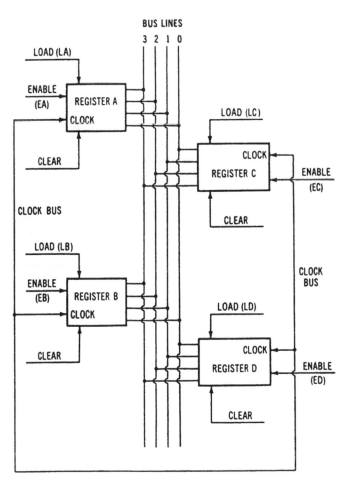

Figure 8-17. Registers connected to a common bus.

The beauty of bus-connected components is the ease with which a data word can be transferred or loaded into registers. In operation, each register has inputs labeled clock, enable, load, and clear. When the load and enable input lines are low or at zero, each register is isolated from the common bus line.

To transfer a word from one register to another, it is necessary to make the appropriate inputs high or at the 1 state. For instance, to transfer the data of register A to register D, place the enable A (EA) and load D (LD) inputs both in the 1 state. This will cause the data of register A to appear on the common bus line. When a clock pulse arrives at the common inputs, the transfer process is completed.

The word length of a bus is based on the number of conductor paths that it uses. Buses for four, eight, and sixteen bits are commonly used in microprocessors. New MPUs tend to use an 8-bit data bus and a 16-bit address bus.

Microprocessor Architecture

The physical layout or *architecture* of a microprocessor is much more complex than the stripped-down version presented in Figure 8-16. In practice, it is not really necessary to know what goes on inside of a microprocessor other than to have some idea of its primary functions. As a rule, the operation is somewhat complex. If one attempts to analyze a specific unit, it is easy to get bogged down in details. An expanded block diagram of a representative microprocessor is shown in Figure 8-18. This can be used to compare its architecture with the simplified unit.

The MC6800 is a member of Motorola's microcomputer family. It is housed in a forty-pin dual in-line package that is shown in Figure 8-19. Two accumulators, an index register, and a condition code register are included in its physical makeup. Only two of these registers differ from those of the simplified model.

Accumulators

The MC6800 MPU has two accumulators instead of one, which are labeled AA and AB. Each accumulator is eight bits wide, which is indicated by the numbers 0 to 7 in Figure 8-19. These numbers are often used to make up a programming model of the chip, showing its capacity. Each accumulator is designed to hold operands or data from the ALU.

Program instructions for each accumulator includes a *mnemonic* of its name and the operation to be performed. Loading accumulator A is

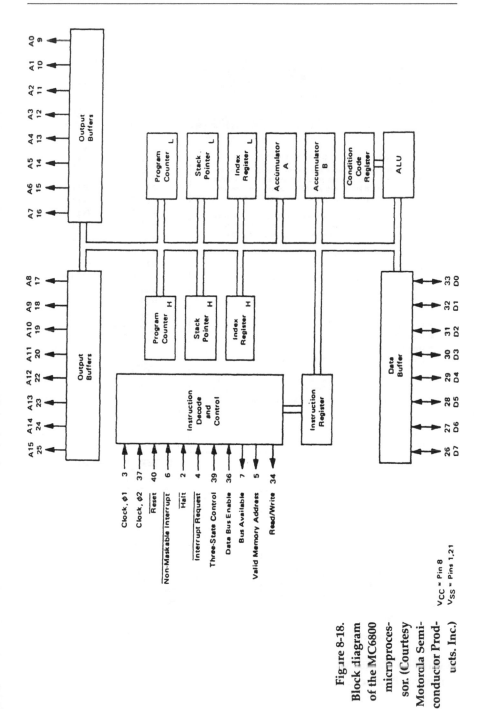

Figure 8-18. Block diagram of the MC6800 microprocessor. (Courtesy Motorola Semiconductor Products, Inc.)

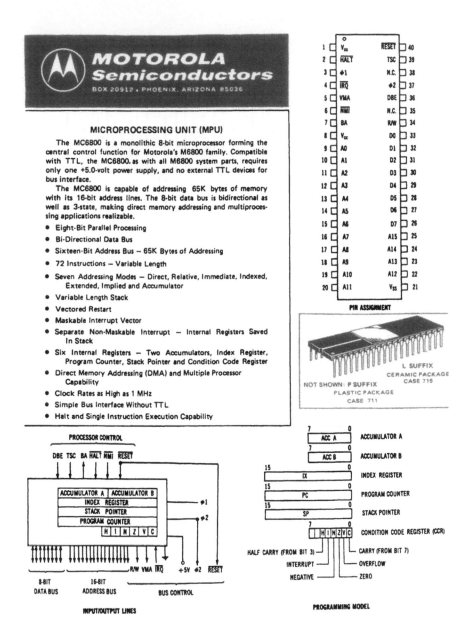

Figure 8-19. Partial MC6800 data sheet and diagrams. (Courtesy Motorola Semiconductor Products, Inc.)

LDAA, and LDAB is the mnemonic for loading accumulator B. Storing data in accumulator A is STAA, with STAB achieving the same operation for accumulator B. With two accumulators in the MPU, arithmetic and logic operations can be performed on two different numbers at the same time without shifting memory.

Index Register

The *index register* of the MC6800 has a 16-bit (two-byte) capacity that is used to store memory addresses. This register has the capability of being loaded from two adjacent memory locations. Through this feature, data can be moved in two-byte groups. The contents value increases by one when given the increment index register instruction (INX). A DEX instruction applied to the unit will cause it to decrease by one. This latter instruction is called *decrementing* the index register.

Program Counter

The *program counter* (PC) of the MC6800 is a 16-bit register that holds the addresses of the next byte to be fetched from memory. It can accommodate 2^{16} or $65,536_{10}$ different memory addresses. Two 8-bit bytes are used for obtaining a specific address location.

Stack Pointer

The *stack pointer* (SP) of the MC6800 is a special 16-bit register. It uses a section of memory which has a last-in, first-out capability which allows the status of the MPU registers to be stored when branch or interrupt subroutines are being performed. An address in the SP is the starting point of sequential memory locations in memory where the status of MPU registers is stored. After the register status has been placed into the stack pointer, it is decremented. When the SP is accessed, the status of the last byte placed on the stack will serve as the first byte to be restored.

Condition Code Register

The *condition code register* (CCR) of the MC6800 is a special 8-bit register that is actuated by the execution of an instruction. The outputs of this register are indicated as 11 HINZVC. The carry-in-borrow condition is indicated by the C output. When C equals 1, it indicates a carry-for addition or a borrow-for subtraction. The C equals 0 condition indicates the reset state when no borrow or carry occurs.

A 1 appearing at the V output location is used to indicate the results of an overflow condition when the twos-complement arithmetic operation has been performed. When V equals 0, the overflow condition does not occur.

The Z or zero output location of the register indicates when the output of an arithmetic operation is zero. If Z equals 0, there is a resulting zero and if Z equals 1, there is a not-zero condition.

The N or negative bit of the condition code register indicates the status of bit 7 after an arithmetic operation. When bit 7 is 1, N equals 1. This is used to denote a negative value for a twos-complement operation. When N equals 0, bit 7 is 0, which indicates a positive status.

When the I bit of the condition code output is equal to 1, the MPU cannot respond to an interrupt request from an outside source. An I equals 0 state permits interrupts to occur.

The H bit of the CCR output is used to indicate a half-carry during adding, adding with carry, or adding accumulator operations. When H equals 1 during an execution of instructions, it indicates there is a resulting carry from bit 3 to bit 4. An H equals 0 state indicates no resulting carry from the bit 3 position.

Bit positions 6 and 7 of the CC register output are not specific indicators in normal operation. They remain in the 1 state during regular operation.

16-BIT MICROPROCESSORS

To improve the data handling and addressing capabilities of their products, microprocessor manufacturers introduced a number of new devices in the early 1980s. These chips were designed to manipulate sixteen bits of data at a time. This brought about a whole new generation of personal computers. Most of the 16-bit PCs are built around a few of the popular microprocessors. Intel's 8088, for example, is found in the IBM personal computer and a dozen or more clones. The data path between the 8088 and the primary storage is only eight bits wide. This permits all operations to and from memory to be done eight bits at a time. Once the data are retrieved, they are processed internally sixteen bits at a time. The 8088 is described as a *8/16-bit microprocessor*. It functions as a fast 8-bit device with an extended set of instructions. Intel's 8086 has internal and external data paths that are all sixteen bits wide. The 8086 is classified as a *16/16-bit microprocessor.* The address bus of this chip has twenty lines, which permits it to identify one megabyte *of* memory.

A simplified block diagram of Intel's 8086 is shown in Figure 8-20 The functional blocks of this unit are similar to those defined for the 8-bit device, but the layout and construction of the chip is significantly different. The internal functions of this chip are divided into two processing units. The left side of the diagram shows the execution unit (EU). It has a register file and a 16-bit ALU, and is interconnected to the control and timing unit. The right side of the diagram shows the bus interface unit (BIU). It has a relocation register file, bus interface unit, and instruction queue, and is connected to the control and timing unit. The chip is housed in a forty-pin dual in-line package.

Motorola's 68000 is another one of the popular 16-bit microprocessors. This chip is used in the Apple Macintosh computer and many other systems. The 68000 uses a 16-bit external bus to move data between storage and the chip. Data processing is done internally, thirty-two bits at a time. This chip is therefore classified as a *16/32-bit microprocessor.* The address bus has twenty-four lines and permits a primary storage capacity of up to sixteen megabytes.

The 68000 family of microprocessors has two modes of operation, called the *user* and *supervisor.* The user mode of operation is reserved for executing programs written by the user. The supervisor mode of operation is designed to handle all other programs presented to the chip. These are grouped under the heading of *exception programs.* They are exceptions to normal user programming. Automatic switching occurs when an exception program is to be executed.

A programming model of the 68000 is shown in Figure 8-21. This model shows that there are no separate accumulators or index registers in the chip. Instead, there are sixteen data registers, identified as D_0 through D_{15}; an address register, numbered A_1 through A_7; and two stack pointer registers, identified as A_8 and A_{23}. The stack pointer registers can be used for most applications as two eight-address registers. The data registers can all be used as accumulators for data handling. In addition, all fifteen data and address registers and the stack pointers can be used as index registers. Pins A_1 through A_{23} form the address bus to interconnect memories and other devices to the microprocessor. Address information always flows from the microprocessor to the devices connected to the bus.

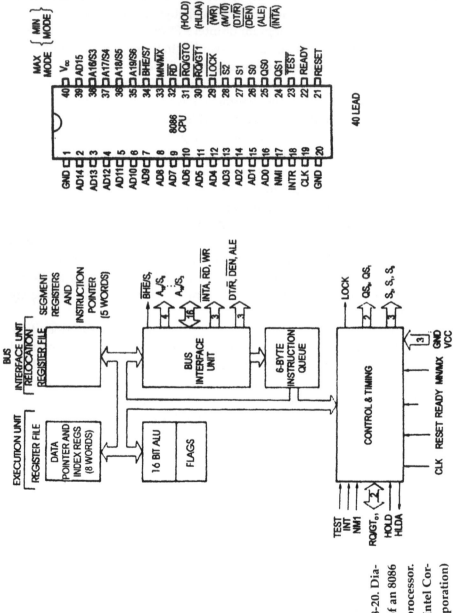

Figure 8-20. Diagram of an 8086 microprocessor. (Courtesy Intel Corporation)

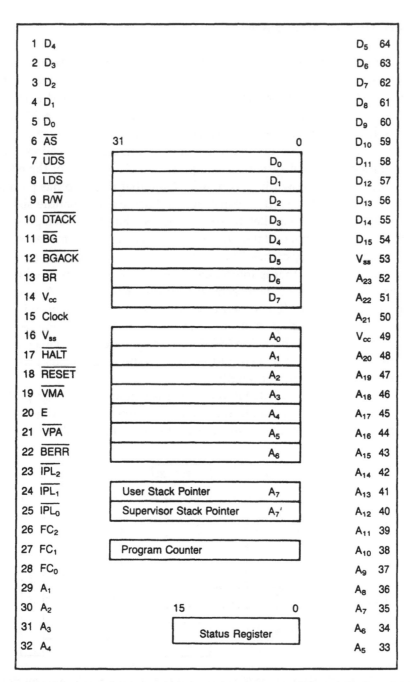

Figure 8-21. Programming model of the 68000. (Courtesy Motorola Semiconductor Products. Inc.)

32-BIT MICROPROCESSORS

The 32-bit microprocessor is primarily an enhancement or extension of the 16-bit devices. Motorola's 68030 is an enhanced version of the 68000 chip. The MC68030 has 300,000 transistors in its circuitry and is classified as a 32/32-bit device. A base operating frequency of up to 20 MHz makes this chip compatible with the most demanding applications in process and numerical control, robotics, workstations, communications, and environmental control systems. A simplified block diagram of the 68030 microprocessor is shown in Figure 8-22. Note that this chip has parallel address and data buses III its construction. Each bus can handle sixteen bits of data. By being parallel, these bus lines can handle sixteen plus sixteen or thirty-two bits of data. The chip also has a memory management unit and two memory caches. The term *cache* refers to a localized block of read/write memory that is internally accessible by the chip in which it resides. The 68030 has a data cache and an instruction cache. A memory management unit (MMU) is used to keep track of where memory is and is not, as well as which portions of physical memory have been used the least or farthest in the past. The 68030 has a paged memory management unit in its construction. Paged addressing has the length of its memory addresses fixed to a specific length.

Intel's 386SX is a full 32-bit microprocessor. It has a very high performance ALU that ensures short average instruction-execution times. This unit is capable of execution at sustained rates of 2.5 to 3.0 million instruc-

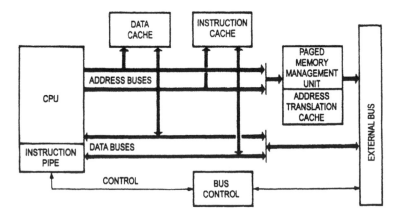

Figure 8-22. Block diagram the 68030. (Courtesy Motorola Semiconductor Products, Inc.)

tions per second. The memory management unit of this chip includes an address translation cache, multitasking hardware, and a four-level hardware-enforced protection mechanism. The circuitry also offers on-chip testability and debugging features.

A block diagram of the internal structure of a 386SX is shown in Figure 8-23. This diagram shows that the chip consists of a central processing unit, memory management unit, and a bus interface. The CPU has execution and instruction units in its construction. The execution unit consists of the register file, multiply/divide, barrel shifter, control, and an instruction decoder to manipulate its operations. The instruction unit decodes the opcode signals and stores them in the decoded instruction queue for immediate use by the execution unit. The MMU has segmentation and paging units. Segmentation is used to isolate and protect operations. The paging unit is transparent to segmentation and manages the address space. This chip also has real and protected modes of operation. This permits the circuit to respond as a fast 8086 by using the paging unit. Finally, the bus interface serves as a direct address pipe for each byte of applied data. The 386 has a 32-bit CPU with a 16-bit external data bus and a 24-bit external address bus that brings high-performance software to midrange systems.

64-BIT MICROPROCESSORS

Progress in microprocessor development has not stopped with the 32-bit chips. Most manufacturers are now producing 64-bit devices. The Intel i860 is a representative 64-bit micro-processor. This unit has balanced integers, floating decimal points, and graphics performance for applications such as engineering workstations, scientific computing, three-dimensional graphics, and networking. Its parallel architecture permits high-speed throughout for graphic techniques, pipeline processing units, wide data paths, on-chip memory caches, over a million transistors in its construction, and 1 μ fast operation. Microprocessors of this type make the small personal computer equivalent to the supercomputer of a few years ago. This innovation permits the computer to serve as the building block of industrial automation.

A block diagram of Intel's i860 microprocessor is shown in Figure 8-24. This chip consists of a core execution unit, floating-point control, floating-point adder, floating-point multiplier, graphics unit, paging, instruction cache, data cache, and bus/cache control.

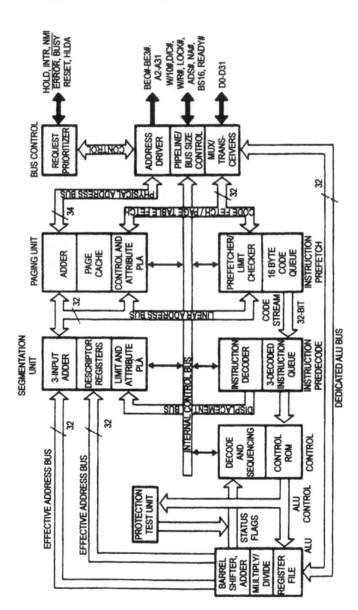

Figure 8-23. Block diagram of the 386SX. (Courtesy Intel Corporation)

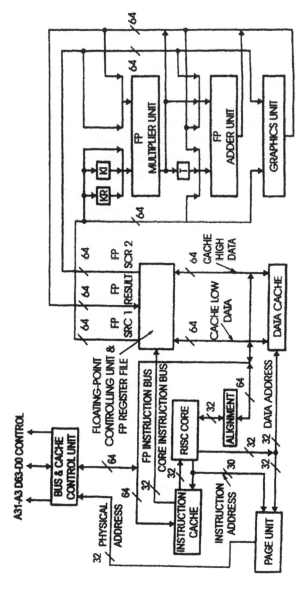

Figure 8-24. Block diagram an i860 microprocessor. (Courtesy Intel Corporation)

The *core execution unit* is designed to control the overall operation of the i860. This part of the chip executes load, store, integer, bit, and control-transfer operations, and fetches instructions for the floating-point unit. A set of 32 × 32 registers are provided to manipulate integer data. The load and store instructions are designed to move data to and from these registers. Its integer, local, and control-transfer instructions give the core unit the ability to execute system programs.

The *floating-point* (FP) function of a microprocessor keeps track of the decimal point in math operations, as contrasted with fixed-point arithmetic. Floating-point hardware is connected to registers which can be easily accessed. The FP control circuit responds to the adder and multiplier when appropriate instructions are issued. The adder performs addition, subtraction, comparison, and conversions of 64-bit data. The multiplier performs integer multiplication and reciprocal operations on 64-bit data values. Addition is executed in three clock cycles, while multiplication is achieved in three to four clock cycles.

The *graphics unit* of the i860 is a special logic circuit that supports three-dimensional drawing in a graphic frame, with color intensity and surface elimination. This part of the chip recognizes the display pixel as 8-, 16-, or 32-bit data. A *pixel* is a definable picture element that forms images on the display screen. The graphics unit computes the individual color intensities within a pixel. The surface of a solid object is drawn with polygon patches whose shapes approximate the original object. The color of the polygon and the distance from the viewer is known, but the other points of the object must be calculated. The graphics unit aids in this calculation.

The *paging unit* of the i860 implements a special memory function of the chip. This unit translates logical address information into a physical address and checks for access violations. This circuit has user and supervisor levels of control.

The i860 has two memory caches in its circuitry. The instruction cache is a two-way memory storage unit. It transfers up to 64 bits of memory per clock cycle. The data cache is a two-way memory unit. It transfers up to 128 bits of data per clock cycle. These caches can be updated or inhibited by software when necessary.

The bus and cache control section of the i860 is responsible for data and instruction accesses for the microprocessor. It receives data externally and serves as an interface to the external bus. This structure supports up to three bus cycles during its operation.

MEMORY

Microcomputer applications range from a number of single-chip microprocessor units to some rather complex networks that employ several auxiliary chips interconnected together in a massive system. As a general rule, the primary difference in this broad range of applications is in the memory capabilities of the system. Single-chip microprocessors are quite limited in the amount of memory they can process because of the large number of essential logic functions needed to make the unit operational. Additional memory, as a rule, can be achieved much more economically through the use of auxiliary chips. The potential capabilities of a microcomputer system are primarily limited by the range of memory that it employs.

Memory, in general terms, refers to the capability of a device to store logical data in such a way that a single bit or group of bits can be easily accessed or retrieved. In practice, memory can be achieved in a variety of different ways. Microprocessor systems are usually concerned with *read/write* memory and *read-only* memory. These two classifications of memory are accomplished by employing numerous semi-conductor circuit duplications on a single IC chip. Read/write memory permits the access of stored memory (reading) and the ability to alter the stored data (writing). Read/write memory is commonly called *random access memory* (RAM). The read-only memory (ROM) function is primarily concerned only with reading. ROM is primarily concerned with storing information that is not subject to change. When operational power is removed from a ROM device, it continues to retain its contents.

Random Access Memory

Random access memory or RAM semiconductor memories are the most widely used form of electronic memory found in microprocessors. Two basic types of RAM have evolved in these systems. *Dynamic RAM* is noted for high capacity, moderate speeds, and low-power consumption. A memory cell of this type of circuit is achieved by charge-storage capacitors with drive transistors. The presence or absence of charge in a capacitor is interpreted by the RAM's sense line as a logical 1 or 0. Because of a natural tendency for a charge to discharge itself into a lower energy-state configuration, dynamic RAM requires periodic charge refreshing in order to maintain data storage. Traditionally, the charge refreshing function of a RAM means that this type of system needs additional circuitry in order to perform its memory function.

Static RAM is an alternative to dynamic RAM. Static RAM is designed to store 1s and 0s using traditional flip-flop gate configurations. This type of memory is faster and requires no refresh circuitry for continuous operation. An operator simply addresses the static RAM and, after a very brief delay, obtains the stored information from a specific location. Static RAM is easier to use and to design circuitry around, but the complexity of a memory cell occupies a great deal more space than its dynamic counterpart. Static memory cells are also classified as *volatile* memory. With the energy source removed or turned off momentarily, a RAM will lose its memory.

A simplification of the memory process is represented by the 8 × 8 state memory unit in Figure 8-25. Memory ICs are usually organized in a rectangular pattern of rows and columns. This particular diagram employs eight rows that can store 8-bit words or sixty-four single bits of memory. To select a particular memory address, a 3-bit binary number

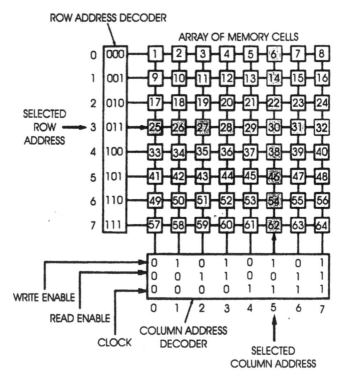

Figure 8-25. Simplification of the memory location process.

is used to designate a specific row location, and three additional bits are used to indicate the column location. In this example, the row address is 3_{10}, or 011_2, and the column address is 5_{10}, or 101_2. The selected memory cell address is at location 30.

Many RAMs employ a single metal oxide semiconductor (MOS) transistor for each memory location (see Figure 8-26). Binary information is stored in the transistor as a charge on a small capacitor. No charge across the transistor gate-channel electrodes indicates a 0 state, and a charge appearing across the two electrodes represents a 1. When a row select line is activated, it energizes the gate of each transistor in the entire row. When a column line is selected, it energizes the source-drain electrodes of each transistor in the column. Simultaneous activation of a row and column energizes a specific transistor memory cell.

A charge placed on the discrete MOS transistor of a memory cell may be restored periodically in order to overcome leakage. Charge regeneration is generally achieved by a special transistor thresholding amplifier. In practice, charge regeneration occurs every few milliseconds on a continuous basis.

Eight-bit word storage is achieved in the memory unit by energizing one row and all eight columns simultaneously. Some of the read/write memory units have capacities of 8×8, 32×8, 128×8, $1{,}024 \times 8$, and $4{,}096 \times 8$.

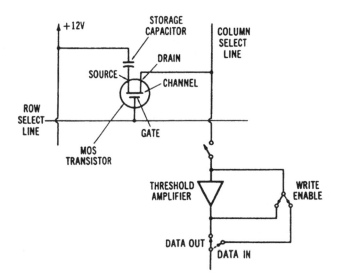

Figure 8-26. MOS transistor memory cell.

To write a word into memory, a specific address is first selected according to the data supplied by the address bus.

This is shown in the block diagram of a read/write memory unit in Figure 8-27. The address decoder, in this case, then selects the appropriate row and column select lines. A high or 1 *write-enable* signal applied to the control unit causes the data bus signal to be transferred to the selected memory address. These data then charge the appropriate memory cells according to the coded 1 or 0 values. Removing the write-enable signal causes the data-charge accumulations to remain at each cell location. The output is disconnected from the data bus after the write operation has been completed.

To read the charge accumulation appearing at each memory cell, the *read-enable* control line must be energized. Selecting a specific memory location will cause charge data to appear at the data bus as a memory output signal. Charge restoration from the thresholding amplifier continuously maintains the same charge at each cell, This means that reading from memory does not destroy the charge data at each cell. All of this takes place as long as the memory unit is energized electrically. A loss of electrical power, or turning the unit off, destroys the data placed at each memory location. Solid-state read/write memory is classified as volatile because of this characteristic.

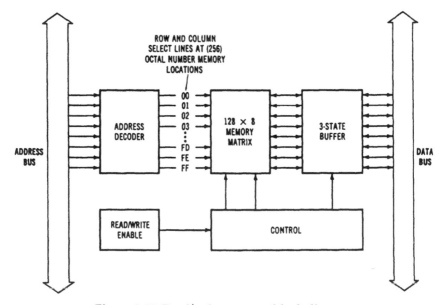

Figure 8-27. Read/write memory block diagram.

Read-only Memory

Most computer systems necessitate memories that contain permanently stored or rarely altered data, A prime example of this would be math tables and permanent program data, Storage of this type of memory unit is provided by read-only memory. Information is often placed in this type of memory unit when the chip is manufactured. ROM data are non-volatile, which means that they are not lost when the power source is removed or turned off.

Read-only memory is achieved in a variety of ways. One process employs *fusible links* built into each memory cell. A data pattern can then be placed into memory by "blowing out" the unwanted fusible links. This action may be used to open interconnecting conductors, place a diode between two connections, or place a small capacitor between two electrodes. A fusible link cannot be reformed after it has been destroyed. The fusible-link principle is used only once to program a read-only memory chip.

Erasable programmable read-only memory (EPROM) can be optically erased by exposure to an ultraviolet light source. Meter light exposure, each cell of the entire unit goes to a zero state. Writing data into the chip again initiates the new program.

Figure 8-28 shows a cross-sectional view of an optically erasable ROM. In this case, the floating gate of the MOS transistor is not electrically connected to anything. Data to be stored in the cell are "written" into the

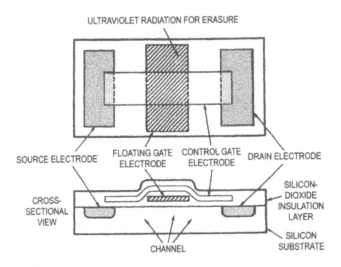

Figure 8-28. Optically erasable ROM using an MOS transistor.

transistor by applying 25 V between the gate and drain electrodes while the source and substrate are at ground potential. This action causes a static field to appear between the gate and source, which in turn causes electron movement with considerable velocity. Electrons that move through the thin silicon-dioxide insulator become trapped on the floating gate. The charged gate drain serves as a small capacitor, A charged condition represents a 1 state in memory and an uncharged condition indicates a 0 state, The charged cell condition produced by each transistor is nonvolatile.

Erasing the charged data of each MOS cell is achieved by exposing the chip to ultraviolet light. This action temporarily makes the silicon-dioxide insulation layer more conductive. As a result, excessive leakage causes the floating gate charge to dissipate, In practice, each cell of the chip is discharged by the exposure process at the same time, Memory can be restored by writing data back into each transistor cell. EPROMs of this type can be altered while in the circuit board if the need arises.

An alternative to EPROM is *electrically erasable programmable read-only* memory (EEPROM), This type of chip permits erasures of individual cells or word locations instead of the entire chip. Cell structure is similar to the EPROM, but the floating gate structure of each cell is altered by having a discrete interface insulation strip between it and the drain, Selective charge and discharge of each cell can be achieved by signals applied to the gate connection. The potential usefulness of EEPROM is quite large. It is presently used in programmable controllers, data loggers, and systems that have user-entered, alterable programs.

MICROCOMPUTER FUNCTIONS

There are certain functions that are basic to almost all microcomputer systems. Included in these operations are *timing, fetch and execution, read* memory; *write* memory; *input/output transfer,* and *interrupts.* An understanding of these functions is an important prerequisite to examining the operation of a specific microcomputer.

Timing

The operational activities of a microcomputer system are achieved by a sequence of cycling instructions. The MPU, for example, fetches an instruction, executes it, and continues to operate in a cycling pattern. This means that all actions occur at or during a precisely defined time interval.

An orderly sequence of operations like this necessitates some type of a precision clock mechanism. In practice, a free-running electronic oscillator or clock is responsible for this function. In some systems, the clock may be built into the MPU, while in others it is provided by an independent unit that feeds the system through a separate clock control bus. The entire sequence of operations is controlled by the timing signal.

Sequential operations such as fetch and execute are achieved within a period called the *MPU cycle*. The fetch portion of this cycle consists of the same series of instructions, and therefore takes the same amount of time for each instruction, The execute phase of an operation, by comparison, may consist of many events and sequences, depending on the specific instruction being performed. This portion of the cycle will vary a great deal with each particular instruction.

The total interval in which a timing pulse passes through a complete cycle from beginning to end is called a *period*. In practice, one or more clock periods may be needed to complete an operational instruction such as fetch. The execute operation, as a rule, may require a larger number of timing periods to cycle through its sequence. Essentially, this means that the machine cycle of a system may be variable, while the period of a timing pulse is consistent.

Fetch and Execute Operations

After programmed information has been placed in memory, its action is directed by a series of fetch and execute operations, This sequence of operations is repeated over and over again until the entire program has cycled to its conclusion. This program tells the MPU specifically what operations it must perform.

The entire operation begins when the start function is initiated, This signal actuates the control section of the MPU, which automatically starts the machine cycle, The first instruction that the MPU receives is to fetch the next instruction from memory. In a normal sequence of events, the MPU issues a read operational code instruction and the contents of the program counter are sent to the program memory. It responds by returning the next instruction word. The first word of this instruction is then placed in the instruction register. If more than one word is included in the instruction, a longer cycling time is needed to complete the instruction.

The execution phase of operation is based on which instruction is to be performed by the MPU, This instruction might call for such things as read memory, write in memory, read the input signal, transfer to output, or anyone of several MPU operations, such as add registers, subtract, or register to register transfer. The magnitude or time of this operation depends on programmed information which is placed in memory.

A 16-bit microprocessor takes a number of clock periods to perform its operations, This group of microprocessors can be operated at a clock rate of 2, 5, 8, 10, or 12.5 MHz. A single cycle of the 2-MHz clock has a period of 1/2,000,000 or 0.0000005 second. Periods this small are best expressed in microseconds (0.5 µs) or nanoseconds (500 ns). The fastest instruction change that can be achieved by this chip requires four clock periods, or 4 times 500 ns (2 µs). Its slowest instruction requires eighteen periods, or 18 times 500 ns (9 µs).

Thirty-two-bit microprocessors operate at a clock rate of 25 to 33 MHz. A single cycle of the 25 MHz clock has a period of 1/25,000,000 s or 40 ns. Sixty-four-bit microprocessors operate at a clock rate of 33 or 40 MHz. A single cycle of the 40 MHz clock has a period of 1/40,000,000 s or 25 ns. The operational time of an MPU is a good measure of its effectiveness and how powerful it is as a functional device.

Read-memory Operation

The read-memory operation is an instruction that calls for data to be read from a specific memory location and applied to the MPU. To initiate this operation, the MPU issues a read-operation code and sends it to the proper memory address. In return, the read/write memory unit sends the data stored at the selected address into the data bus. This 8-bit number is ultimately fed to the MPU, where it is placed in the accumulator after an appropriate timing pulse has been initiated.

Figure 8-29 shows a simplification of the read-memory operation in a series of sequential steps. In Step 1, the address bus supplies an 8-bit memory select address to the decoder. These data come from the programmed instructions and are initiated by the MPU. Step 2 is performed by the address decoder, which generates the appropriate row and column select lines that will actuate the memory location. These address input lines will allow anyone of the 1,024 locations to be addressed. When several memory chips are used, a specific chip is selected first, and then the desired address is selected, This operation involves two steps, called *chip selection* and *address selection*.

Step 3 is a control function. When the read-memory signal is applied to the read/write control line, the data transfer direction of the three-state buffer is decided. The reading operation directs data from

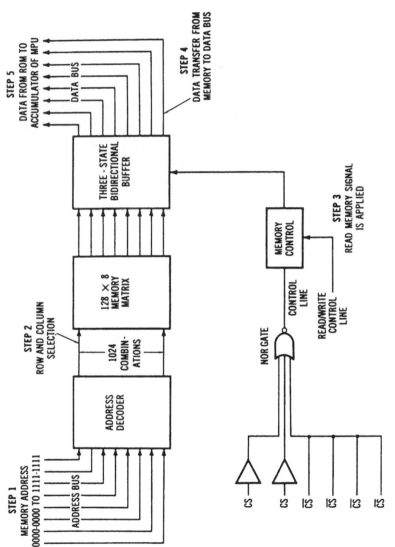

Figure 8-29. Operational sequence steps for read memory.

the memory to the data bus line. Step 4 deals with the chip-select control signal. An appropriate 6-bit binary code will energize a logic gate in the control line. This action causes the three-state buffer to change from a high-impedance condition to one of low impedance, As a result, data are transferred from memory into the data bus. Step 5 shows data being transferred to the accumulator of the MPU, which completes the read-memory operation.

Write-memory Operation

The write operation is an instruction that calls for data to be placed into or stored at a specific memory location. This function is initiated when the MPU issues a write-operation code and sends it to a selected read/write memory unit. Data from the data bus are then placed into the selected memory location for storage. In practice, 8-bit numbers are usually stored as words in the memory unit.

Figure 8-30 shows a simplified block diagram of the write-memory operation in a series of sequential steps. In Step 1, the MPU places an 8-bit address location into the address bus. In Step 2, these data are applied to the address decoder, which generates the desired row and column select lines that actuate the memory location. Anyone of $1,024._{10}$ locations can be selected in this particular example. If more than one memory chip is used, the operation is preceded by a chip-select signal applied to the chip-select (CS) inputs, The complete operation would then involve chip and address selection within the chip.

Step 3 is a control operation that prepares the chip to receive data. When the write-in memory signal is applied to the read/write control line, the data transfer direction of the three-state buffer is decided. The writing operation directs data from the data bus into the buffer (notice the direction of the data flow arrows), The data may come from the MPU, an alternate memory source, or through the system input.

Step 4 deals with the chip-select control signal. A 6-bit binary code will energize the NOR gate control line. This action causes the three-state buffer to change so that it is low impedance from the data bus to the memory matrix.

Data are then transferred into the matrix, where they actuate specific memory cells according to a prescribed level. Step 5 opens the path through the three-state buffer, which prevents any further change in the stored data at this location.

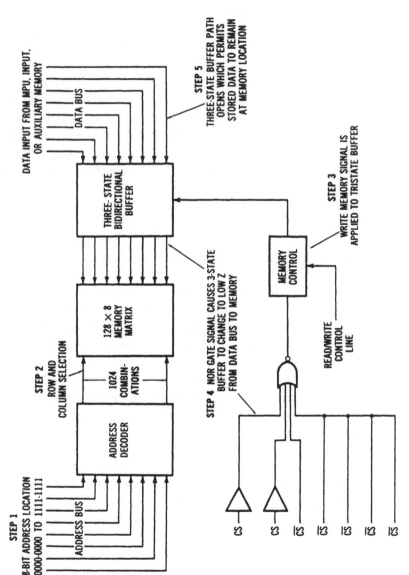

Figure 8-30. Operational sequence steps for write memory.

Input/Output Transfer Operations

In a microcomputer system, *input/output* (I/O) transfer operations are similar to the read/write operations just discussed. The major difference is the opcode data number used to call up the operation. When the MPU issues an input/output opcode, it actuates the appropriate I/O port, which either receives data from the input or sends them to the output device according to the coded instructions.

A simplification of a microcomputer system with a read/write memory. read-only memory, output, and input connected to a common address bus and data bus is shown in Figure 8-31. In this type of system, all data are acted upon by the MPU. As noted by the data bus arrows, data move in either direction to the read/write memory and flow only from the ROM or input into the data bus. The output only flows from the data bus to the output device.

In a bus-controlled system, data from a specific source must be transferred independently of the others, This is accomplished by assigning each destination or potential data source a different address, The chip-enable pin of each data source is selected by an activating code that energizes the desired data source.

The address assigned to each data source is primarily dependent on the capacity of the memory being employed. If a $512._{10}$-byte read/write memory is used, it might find itself assigned memory locations from $000._{16}$ to $01FF._{16}$. A $1,024._{10}$-byte ROM, if used, is usually assigned locations near the upper extreme of the data-bus capacity. In this case, locations might be $FC00._{16}$ to $FFFF._{16}$. When any of these memory address locations appear on the address bus, their data can be transferred to the data bus when the appropriate chip-enable code is applied. The latch circuit of the output and the buffer of the input are also assigned specific address locations. When these address locations are selected, data will be transferred to the output or brought into the data bus via the input device. This action occurs only when a chip-enable control signal is received by the appropriate component.

The output of each data source must not interfere with or upset the data appearing at the other source when it is placed on the data bus. To prevent cross-interference, each data source is fed through a three-state logic device. When a specific data source is disabled, it automatically assumes a high-impedance state with the data bus, and when enabled, it immediately changes to a low-impedance state, which provides connection to the data bus. Through this kind of circuitry, only one data source appears on the data bus at a particular time.

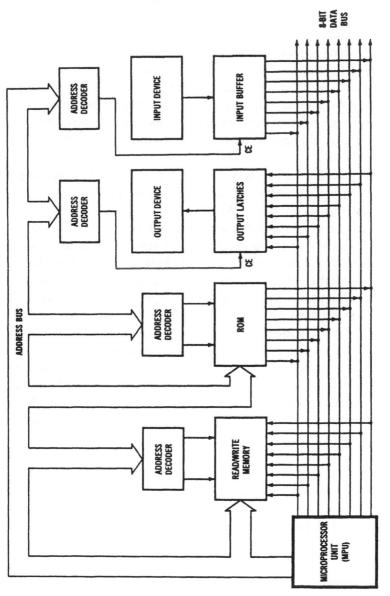

Figure 8-31. Simplification of MPU address bus and data bus connections.

Interrupt Operations

Interrupt operations are often used to improve the efficiency of a microcomputer system. Interrupt signals are generated by peripheral equipment such as keyboards, displays, printers, or process control devices. These signals are applied to the MPU to inform it that a particular peripheral device needs attention.

Consider a microcomputer system that is designed to process a large volume of data, some of which are to be the output to a line printer, The MPU, in this case, must output data at a high rate compared to the time needed to actuate a character representing a specific data byte. This means that the MPU would have to remain idle while waiting for the printer to complete its task.

If an interrupt capability is used by the MPU of a microcomputer system, it can output a data byte and then return to processing while the printer completes its operation, When the printer is ready for the next data byte, it simply requests an interrupt. Upon acknowledging the interrupt, the MPU suspends program execution and automatically branches to a subroutine that will output the next data byte to the printer. The MPU then continues the main program execution where it was interrupted. Through this procedure, high-speed MPU operation is not restricted by slow-speed peripheral equipment.

In practice, the interrupt operation is achieved by using registers within the MPU in conjunction with instructions stored in memory. Upon receipt of an interrupt request, it is necessary to freeze the contents of the MPU internal registers. This is usually achieved by storing them in the memory locations of the stack pointer register, which has last-in, first-out memory capabilities. When the program is resumed again, data at the interrupted point can be lifted from the stack pointer, which calls for execution to be restored,

Interrupt operations are only needed in systems where the MPU has a task to perform that must occur while a particular peripheral device is progressing its sequence. In industrial process control applications, where a limited amount of data is being moved, interrupt operations are generally avoided.

PROGRAMMING

The term *program* generally refers to a series of acceptable instructions developed for a computer that will permit it to perform a prescribed operation or function, In microcomputer systems, some programs are *hard-wired*, some are in ROM called *firmware*, and others are described as *software*, A dedicated system that is designed to achieve a specific function is generally *hardware programmed*, which means that it cannot be adapted for other tasks without a physical circuit change. Firmware systems have programmed material placed on ROM chips. Program changes for this type of system can be achieved by changing the ROM, A software type of program has the greatest level of flexibility. This type of program is created on paper and transferred to the system by a keyboard, floppy disk, or hard disk-drive unit. Instructions are stored in RAM locations and performed by calling for these instructions from memory.

Most microprocessor systems used in industrial applications are replacements for hard-wired circuits or systems. The design of the system and its purpose usually dictate the type of programming method employed. Combined firmware- and software-programmed systems, with a keyboard, are very common today, and these systems often adapt personal computers for specific applications. Computers of this type can be adapted to systems through hardware, by plug-in circuit boards, and through software, by programming. Personal computers are widely used in industrial systems because of the ease with which they can be adapted.

The instructions for a microcomputer system normally appear as a set of characters or symbols which are used to define a specific operation. These symbols, which may appear alone or with other characters, are similar to those that appear on an ordinary typewriter, including decimal digits (0 through 9), letters (A through Z), and, in some cases, punctuation marks and specialized keyboard characters. The presentation of symbolized instructions may appear as binary numbers, hexadecimal numbers, or mnemonic codes.

Microcomputers may use one of several different codes in a program. As a general rule, each type of MPU has a unique *instructional set* or repertoire of instructions which it is designed to understand and obey. These instructions primarily appear as binary data symbols and words. Machine instructions of this type are usually held in a read-only memory unit that is address selected and connected to the MPU through the common data bus. Instructional sets of this type are firmware because they are fixed at a specific memory location and cannot be changed by program material.

Microcomputer instructions usually consist of one, two, or three bytes of data. These type of data must follow the instruction commands in successive memory locations. The instructions are called *addressing modes*.

There are several distinct types of addressing modes in a typical micro-computer system.

Inherent Mode Addressing. One-byte instructions are often described as inherent mode instructions. These instructions are primarily designed to manipulate data to the accumulator registers of the ALU. As a rule, no address code is needed for this type of instruction because it is an implied machine instruction. The instruction *CLA*, for example, is a one-byte opcode that clears the contents of accumulator A. No specific definition of data is needed for further data manipulation. The instruction simply clears the accumulator register of its data.

Inherent mode instructions differ a great deal between manufacturers. Some representative opcode instructions are shown in Chart 8-1. Note particularly that the meaning of the instruction, its code in hexadecimal form, and a mnemonic are given for each instruction. Each microcomputer system has a number of inherent instructions of the one-byte type which contain only an opcode.

Immediate Addressing. Immediate addressing is accomplished by a two-byte instruction that contains an opcode and an operand. In this addressing mode, the opcode appears in the first 8-bit byte followed by an 8-bit operand. A common practice is to place intermediate addressing instructions in the first 256 memory locations. Through this procedure, these instructions can be retrieved quickly, making this the fastest mode of operation.

An example of an intermediate address instruction would be "load accumulator A with the number $53._{16}$." This instruction would appear as

Memory Location	Hexadecimal-Binary	Function
0010	$86._{16}$ 1000-0110	LDA (opcode)
0011	$53._{16}$ 0101-0011	(Data)

The number 86_{16} in this instruction is an indication of the hexadecimal opcode LDA A located at memory 0010. The number 53_{16} is the data to be manipulated, which appears at the next consecutive memory location. Ultimately, this means that the number 53_{16} ($0101\text{-}0011_2$) has been loaded into accumulator A.

Relative Addressing. Relative addressing instructions are primarily designed to transfer program control to a location other than the next consecutive memory address. In this type of addressing, two 8-bit bytes are used for the instruction. Transfer operations of this type are often limited

Mnemoni	Opcode	MEANING
ABA	1B	Add the contents of accumulators A and B. The result is stored in accumulator A. The contents of B are not altered.
CLA	4F	Clear accumulator A to all zeros.
CLB	5F	Clear accumulator B.
CBA	11	Compare accumulators: Subtract the contents of ACCB from ACCA. The ALU is involved but the contents of the accumulators are not altered. The comparison is reflected in the condition register.
COMA	43	Find the ones complement of the data in accumulator A, and replace its contents with its ones complement. (The ones complement is simple inversion of all bits.)
COMB	53	Replace the contents of ACCB with its ones complement.
DAA	19	Adjust the two hexadecimal digits in accumulator A to valid BCD digits. Set the carry bit in the condition register when appropriate. The correction is accomplished by adding 06, 60, or 66 to the contents of ACCA.
DECA	4A	Decrement accumulator A. Subtract 1 from the contents of accumulator A. Store result in ACCA.
DECB	5A	Decrement accumulator B. Store result in accumulator B.
LSRA	44	Logic shift right, accumulator A or B.
LSRB	54	

$$0 \rightarrow \boxed{b_7\ b_6\ b_5\ b_4\ b_3\ b_2\ b_1\ b_0} \rightarrow \boxed{C}$$

SBA	10	Subtract the contents of accumulator B from the contents of accumulator A. Store results in accumulator A.
TAB	16	Transfer the contents of ACCA to accumulator B. The contents of register A are unchanged.
TBA	17	Transfer the contents of ACCB to accumulator A. The contents of ACCA are unchanged.
NEGA	40	Replace the contents of ACCA with its twos complement. This operation generates a negative number.
NEGB	50	Replace the contents of ACCB with its twos complement. This operation generates a negative number.
INCA	4C	Increment accumulator A. Add 1 to the contents of ACCA and store in ACCA.
INCB	5C	Increment accumulator B. Store results in AACB.
ROLA	49	Rotate left, accumulator A or B.
ROLB	59	

$$\hookleftarrow \boxed{C} \leftarrow \boxed{b_7\ b_6\ b_5\ b_4\ b_3\ b_2\ b_1\ b_0} \hookleftarrow$$

Carry/borrow in condition register.

RORA	46	Rotate right, accumulator A or B.
RORB	56	

$$\hookrightarrow \boxed{C} \rightarrow \boxed{b_7\ b_6\ b_5\ b_4\ b_3\ b_2\ b_1\ b_0} \rightarrow$$

ASLA	48	Shift left, accumulator A or B (arithmetic).
ASLB	58	

$$\boxed{C} \leftarrow \boxed{b_7\ b_6\ b_5\ b_4\ b_3\ b_2\ b_1\ b_0} \leftarrow \boxed{0}$$

In condition code register   Always enters zeros

ASRA	47	Shift right, accumulator A or B (arithmetic).
ASRB	57	

$$\boxed{b_7\ b_6\ b_5\ b_4\ b_3\ b_2\ b_1\ b_0} \rightarrow \boxed{C}$$

Retains sign bit

Chart 8-1. Opcode instruction explanations.

to a specific number of memory locations. An example of this would be 127 forward and 128 reverse locations.

The two-byte instruction contains an opcode in the first byte and an 8-bit memory location in the second byte. The second byte points to the location of the next instruction to be executed. This type of instruction is designed to achieve branching operations.

Indexed Addressing. Indexed addressing is achieved by a two-byte instruction and is similar to the addressing mode of instruction. In this type of addressing, the second byte of instruction is added to the contents of the index register to form a new effective address. This address is obtained during program execution rather than being held at a predetermined location as with the other addressing modes. The newly created effective address is also held in a temporary memory address register so that it will not be altered or destroyed during the processing operation.

Direct Addressing. Direct addressing is the most common of all modes of instruction. In this type of instruction, the address is located in the next byte of memory following the opcode. This permits the first 256 bytes of memory, from 0000_{16} to $D0FF_{16}$, to be addressed.

An example of direct addressing would be

Memory Location	Contents	Function
0010	$96._{16}$	LDAA (opcode)
0011	$62._{16}$	Location of Data

The number 96_{16} is used as the opcode command for direct addressing of accumulator A. The mnemonic LDAA describes this operation. The number 62_{16} is an indication of the address where data are to be fetched from memory. Whatever data appear at this location are then transferred through the data bus to accumulator A. In direct addressing: the second byte of the instruction is considered to be the absolute memory address.

Extended Addressing. Extended addressing, as its name implies, is a method of increasing the capability of direct addressing so that it can accommodate more data. This mode of addressing is used for memory locations above $00FF_{16}$ and requires three bytes of data for the instruction. The first byte is a standard 8-bit opcode. The second byte is an address location for the most significant or highest order of the data in eight bits. This is followed by the third byte, which holds the address of the least significant or lowest eight bits of the data number being processed.

A representative extended-memory instruction would appear as

Memory Location	Contents	Function
0010	$B6._{16}$	LDA A (opcode)
0011	$40._{16}$	Address of highest eight bits of data
0100	$53._{16}$	Address of lowest eight bits of data

The number $B6_{16}$ in this instruction indicates the opcode command for extended loading of accumulator A. The mnemonic LOA A describes the operation. The number 40_{16} is an indication of the address of the most significant eight bits of data. The second half of the number, the least significant eight bits of data, is stored at location 53_{16}. After the instruction has been executed, the combined data at the designated locations would be transferred to the data bus, where it is loaded into accumulator A.

A Programming Example

In order to demonstrate the potential capabilities of a microcomputer system, the ability to solve a straight-line computation problem will be discussed. In this problem, the system will be used to simply add two numbers and then indicate the resulting sum. This type of problem is obviously quite simple and could be easily solved without the help of a microcomputer system, but it is used to demonstrate a principle of operation and to show a procedure. A microcomputer system could not solve even the simplest type of problem without the help of a defined program that works out everything right down to the smallest detail. After the program has been developed, the system follows this procedure to accomplish the task. Programming is an essential part of nearly all computer-system applications.

Before a program can be effectively prepared for a microcomputer system, the programmer must be fully aware of the specific instructions that can be performed by the system. In general, each microcomputer system has a unique list of instructions that are used to control its operation. The instruction set of a microcomputer is the basis of all program construction. Figure 8-32 shows an instruction set for the Motorola MC6800 microprocessor system.

Assume that the programmer is familiar with the instruction set of the system being used to solve the straight-line computation problem. The next step in this procedure is to decide what specific instructions are needed to solve the problem. A limited number of operations can generally be developed without the aid of a diagrammed plan of procedure. Complex problems, by comparison, usually require a specific plan to reduce confusion and avoid the loss of an important operational step. *Flowcharts* are sometimes used to aid the programmer in this type of planning. Figure 8-33 shows some of the flowchart symbols that are commonly used in program planning.

The first step in preparing a program to solve a problem is to make a

OPERATIONS	MNEMONIC	IMMED OP	~	=	DIRECT OP	~	=	INDEX OP	~	=	EXTND OP	~	=	IMPLIED OP	~	=	BOOLEAN/ARITHMETIC OPERATION (All register labels refer to contents)	H	I	N	Z	V	C
Add	ADDA	8B	2	2	9B	3	2	AB	5	2	BB	4	3				$A + M \to A$	t	•	t	t	t	t
	ADDB	CB	2	2	DB	3	2	EB	5	2	FB	4	3				$B + M \to B$	t	•	t	t	t	t
Add Acmltrs	ABA													1B	2	1	$A + B \to A$	t	•	t	t	t	t
Add with Carry	ADCA	89	2	2	99	3	2	A9	5	2	B9	4	3				$A + M + C \to A$	t	•	t	t	t	t
	ADCB	C9	2	2	D9	3	2	E9	5	2	F9	4	3				$B + M + C \to B$	t	•	t	t	t	t
And	ANDA	84	2	2	94	3	2	A4	5	2	B4	4	3				$A \cdot M \to A$	•	•	t	t	R	•
	ANDB	C4	2	2	D4	3	2	E4	5	2	F4	4	3				$B \cdot M \to B$	•	•	t	t	R	•
Bit Test	BITA	85	2	2	95	3	2	A5	5	2	B5	4	3				$A \cdot M$	•	•	t	t	R	•
	BITB	C5	2	2	D5	3	2	E5	5	2	F5	4	3				$B \cdot M$	•	•	t	t	R	•
Clear	CLR							6F	7	2	7F	6	3				$00 \to M$	•	•	R	S	R	R
	CLRA													4F	2	1	$00 \to A$	•	•	R	S	R	R
	CLRB													5F	2	1	$00 \to B$	•	•	R	S	R	R
Compare	CMPA	81	2	2	91	3	2	A1	5	2	B1	4	3				$A - M$	•	•	t	t	t	t
	CMPB	C1	2	2	D1	3	2	E1	5	2	F1	4	3				$B - M$	•	•	t	t	t	t
Compare Acmltrs	CBA													11	2	1	$A - B$	•	•	t	t	t	t
Complement, 1's	COM							63	7	2	73	6	3				$\overline{M} \to M$	•	•	t	t	R	S
	COMA													43	2	1	$\overline{A} \to A$	•	•	t	t	R	S
	COMB													53	2	1	$\overline{B} \to B$	•	•	t	t	R	S
Complement, 2's (Negate)	NEG							60	7	2	70	6	3				$00 - M \to M$	•	•	t	t	①	②
	NEGA													40	2	1	$00 - A \to A$	•	•	t	t	①	②
	NEGB													50	2	1	$00 - B \to B$	•	•	t	t	①	②
Decimal Adjust, A	DAA													19	2	1	Converts Binary Add. of BCD Characters into BCD Format	•	•	t	t	t	③
Decrement	DEC							6A	7	2	7A	6	3				$M - 1 \to M$	•	•	t	t	④	•
	DECA													4A	2	1	$A - 1 \to A$	•	•	t	t	④	•
	DECB													5A	2	1	$B - 1 \to B$	•	•	t	t	④	•
Exclusive OR	EORA	88	2	2	98	3	2	A8	5	2	B8	4	3				$A \oplus M \to A$	•	•	t	t	R	•
	EORB	C8	2	2	D8	3	2	E8	5	2	F8	4	3				$B \oplus M \to B$	•	•	t	t	R	•
Increment	INC							6C	7	2	7C	6	3				$M + 1 \to M$	•	•	t	t	⑤	•
	INCA													4C	2	1	$A + 1 \to A$	•	•	t	t	⑤	•
	INCB													5C	2	1	$B + 1 \to B$	•	•	t	t	⑤	•
Load Acmltr	LDAA	86	2	2	96	3	2	A6	5	2	B6	4	3				$M \to A$	•	•	t	t	R	•
	LDAB	C6	2	2	D6	3	2	E6	5	2	F6	4	3				$M \to B$	•	•	t	t	R	•
Or, Inclusive	ORAA	8A	2	2	9A	3	2	AA	5	2	BA	4	3				$A + M \to A$	•	•	t	t	R	•
	ORAB	CA	2	2	DA	3	2	EA	5	2	FA	4	3				$B + M \to B$	•	•	t	t	R	•
Push Data	PSHA													36	4	1	$A \to M_{SP}, SP - 1 \to SP$	•	•	•	•	•	•
	PSHB													37	4	1	$B \to M_{SP}, SP - 1 \to SP$	•	•	•	•	•	•
Pull Data	PULA													32	4	1	$SP + 1 \to SP, M_{SP} \to A$	•	•	•	•	•	•
	PULB													33	4	1	$SP + 1 \to SP, M_{SP} \to B$	•	•	•	•	•	•
Rotate Left	ROL							69	7	2	79	6	3				M ⎫ (see diagram)	•	•	t	t	⑥	⑥
	ROLA													49	2	1	A ⎬	•	•	t	t	⑥	⑥
	ROLB													59	2	1	B ⎭	•	•	t	t	⑥	⑥
Rotate Right	ROR							66	7	2	76	6	3				M ⎫ (see diagram)	•	•	t	t	⑥	⑥
	RORA													46	2	1	A ⎬	•	•	t	t	⑥	⑥
	RORB													56	2	1	B ⎭	•	•	t	t	⑥	⑥
Shift Left, Arithmetic	ASL							68	7	2	78	6	3				M ⎫ (see diagram)	•	•	t	t	⑥	⑥
	ASLA													48	2	1	A ⎬	•	•	t	t	⑥	⑥
	ASLB													58	2	1	B ⎭	•	•	t	t	⑥	⑥
Shift Right, Arithmetic	ASR							67	7	2	77	6	3				M ⎫ (see diagram)	•	•	t	t	⑥	⑥
	ASRA													47	2	1	A ⎬	•	•	t	t	⑥	⑥
	ASRB													57	2	1	B ⎭	•	•	t	t	⑥	⑥
Shift Right, Logic	LSR							64	7	2	74	6	3				M ⎫ (see diagram)	•	•	R	t	⑥	⑥
	LSRA													44	2	1	A ⎬	•	•	R	t	⑥	⑥
	LSRB													54	2	1	B ⎭	•	•	R	t	⑥	⑥
Store Acmltr	STAA				97	4	2	A7	6	2	B7	5	3				$A \to M$	•	•	t	t	R	•
	STAB				D7	4	2	E7	6	2	F7	5	3				$B \to M$	•	•	t	t	R	•
Subtract	SUBA	80	2	2	90	3	2	A0	5	2	B0	4	3				$A - M \to A$	•	•	t	t	t	t
	SUBB	C0	2	2	D0	3	2	E0	5	2	F0	4	3				$B - M \to B$	•	•	t	t	t	t
Subtract Acmltrs	SBA													10	2	1	$A - B \to A$	•	•	t	t	t	t
Subtr. with Carry	SBCA	82	2	2	92	3	2	A2	5	2	B2	4	3				$A - M - C \to A$	•	•	t	t	t	t
	SBCB	C2	2	2	D2	3	2	E2	5	2	F2	4	3				$B - M - C \to B$	•	•	t	t	t	t
Transfer Acmltrs	TAB													16	2	1	$A \to B$	•	•	t	t	R	•
	TBA													17	2	1	$B \to A$	•	•	t	t	R	•
Test, Zero or Minus	TST							60	7	2	7D	6	3				$M - 00$	•	•	t	t	R	R
	TSTA													4D	2	1	$A - 00$	•	•	t	t	R	R
	TSTB													5D	2	1	$B - 00$	•	•	t	t	R	R

Condition Code Register columns: 5 4 3 2 1 0 = H I N Z V C

LEGEND:

OP Operation Code (Hexadecimal).
~   Number of MPU Cycles.
=   Number of Program Bytes.
+   Arithmetic Plus.
-   Arithmetic Minus.
·   Boolean AND.
M_SP Contents of memory location pointed to be Stack Pointer.

+ Boolean Inclusive OR.
⊙ Boolean Exclusive OR.
M̄ Complement of M.
→ Transfer Into.
0 Bit = Zero.
00 Byte = Zero.

CONDITION CODE SYMBOLS:

H Half-carry from bit 3.
I Interrupt mask
N Negative (sign bit)
Z Zero (byte)
V Overflow, 2's complement
C Carry from bit 7
R Reset Always
S Set Always
: Test and set if true, cleared otherwise
• Not Affected

Note - Accumulator addressing mode instructions are included in the column for IMPLIED addressing

Figure 8-32. MC6800 Instructional set (Courtesy Motorola Semiconductor Products, Inc.)

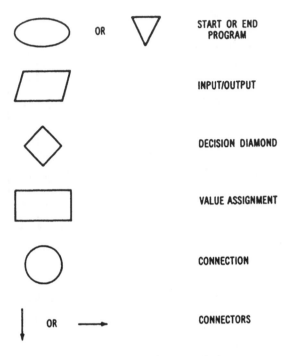

OR   START OR END
                     PROGRAM

                     INPUT/OUTPUT

                     DECISION DIAMOND

                     VALUE ASSIGNMENT

                     CONNECTION

OR   CONNECTORS

Figure 8-33. Flowchart symbols.

flowchart that shows the general plan of procedure to be followed. In this
case, the program will be set to accomplish the following:

Find the number N (a decimal value)

Use the equation $\qquad$ $x + y = N$

Let $\qquad\qquad\qquad$ $x = 0A._{16}$ (operand)

Let $\qquad\qquad\qquad$ $y = 07._{16}$ (operand)

Figure 8-34 shows a simple flowchart plan of procedure to be followed by this program to solve the problem. The first step, Step 0, is a simple statement of the problem. Steps 1 and 2 are then used as the operands
of the problem and indicate the values of x and y. After these values have
been obtained, each is placed into a specific accumulator. Step 3 is then
responsible for the addition operation. Since this system deals with binary
numbers, conversion to decimal values is also necessary. Step 4 is a conversion operation that changes binary numbers to bcd values. These values can then be used to energize an output to produce a decimal readout.
The fifth and final step of the program is an implied halt opcode that stops

the program. A programming sheet for the example problem is shown in Chart 8-2. Notice that this sheet indicates the step number, a representative memory address in hexadecimal values, instruction bytes in hexadecimal values, opcode mnemonics, binary equivalents, addressing modes, and a description of each unction. The memory address locations employed in the program begin with 66_{16}. In practice, it is a common procedure to reserve the first 100 memory locations, or $00._{16}$ to $63._{16}$, for branch instructions. Location $66._{16}$ ($102._{10}$) has been arbitrarily selected to avoid those addresses being reserved for branching operations.

The example program shown here is only one of literally thousands of programs that can be employed by a microcomputer to perform useful industrial operations. In this case, only one simple problem has been shown, but the potential capabilities of this type of system are unlimited. The type of microprocessor used by a system and its individual instructional set are the factors that govern its operation in program planning.

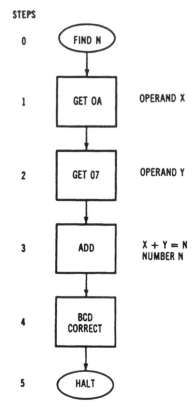

Figure 8-34. Flowchart of a problem.

COMPUTER AUTOMATION

Probably the single most influential technological development that has taken place in this century is the computer. The computer is now finding its way onto factory floors to control industrial processes and is widely used as a tool to achieve automation. *Computer integrated manufacturing* (CIM) is a buzzword used to describe this computer application. In general, there is no explicit definition of CIM. One common way of describing this term refers to CIM as a totally automated factory in which all manu-

Chart 8-2. Programming sheet.

Steps	Memory Address (Hexadecimal)	Instruction			Opcode	Addressing Mode	Operand (Binary)	Description
		Byte 1	Byte 2	Byte 3				
1	66	86	–	–	LDA A	Immediate		Load accumulator A
	67	–	0A	–			0000 1010	Use data 0A₁₆
2	68	C6	–	–	LDA B	Immediate		Load accumulator B
	69	–	08	–			0000 1000	Use data 08₁₆
3	6A	1B	–	–	ABA	Inherent		Add the contents of accumulators A and B
4	6B	19	–	–	DAA	Inherent		Correct for bcd output
5	6C	3E	–	–	HLT	Inherent		Stop all operations

Title: X + Y = N Purpose: Find N Date: ___ Time: ___

facturing processes are integrated and controlled by a computer system. This system enables production planners, schedulers, foremen, designers, accountants, and engineers to use the same database system. CIM is a philosophy for integrating hardware and software in such a way as to achieve total automation, but each company has its own idea of what CIM really means. Many of them follow a pattern similar to that in Figure 8-35. In this diagram, dedicated processing tasks are distributed around the factory. When computers are removed from the actual manufacturing area, their function shifts from real-time control to supervision.

It is generally agreed that a CIM operation contains different departments, areas, or levels of operation that are connected through a common computer network. Each level has certain tasks within its range of responsibility. At least three levels of integration are found in a CIM operation.

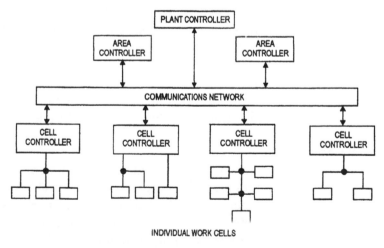

INDIVIDUAL WORK CELLS

Figure 8-35. CIM organizational pattern.

These are described as *cell level, area level,* and *plant level.*

Individual work cells actuate microprocessor controllers or personal computers. This level of operation is primarily responsible for data acquisition and direct machine control.

Individual cells are usually actuated by sensors and transducers. Area controllers are responsible for machine management, assembly operations, maintenance, material handling, simulation, engineering, and design procedures. Plant-level operation is generally considered to be *soft technology.* Plant-level computers are responsible for purchasing, accounting, materials management, resource planning, report generation, and system supervision. These three levels of operation are all interconnected through a common communications network.

Plant-level Computers

Plant-level computers are designed to supervise and control the entire operation of a CIM system. In large systems, *mainframe* computers are often used for this function. In some cases, these computers may have already been in operation to perform payroll, accounting, inventory, and general supervisory operations. If a computer has the capacity to perform additional operations, it may be connected or *interfaced* with the communications network. Interfacing generally necessitates some type of circuitry to make the computer compatible with the network. As a rule, some form of software is also needed to make the system operational. The mainframe computer is usually housed in a clean, temperature-controlled environment, located some distance from the main factory floor.

A number of companies are now producing customized assemblies that are designed to function as plant-level computers. The hierarchical structure of CIM makes this type of computer desirable. Customized assemblies of this type reduce the need for auxiliary interfacing and updating older technologies. As a rule, customized units are more efficient, operate at a faster rate, and permit the system to be very flexible.

Area-level Computers

Area-level computers are usually located on the plant floor or near actual manufacturing operations. These computers are responsible for machine management, maintenance, material handling, simulation, design, engineering, and intermediate supervisory operations. The area computer serves as an intermediary between the plant computer and the working-cell units. A special breed of industrial computers have been developed for this application. In general, this type of computer is designed for operation in a harsh environment. The primary difference between an area computer and a conventional personal computer is in its physical construction. Industrial computers are built to withstand high temperatures, vibration, electromagnetic noise, and rough handling. Since the primary role of the area computer is supervision, its hardware does not differ from that of a conventional PC, so many personal computers are now being used to perform this operation. The PC is inexpensive, readily available, and has an abundance of software. As a rule, this type of computer can be readily adapted to a CIM operation as long as the environment is not extremely harsh. This tends to enhance the flexibility of CIM.

Work-cell Controllers

The work-cell controller of a CIM operation is a specialized computer. It is used to control the operation of a group of work cells. A work cell is defined as a group of machine tools or equipment that is integrated into the system to perform different manufacturing processes. The work-cell computer coordinates the operation of each cell. This assembly necessitates special hardware and software requirements in order to be functional. The computer must have multiple data paths or input/output ports through which it communicates with respective cell devices.

Traditionally, cell control has been achieved by *programmable logic controllers* (PLCs), which were specifically designed for this type of control operation. A PLC is a microprocessor-based sequential controller that is used to control automated processes such as assembly, material handling, packaging, painting, welding, and machining-operations that used to be accomplished by electromechanical relays, timers, counters, and controllers. A PLC has a user-programmable memory for the storage of instructions, which permits it to implement specific instructions such as arithmetic, counting, data manipulation, input/output control, logic, and timing, PLCs consist of a processor, I/O interface, memory, and a programming device that uses relay logic symbols. Figure 8-36 shows a block diagram of a PLC that is designed to control the operation of an instrument, machine, or manufacturing process.

Figure 8-37 shows an equivalent PLC diagram that achieves control of two loads. The input source, which is isolated from the rest of the assembly, controls the energy applied to the input module. Control signals developed by the input module are applied as input data for the processor, which then directs the output module to conform with programmed

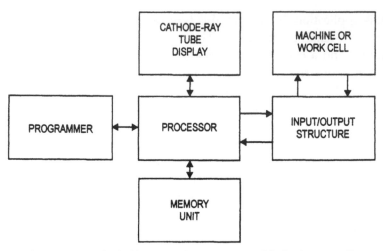

Figure 8-36. Block diagram of a programmable logic controller.

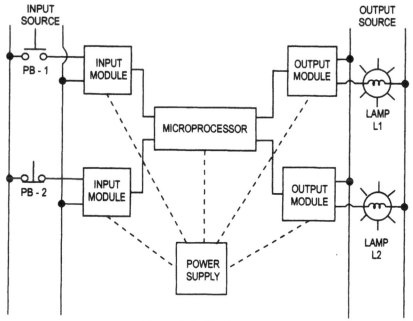

Figure 8-37. PLC system.

information that controls power to the load. Respective, loads are controlled by the output modules, which isolate the load energy source from the processor. This means that the energy sources of the load and input are completely independent and isolated from the processor. Isolation and independent energy sources makes this PLC a very flexible cell controller.

In a CIM operation, cell controllers must communicate with other cells, area-level computers, and the main plant computer through a communications network. Today, this is somewhat difficult for the PLC to accomplish, as the language of a PLC is somewhat restrictive. It does not work well in record keeping, high-speed communication, and data acquisition analysis. Manufacturers are now developing PLCs that have greater software capabilities. The *cell computer* is a recent development that confronts this problem. In this assembly, standard PLCs are attached to the input/output of the computer, permitting the cell controller to be completely integrated into the CIM operation. A number of inexpensive personal computers are also finding their way into this type of automation. This tends to make CIM even more flexible for future applications.

Communications Network

The key to a successful CIM operation is its communications network. The network must be compatible with each computer and the processes being performed in ,the manufacturing operation, which tends to cause some problems. Each computer, for example, may have a different set of communication standards. As the level of communication becomes more sophisticated, these standards change. In anticipation of this problem, the International Standards Organization (ISO) has developed a communications-network architecture model. Figure 8-38 shows a layer diagram of the ISO communications network. Each block of this diagram represents a different level of sophistication. Starting at the bottom, each layer provides the necessary support for the layers above it. Implementing this type of structure permits the user to connect any data-communication device to the network and be assured of compatibility.

Each layer of the ISO network structure has built-in communication protocols. A *protocol* is a set of rules that governs the exchange of information between computers. These protocols permit handshaking between computers. Protocols exist at all levels of communication and are usually assigned to the appropriate network layer according to its level of sophistication.

The need for protocols at different levels of a communications network is illustrated by a *FAX* transmission. The term *FAX* is an acronym for *facsimile*. It is the equipment that facilitates the transmission of images over a common carrier network, such as telephone lines. In this application, there must be some convention for the way wires are connected,

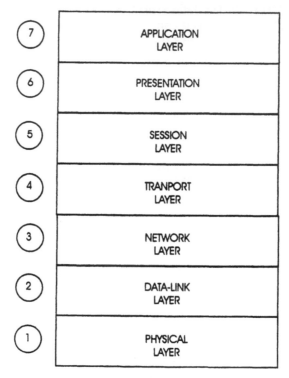

Figure 8-38. International Standards Organization network model.

the allowable bandwidth of the communication channel, and the signal levels being used. The sequence of signal exchange must then be established, This involves dial-tone generation and transmission, dial-pulse or touch-tone detection, station ringing, and busy-signal detection and transmission. Each of these must be clearly defined and established before the system can be functional. Finally, when the remote station connection has been established, image transmission or user signal development goes into effect. There are three protocol levels in this example. These are physical, data-link connection, and information dissemination. Each protocol level must effectively support the next level. Any change in the physical level should not affect the other protocols, which means that changes in the first layer are *transparent* to the other layers. This is the operational premise of the ISO's network architecture.

Presently, only Layers 1, 2, and 3 of the ISO network architecture model have been defined and successfully implemented. Layers 4 through 7 are for future developments. They are available for more complex protocols that are being developed.

Layer 1 deals with physical-link protocols, such as signal levels, wiring conventions, and cable standards. Signal information applied to a communications network is in the form of digital data. It is transmitted in either a serial or parallel format. Parallel communications are used for connections between process instruments, PCs, and printers. Parallel communication has a line for each bit of data being transmitted. Transmission of an 8-bit word needs a minimum of eight lines. Serial connections are made between computers and other peripherals. Serial transmission sends data one bit at a time over the communication line. Serial transmission is used for long-distance communication, while parallel transmission is used for short distances where high volumes of data are moved.

A Layer-1 protocol of the ISO model deals with cable standards. These are called *RS-232C, RS-422,* and *RS-485 options.* The RS-232C cable is available with four, nine, or twenty-five-pin wiring. The twenty-five-pin cable connects every pin of the cable, The RS-232C is a serial data standard for communication between terminal devices, printers, computers, and modems, and supports transmission distances up to fifty feet. Most microcomputers provide RS-232C interfaces. The RS-422 is a recently adopted standard for a very high-speed serial port. It overcomes some of the distance problems of the RS-232C, and can support a transmission distance up to 3,600 feet. The RS-232C and the RS-422 are both designed for connection between only two devices at one time. This is normally a computer and an instrument.

The RS-485 supports similar transmission distances as the RS-422, but it has multidrop capabilities so that up to thirty-two devices may be connected at one time. The RS-422 uses a four-wire design: transmit+, transmit−, receive+, and receive−. The RS-485 is normally a two-wire system, with a plus and a minus signal line. When a device needs to send data, it connects to the transmit signal lines. Receiving data is accomplished by connection to the receive data lines. Some serial networks combine the four wires of the RS-422 and the multidrop possibility of the RS-485 and call it an *RS-422/485 protocol.* The physical properties of communication networks are classified as *wide-area networks* (WANs) or *local-area networks* (LANs). The difference between these two groups lies in the transmission distance between stations. CIM and nearly all factory computer operations are connected to LANs. When two or more factories of a corporation are connected, they communicate over a WAN.

LANs are characterized by their method of station access. Two methods are in common use today: *random access* and *token access.* In a random access network, all stations listen to the transmission signal and wait until the line is free before attempting a transmission. If a collision occurs, the transmitting stations must shut down and wait for a new opportunity to access the network. This method of station access is called *carrier-sense multiple access with collision detection* (CSMA/CD). A number of protocols apply to this procedure. This type of access occurs at the transport level of the ISO model.

Token access is a special method of network accession that requires a token to enter the network. The token is a special code that precedes a transmission, When the line receives the token code, it accesses the network and permits the station to transmit. The token then passes between different stations until it reaches one with a message to transmit. That station then removes the token and replaces it with the data being transmitted. After the transmission has been completed, the station returns the token to the transmission line and the token proceeds to the next station. This procedure permits each station an equal opportunity to access data to the network. General Motors Corporation calls this its *Manufacturing Automation Protocol* (MAP). A number of other corporations use the token protocol for network access.

SUMMARY

Microcomputer technology has virtually revolutionized the industrial process field by providing small-capacity computer systems that can be built into instruments.

Computer systems use two-state data, such as 1s and 0s, in their operations. A single piece of two-state data is called a bit and a group of bits forms a word. A byte is an 8-bit word that finds widespread usage in microcomputers.

A digital computer system has input/output, an arithmetic/logic unit, control circuitry, and memory. Large-scale computers employ thousands of components, have enormous memory capacities, and operate at high speed. Microcomputer systems are tiny units built on a single circuit board or chip and are designed to handle specific applications similar to those of a central computer.

Microcomputers use decimal, binary, octal, and hexadecimal numbering systems, The base or radix of the system indicates the number of discrete digits it uses. Nearly all numbering systems have place value.

A microprocessor is the primary control section of a computer, scaled down so that it fits on a single IC chip, The ALU part of the chip achieves the arithmetic function. Accumulators are temporary registers that store operands that are to be processed by the ALU. Data registers are used to temporarily store the address of a memory location that can be accessed for data. A program counter is used to hold the address of the next instruction to be executed in a program. Instruction decoders are used to decipher an instruction after it has been pulled from memory. Sequence controllers maintain the logical order in which events are performed by the MPU. Buses are a group of conductor paths that supply words to registers.

Most microcomputer systems employ auxiliary memory units to extend operating capabilities. Memory permits data to be accessed or retrieved. Read/write memory permits data to be placed or stored into specific memory cells and retrieved at a later time when it is needed. Read-only memory contains permanently stored or rarely altered data. Permanent program data is an example of a ROM application. EPROM is erasable programmable read-only memory, in which electrical energy or ultraviolet light is used to erase stored data.

There are certain functions that are basic to almost all microcomputer systems, such as timing, fetch and execute, read memory, write memory, input/output transfer, and interrupt operations.

Programming refers to a series of acceptable instructions developed for a computer that will permit it to perform a prescribed operation or function. A hard-wired program is achieved by electrical-circuit construction. Firmware systems have programmed material placed on read-only memory chips. Software programming is created on paper and transferred to the system through a keyboard, Most microcomputer systems combine both firmware and software instructions in their programming material.

An instruction set is the individual set of commands designed to control the internal operations of a microcomputer system. Inherent instructions are one-byte opcodes that are designed to manipulate data to the accumulator registers. Immediate addressing is achieved by two-byte instructions containing an opcode and an operand. Relative-addressing instructions are designed to transfer program-control data to a location other than the next consecutive memory address. Direct addressing is a two-byte instruction that transfers data directly through the data bus to an accumulator. Extended addressing increases the capability of direct addressing so that it can accommodate more data.

When programming a microcomputer system, the programmer must be aware of specific unit instructions, decide on what instructions are needed to solve a problem, plan a flowchart, develop a programming sheet, initiate the program, correct it if needed, and execute its operation.

Computer integrated manufacturing is an important application of the computer that permits manufacturing to be achieved automatically. It has three levels of operation: cell, area, and plant. Each level has certain tasks within its range of responsibility: Work cells are responsible for data acquisition and direct machine control; area-level controllers are responsible for machine management, design, maintenance, and material handling; and plant-level computers are responsible for soft technology such as purchasing, accounting, planning, and supervision. These three levels are all interconnected through a common communications network.

Chapter 9

Automated Processes and Robotic Systems

OBJECTIVES

Upon completion of this chapter, you will be able to

1. Define hard automation and flexible automation processes.
2. Identify and describe the five major subsystems of an industrial robot.
3. Describe servo and non-servo robotic systems.
4. Discuss degrees of freedom in regard to automated processes performed by industrial robots.
5. Describe revolute, Cartesian, cylindrical and spherical robotic configurations.
6. List and describe at least eight automated robotic processes used in industry.

KEY TERMS

Automation— Processes that occur automatically, usually with electronic/computer control to manufacture parts or products.

Cartesian configuration—Geometric description of the arm movement of a robot along three intersecting perpendicular axes (straight lines) that can start or stop simultaneously.

Closed loop system—A system that allows for feedback that affects the output of a system.

Controller—The part of a robot that coordinates all movements of the mechanical system, usually a microprocessor and various sensors

Cylindrical configuration—Geometric description of a robot in which its range of motion assumes a cylindrical shape.

Degrees of freedom—A term used to describe a robot's dexterity or freedom of motion, for each degree of freedom an axis of motion is required.

Die casting—The pumping of hot metals into closed dies to form shapes when the metal solidifies and the casting is removed.

Electric drive—An actuator driven by an electric motor.

End of arm tooling—An end effector or other device attached to the wrist of a manipulator to perform a process operation.

Feedback—The interaction between the controlled element and the control unit of a closed loop system.

Flexible automation—Machines that can be programmed to perform different tasks.

Gripper—An end effector that is designed to grasp an object and move it.

Manipulator—The part of a robot consisting of segments that may be joined and moved to allow the robot to do work (the robot's arm).

Open loop system—A system in which no feedback mechanism is used to compare programmed positions to actual positions.

Palletizing—Placement of parts in a uniform series of positions, performed by pick and place robots

Pick and place motion—The motion of following points along each axis in a fixed pattern, usually one at a time.

Pitch—The up and down movement of a robot's wrist.

Point to point motion—The movement of a robot along a number of fixed points to position the end effector at a desired point.

Spherical configuration—Geometric description of the shape of a robot's work envelope that resembles a sphere.

System—An organization of parts that work together to form a functional unit.

Teach pendant—A device used to record movements into a robot's memory.

Work envelope—The area within which a robot's end effector can reach.

Yaw—The side to side movement of a robot' wrist.

INTRODUCTION

Today's industries use various types of automated processes to manufacture parts and products. Two common classifications are hard automation and flexible automation. Hard automation refers to process machinery that has been specifically designed and built to perform one particular task within an assembly line. After the item that the machine was designed to make is no longer needed, the machine must be discarded. This kind of automation can be very costly. The demand for new products and new models of existing products means shorter and shorter product life spans. Today, new hard automation equipment is cost-effective only where volume is high and the machine will be used for a long time.

Flexible automation includes machines that can perform different tasks. Robotic systems belong in this category. As new products or new models are needed, the flexible machine can be reprogrammed to make the parts required. This flexibility saves money because the equipment does not have to be discarded or rebuilt. In addition, it takes much less time to reprogram the same machine than to install a new one. Figure 9-1 shows a work cell for flexible automation.

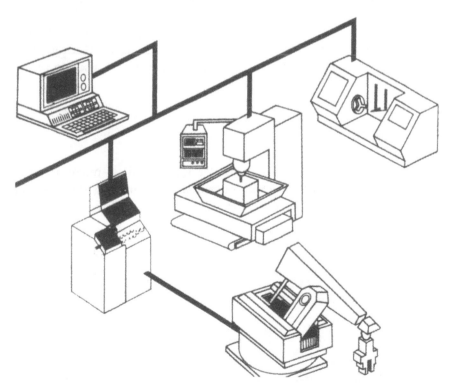

Figure 9-1. A flexible manufacturing workcell can be programmed for more than one task.

Robots are the most flexible type of automation available today. They be reprogrammed easily, quickly, and economically, and they can also be moved from one location in a plant to another. This chapter will primarily discuss the use of robotic systems in automated industrial processes.

Robots come in many shapes and sizes. Typical industrial robots resemble an inverted human arm mounted on a base. Industrial robots consist of a number of subsystems that work together. These subsystems are the controller, the manipulator, the power supply, the end effector, and a means for programming. Figure 9-2 is a diagram of the relationships among these five major systems. The controller is the part of a robot that coordinates all movements of the mechanical system. It also receives input from the immediate environment through various kinds of sensors. The manipulator consists of segments that may be jointed and that move about, allowing the robot to do work. For example, the manipulator may be the robot's arm. The power supply converts ac line voltage to the dc voltages required by the robot's internal circuits, or it may be a pump or compressor providing hydraulic or pneumatic power. The end effector is the robot's hand. The area which the robot's end effector can reach is called its work envelope. The programming system is used to record movements into the robot's memory.

CONTROLLERS

The heart of the robot's controller is generally a microprocessor (computer chip) linked to input/output and monitoring devices. The commands issued by the controller activate the motion control mecha-

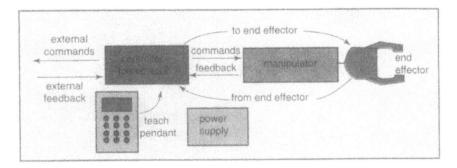

Figure 9-2. The relationships among the five major systems that make up an industrial robot are shown in this diagram.

nism, consisting of various controllers, amplifiers, and actuators. An actuator is a motor or valve that converts power into robot movement. This movement is initiated by a series of instructions, called a program, stored in the controller's memory.

The controller has three levels of hierarchical control. A hierarchical arrangement is one in which a given level is lower (more elemental) than the one above it and is dependent on the level above it for its instructions. Figure 9-3 illustrates the relationships among the basic levels of hierarchical control. The three levels are:

Level I: Actuator Control. This is the most elementary level, at which the separate movements of the robot along various planes, such as the X, Y, and Z axes are controlled. These movements will be explained in detail in the section of this chapter that covers robot configurations.

Level II: Path Control. The path control (intermediate) level coordinates the separate movements along the planes determined in Level I into the desired trajectory (path).

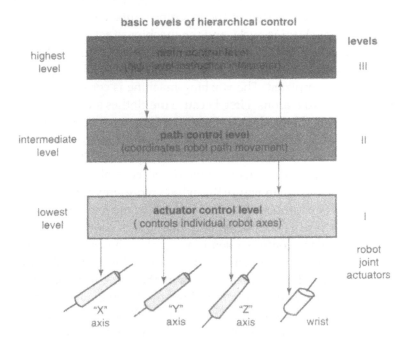

Figure 9-3. The three basic levels of hierarchical control.

Level III: Main Control. The primary function of this highest control level is to interpret the written instructions from the human programmer as to the tasks required. The instructions are then combined with various environmental signals and translated by the controller into the more elementary instructions that Level II can understand.

Robots can be classified according to the type of control system used. The nonservo robot is a nonintelligent robot. Servo robots are classified as either intelligent or highly intelligent. The primary difference between an intelligent and highly intelligent robot is its level of awareness of its environment.

Nonservo Robots

Nonservo robots are the simplest robots, and are often referred to as limited sequence, pick-and-place, or fixed-stop robots. The nonservo robot is an open-loop system. That is, no feedback mechanism is used to compare programmed positions to actual positions. A good example of an open-loop system is the operating cycle of a washing machine used in the home. Figure 9-4 is a block diagram of the steps performed by a typical washing machine. At the beginning of the operation, the dirty clothes and the detergent are placed in the machine's tub. The timer/control is set for the proper cleaning cycle, and the machine is activated by the start button. The machine fills with water and begins to go through the various washing, rinsing, and spinning cycles. The machine finally stops after the set sequence is completed. The washing machine is considered an open-loop system for two reasons. First, because the clothes are never examined by sensors during the washing cycle to see if they are clean. Second, the length of the cycle is not automatically adjusted to compensate for the amount of dirt remaining in the clothes. The cycle and its time span are determined by the fixed sequence of the timer/control.

Study the diagram in Figure 9-5, which represents a three-axis pneumatic (air-controlled) robot. The three axes allow movement along certain planes. For the sake of simplicity, only one axis is shown. At the beginning of the cycle, the controller sends a signal to the control valve of the manipulator. As the valve opens, air passes into the air cylinder, causing the rod in the cylinder to move. As long as the valve remains open, this rod continues to move until it is restrained by the end stop. After the rod reaches the limit of its travel, a limit switch tells the controller to close the control valve, and the controller sends the valve a signal to close. The controller then moves to the next step in the program and initiates the

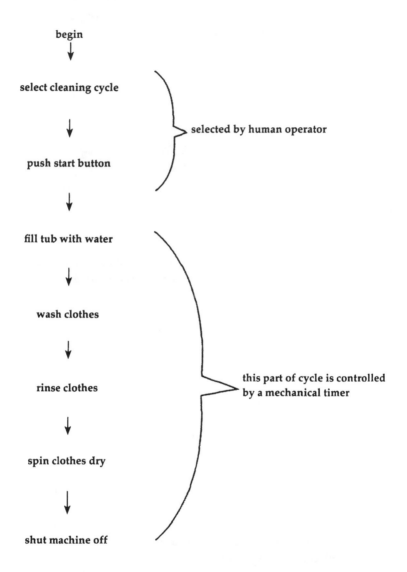

Figure 9-4. In this block diagram depicting the sequence of steps performed by a washing machine, no feedback is used. In such an open-loop control system, the condition of the clothes during the washing operation is not monitored and used to alter the process.

necessary signals. If the signals go to the end effector, for example, they might cause the gripper to close in order to grasp an object. The process is repeated until all the steps in the program have been completed.

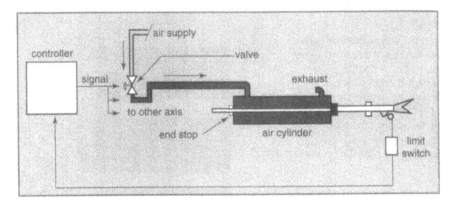

Figure 9-5. In a nonservo system, movement is regulated by such devices as a limit switch, which signals the controller when it is activated.

Nonservo robots have the following characteristics:
- Relatively inexpensive compared to servo robots.
- Simple to understand and operate.
- Precise and reliable.
- Simple to maintain.
- Capable of fairly high speeds of operation.
- Small in size.
- Limited to relatively simple programs.

Servo Robots

The servo robot is a closed-loop system because it allows for feedback. The feedback signal sent to the servo amplifier affects the output of the system. In a sense, a servomechanism is a type of control system that detects and corrects for errors. Figure 9-6 shows a block diagram of a servo-controlled robotic system.

The principle of servo control can be compared to many tasks performed by human beings. One example is cutting a circle from a piece of stock on a power bandsaw, shown in Figure 9-7. The machine operator's eye studies the position of the stock to be cut in relation to the cutting edge of the blade. The eye transmits a signal to the brain. The brain compares the actual position to the desired position. The brain then sends a signal to

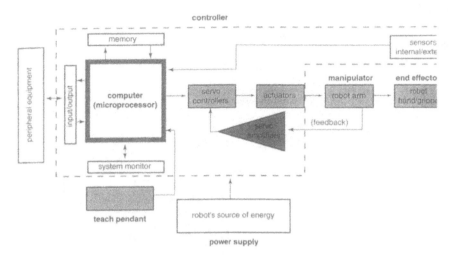

Figure 9-6. A servo-controlled industrial robotic system such as the one depicted in this block diagram might be classified "intelligent" or "highly intelligent," depending on the level of sensory data it can interpret.

the arms to move the stock beneath the cutting edge of the blade. The eye is used as a feedback sensing device, while the brain compares desired locations with actual locations. The brain sends signals to the arms to make necessary adjustments. This process is repeated as the operator follows the scribed line during the sawing operation.

The diagram in Figure 9-8 helps to explain the operation of a six-axis hydraulic robot. Only one axis is shown in detail. When the cycle begins, the controller searches the robot's programming for the desired locations along each axis. By means of feedback signals, the controller determines the actual locations on the various axes of the manipulator. The desired locations and actual locations are compared. Error signals are fed back to the servo amplifier. The greater the error, the higher the intensity of the signal. These error signals are amplified (increased) by the servo amplifier and applied to the servo control valve on the appropriate axis. The valve opens in proportion to the level of the command signal received. The opened valve admits fluid to the proper actuator to move the various segments of the manipulator.

New signals are generated as the manipulator moves. When there are no more error signals, the servo control valves close, shutting off the flow of fluid. The manipulator comes to rest at the desired position. The controller then addresses the next instruction in the program. It may be to move to another location, or to operate some peripheral equipment.

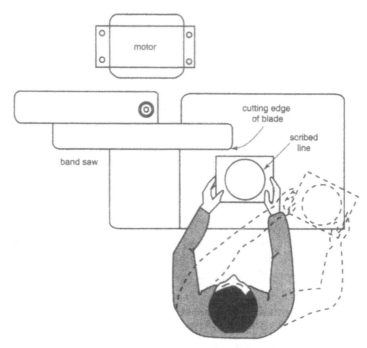

Figure 9-7. Human beings employ the principle of the servomechanism for many tasks, such as cutting a circle on a bandsaw.

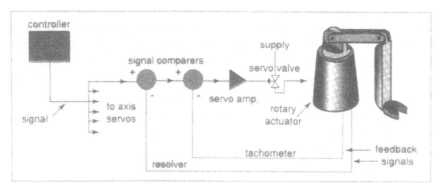

Figure 9-8. Feedback signals from the tachometer and resolver sensing systems allow the system to make corrections whenever the actual speed or position of the robot does not agree with the values contained in the robot's program.

The process is repeated until all steps of the program are completed. The output of a tachometer, a speed measuring device, is used to control acceleration and deceleration of the manipulator's movements.

Servo robots have the following characteristics:

- Relatively expensive to buy and cost more than nonservo robots to operate and maintain.
- Use a sophisticated, closed-loop controller.
- Have a wide range of capabilities.
- Can perform multiple point-to-point transfer as well as transfer along a controlled, continuous path.
- Can respond to very sophisticated programming.
- Use a manipulator arm that can be programmed to avoid obstructions within the work envelope.

Manipulators

The manipulator must move materials, parts, tools, or special devices through various motions to provide useful work. The manipulator is the arm of the robot, and is made up of a series of segments and joints much like those found in the human arm. Joints connect two segments together and allow them to move relative to one another. The joints provide either linear (straight line) or rotary (circular) movement. Figure 9-9 shows simple linear and rotary joints commonly found in manipulator arms.

The muscles of the human body supply the driving force that moves the various body joints. Similarly, a robot uses actuators to move its arm along programmed paths and then to hold its joints rigid once the correct position is reached.

There are two basic types of motion provided by actuators. Linear actuators provide motion along a straight line. They either extend or retract their attached loads. Angular actuators provide rotation. They move their loads in an arc or circle. Rotary motion can be converted into linear motion using a lead screw or other mechanical means of conversion.

TYPES OF ACTUATOR DRIVES

One common method of classifying robots is by the type of drive required by the actuators. Electrical actuators use electric power. Pneumatic actuators use pneumatic (air) power. Hydraulic actuators, shown in Figure 2-14, use hydraulic (fluid) power.

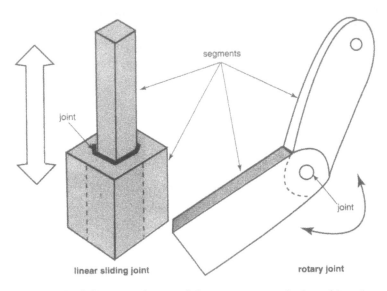

Figure 9-9. Both linear and rotary joints are commonly found in robots.

Electric Drive

Three types of motors are commonly used for electric drive actuators: ac servo motors, dc servo motors, and stepper motors. Both ac and dc servo motors have built-in methods for controlling exact position. Many newer robots use dc servo motors rather than hydraulic or pneumatic ones. Dc servo motors are commonly found on small and medium-size robots. Ac servo motors are found in heavy-duty robots because of their high torque capabilities. A stepper motor is a type of incrementally controlled dc motor.

Conventional electric-drive motors have several advantages. They are quiet and simple and can be used in clean-air environments. Electric robots generally require less floor space, and their energy source is readily available. However, the conventionally geared drive causes problems of backlash, friction, compliance, and wear. These problems cause inaccuracy, poor dynamic response, need for regular maintenance, poor torque control capability, and limited maximum speed on longer moves. Loads that are heavy enough to stall (stop) the motor can cause damage. Conventional electric motors also have poor output power compared to their weight. This means that a larger, heavier motor must be mounted on the robot arm when a large amount of torque is needed.

The rotary motion of most electric drive motors must be geared down (reduced) to provide the speed or torque required by the manipulator. However, manufacturers also offer robots that use direct-drive electric motors, which eliminate some of these problems. These high-torque motors drive the arm directly, without the need for reducer gears.

Applications performed by electric direct-drive robots are mechanical assembly, electronic assembly, and material handling. These robots will increasingly meet the demands of advanced, high-speed, precision applications. Such applications include laser cutting of sheet metal, which requires speeds of over one meter per second and tolerances of less than 0.1 millimeter, and fiber optic assembly, which requires accuracy within 0.1 to 0.5 microns (millionths of a meter).

Hydraulic Drive

Many earlier robots were driven by hydraulic drive systems. A hydraulic system consists of an electric pump connected to a reservoir tank, control valves, and an hydraulic actuator. Hydraulic systems provide both linear and rotary motion using a much simpler arrangement than non-direct-drive electrical systems. The storage tank supplies large amounts of instant power not available from electric systems.

Hydraulic-drive units have several advantages. They provide precise motion control over a wide range of speeds. They can handle heavy loads on the end of the manipulator arm, can be used around high explosives, and are not easily damaged when quickly stopped while carrying a heavy load. However, they are expensive to purchase and maintain, and are not energy efficient. They are also noisier than electric units. Because of hydraulic fluid leaks, they are not recommended for clean environments.

Pneumatic Drive

Pneumatic systems make use of air-driven actuators. Since air is also a fluid, many of the same principles that apply to hydraulic systems are applicable to pneumatic systems. Pneumatic and hydraulic motors and cylinders are very similar. Since most industrial plants have a compressed air system running throughout assembly areas, air is an economical and readily available energy source. This makes the installation of pneumatic robots easier and less costly than that of hydraulic robots.

Pneumatic-drive units work at high speeds and are most useful for small-to-medium loads. They are economical to operate and maintain, and can be used in explosive atmospheres. However, since air is compressible, precise placement and positioning is more difficult to control. It is also difficult to keep the air as clean and dry as needed by the control system. Pneumatic robots are noisy and vibrate as the air cylinders and motors stop.

Power Supplies

The robot power supply provides the energy to drive the controller and actuators. The three basic types of power supplies are electrical, hydraulic, and pneumatic. The most common energy source available where industrial robots are used is electricity. The second most common is compressed air, and the least common is hydraulic power. These primary sources of energy must be converted into the form and amount required by the type of robot being used. The electronic part of the control unit and any electric motor actuators require electrical power. A robot containing hydraulic actuators requires the conversion of electrical power into hydraulic energy through the use of an electric motor-driven hydraulic pump. A robot with pneumatic actuators requires compressed air, which is usually supplied by a compressor driven by an electric motor.

End Effectors

End effector is the technical term for the end-of-arm tooling on the robot. It is often referred to as the hand or gripper. An end effector is better-defined as a device attached to the wrist of the manipulator for the purpose of grasping, lifting, transporting, maneuvering, or performing operations on a workpiece.

Means of Programming

A robot may be programmed using any of several different methods. The teach pendant, also called a teach box or hand-held programmer, is one commonly used device. It teaches a robot the movements required to perform a useful task. The human operator uses a teach pendant to move the robot through the series of points that describe its desired path. The points are recorded by the controller for later use.

Degrees of Freedom

Although robots have a certain amount of dexterity, theirs is limited compared to human dexterity. The movements of the human hand are controlled by 35 muscles. Fifteen of these muscles are located in the forearm. The arrangement of muscles in the hand provides great strength to the fingers and thumb for grasping objects. Each finger can act alone or together with the thumb. This enables the hand to do many intricate and delicate tasks. In addition, the human hand has 27 bones. Figure 9-10 shows the bones found in the hand and wrist. This bone, joint, and muscle arrangement gives the hand its dexterity.

Degrees of freedom is a term used to describe a robot's dexterity, or freedom of motion. For each degree of freedom, a joint is required. A robot requires six degrees of freedom to be completely versatile. Its movements are clumsier than those of a human hand, which has 22 degrees of freedom.

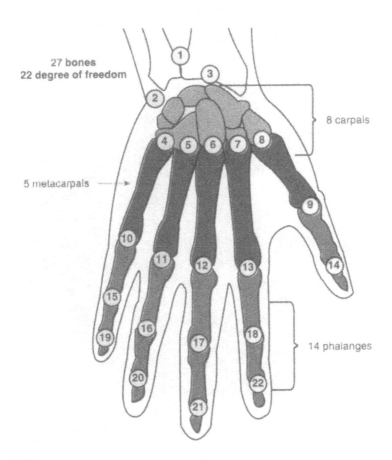

Figure 9-10. The arrangement of bones and joints found in the human hand provides dexterity. Each joint represents a degree of freedom; there are 22 joints, and thus, 22 degrees of freedom in the human hand.

The number of degrees of freedom defines the robot's configuration. For example, many simple applications require movement along three axes: X, Y, and Z. See Figure 9-11. These tasks would require three joints, or three degrees of freedom. The three degrees of freedom in the robot arm are the rotational traverse, the radial traverse, and the vertical traverse. The **rotational traverse** is movement about a vertical axis. This is the side-to-side swivel of the robot's arm about its base. The **radial traverse** is the extension and retraction of the arm, creating in-and-out motion relative to the base. The **vertical traverse** provides up-and-down motion.

For applications that require more freedom, additional degrees can be obtained from the wrist, which gives the end effector its flexibility. See Figure 9-12. The three degrees in the wrist bear aeronautical names: pitch, yaw, and roll. The **pitch**, or bend, is the up-and-down movement of the

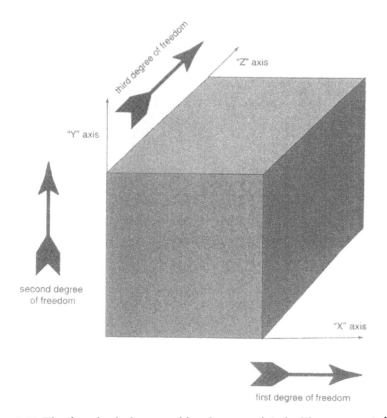

Figure 9-11. The three basic degrees of freedom associated with movement along the X, Y, and Z axes of the Cartesian coordinate system.

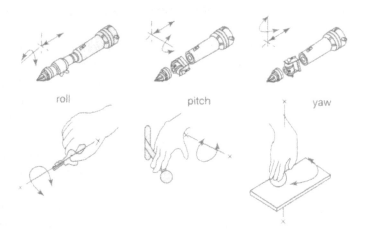

Figure 9-12. The three additional degrees of freedom—roll, pitch, and yaw—are associated with the robot's wrist.

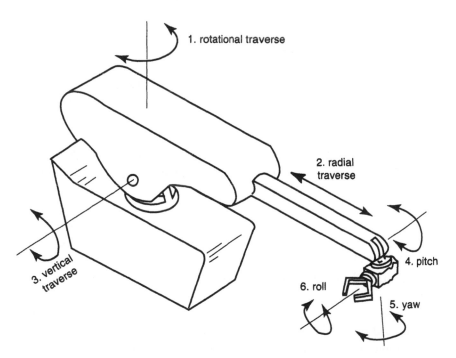

Figure 9-13. Six degrees of freedom provide maximum flexibility for an industrial robot.

wrist. The **yaw** is the side-to-side movement, and the **roll**, or swivel, involves rotation.

A robot requires a total of six degrees of freedom to locate and orient its hand at any point in its work envelope, Figure 9-13. Although six degrees of freedom are required for maximum flexibility, most applications require only three to five. The more degrees of freedom required, the more complex must be the robot's motions and controller design. Some industrial robots have seven or eight degrees of freedom. These additional degrees are achieved by mounting the robot on a track or moving base. The track-mounted robot would have a total of seven. This addition also increases the robot's reach.

Although the robot's freedom of motion is limited in comparison with that of a human, the range of movement in each of its joints is considerably greater. For example, the human hand has a bending range of only about 165 degrees. Figure 9-14 shows the six major degrees of freedom by comparing those of a robot to a person using a spray gun.

ROBOT CONFIGURATIONS

Robots come in many sizes and shapes. They also vary as to the type of coordinate system used by the manipulator. The type of coordinate system, the arrangement of the joints, and the length of the manipulator's segments all help determine the shape of the work envelope. To identify this maximum area, a point on the robot's wrist is used, rather than the tip of the gripper or the end of the tool bit. As a result, the work envelope is slightly larger when the tip of the tool is considered.

Work envelopes vary from one manufacturer to another, depending on the exact design of the manipulator arm. Combining different configurations in a single robot can result in another set of possible work envelopes. Before choosing a particular robot configuration, the application must be studied carefully to determine the precise work envelope requirements.

Some work envelopes have a geometric shape; others are irregular. One method of classifying a robot is by the geometric configuration of its work envelope. Some robots may employ more than one configuration. The four major ones are revolute, Cartesian, cylindrical, and spherical. Each configuration is used for specific applications.

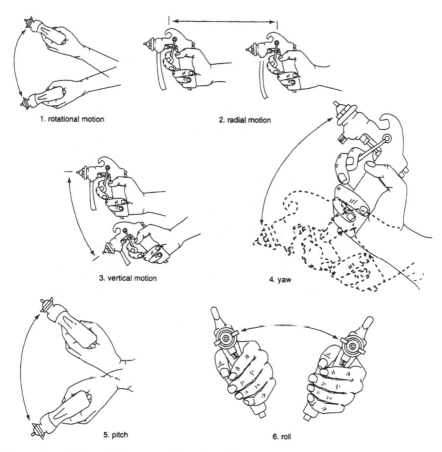

1. rotational motion   2. radial motion

3. vertical motion   4. yaw

5. pitch   6. roll

Figure 9-14. The six degrees of freedom, illustrated by a person using a spray gun. Degrees 1, 2, and 3 are arm movements, 4, 5, and 6 are wrist movements.

Revolute Configuration

The jointed-arm or **revolute configuration** is the most common. These robots are often referred to as **anthropomorphic** (humanlike) in form because their movements closely resemble those of the human body. Rigid segments resemble the human forearm and upper arm. Various joints mimic the action of the wrist, elbow, and shoulder. A joint called the **sweep** represents the waist.

A revolute coordinate robot performs in an irregularly shaped work envelope. There are two basic revolute configurations: vertically articulated and horizontally articulated. The vertically articulated configuration shown in Figure 9-15 has three revolute (rotary) joints.

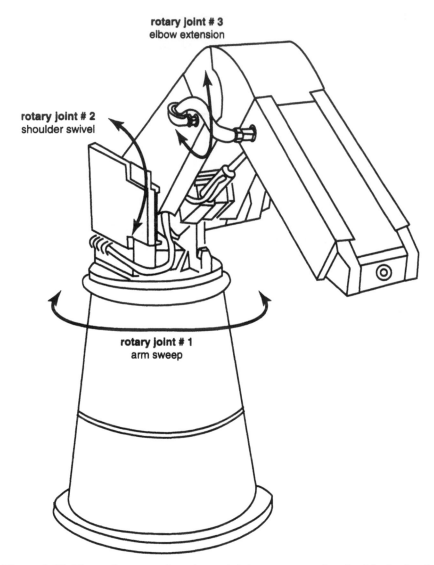

Figure 9-15. These three revolute (rotary) joints are associated with the basic manipulator movements of a vertically articulated robot.

The revolute configuration has several advantages. It is by far the most versatile configuration and provides a larger work envelope than the Cartesian, cylindrical, or spherical configurations. It also offers a more flexible reach than the other configurations, making it ideally suited to welding and spray painting operations.

However, there are also disadvantages to the revolute configuration. It requires a very sophisticated controller, and programming is more complex than for the other three configurations. Different locations in the work envelope can affect accuracy, load-carrying capacity, dynamics, and the robot's ability to repeat a movement accurately. This configuration also becomes less stable as the arm approaches its maximum reach.

Typical applications include:

- Automatic assembly.
- Parts and material handling.
- Multiple-point light machining operations.
- In-process inspection.
- Palletizing.
- Machine loading/unloading.
- Machine vision.
- Material cutting.
- Material removal.
- Thermal coating.
- Paint and adhesive application.
- Welding.
- Die casting.

Cartesian Configuration

The arm movement of a robot using the Cartesian configuration can be described by three intersecting perpendicular straight lines, referred to as the X, Y, and Z axes. See Figure 9-16. Because movement can start and stop simultaneously along all three axes, motion of the tool tip is smoother. This allows the robot to move directly to its designated point, instead of following trajectories parallel to each axis. See Figure 9-17.

Figure 9-18 illustrates the rectangular work envelope of the typical Cartesian configuration. One advantage of robots with a Cartesian configuration is that their totally linear movement allows for simpler controls. They also have a high degree of mechanical rigidity, accuracy, and repeatability. They can carry heavy loads, and this weight lifting capacity does not vary at different locations within the work envelope. As to disadvantages, Cartesian robots are generally limited in their movement to a small, rectangular work space.

Typical applications for Cartesian robots include:

- Assembly.
- Machining operations.

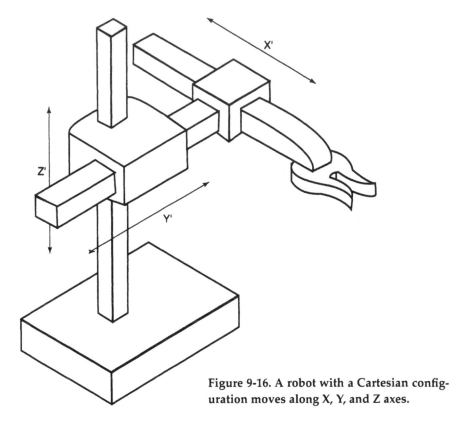

Figure 9-16. A robot with a Cartesian config-uration moves along X, Y, and Z axes.

- Adhesive application.
- Surface finishing.
- Inspection.
- Waterjet cutting.
- Welding.
- Nuclear material handling.
- Robotic X-ray and neutron radiography.
- Automated CNC lathe loading and operation.
- Remotely operated decontamination.
- Advanced munitions handling.

Cylindrical Configuration

A cylindrical configuration consists of two orthogonal (at a 90-degree angle) slides mounted on a rotary axis, Figure 9-19. Reach is accomplished as the arm of the robot moves in and out. For vertical movement, the car-

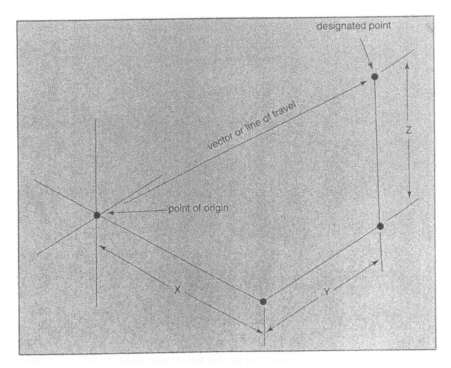

Figure 9-17. With a Cartesian configuration, the robot can move directly to a designated point, rather than moving in lines parallel to each axis. In this example, movement is along the "vextor" connecting the point of origin and the designated point, rather than moving first along the X axis, then Y, then Z.

riage moves up and down on a stationary post, or the post can move up and down in the base of the robot. Movement along the three axes, as shown in Figure 9-20, traces points on a cylinder.

A cylindrical configuration generally results in a larger work envelope than a Cartesian configuration. These robots are ideally suited for pick-and-place operations. However, cylindrical configurations have some disadvantages. Their overall mechanical rigidity is lower because robots with a rotary axis must overcome the inertia of the object when rotating. Their repeatability and accuracy is also lower in the direction of rotary movement. The cylindrical configuration requires a more sophisticated control system than the Cartesian configuration.

Typical applications include:
- Machine loading/unloading.
- Investment casting.
- Upsetter loading.

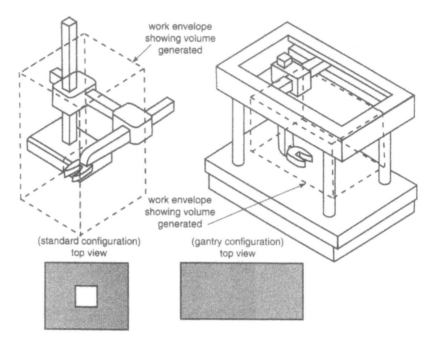

Figure 9-18. In either the standard or gantry construction, a Cartesian configuration robot creates a rectangular work envelope.

- Conveyor pallet transfers.
- Foundry and forging applications.
- General material handling and special payload handling and manipulation.
- Meat packing.
- Coating applications.
- Assembly.
- Injection molding.
- Die casting.

Spherical Configuration

The spherical configuration, sometimes referred to as the polar configuration, resembles the action of the turret on a military tank. A pivot

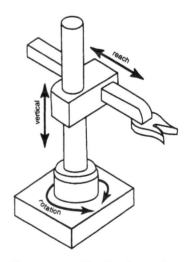

Figure 9-19. The basic configuration of a cylindrical robot.

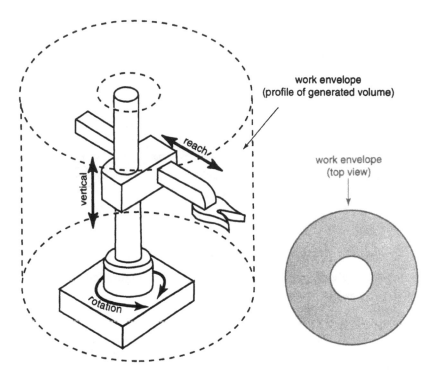

work envelope
(profile of generated volume)

work envelope
(top view)

reach

vertical

rotation

Figure 9-20. Motion along the three axes traces points on a cylinder to form the work envelope.

point gives the robot its vertical movement, Figure 9-21. Reach is accomplished, for example, through use of a telescoping boom that extends and retracts. Rotary movement may occur around an axis perpendicular to the base. Figure 9-22 illustrates the work envelope profile of a typical spherical configuration robot.

The spherical configuration generally provides a larger work envelope than the Cartesian or cylindrical configurations. The design is simple and gives good weight lifting capabilities. This configuration is suited to applications where a small amount of vertical movement is adequate, such as loading and unloading a punch press. Its disadvantages include lower mechanical rigidity and the need for a more sophisticated control system than either the Cartesian or cylindrical configurations. The same problems occur with inertia and spatial resolution in this configuration as they do in the cylindrical configuration. Vertical movement is limited, as well.

Typical applications include:

- Die casting.

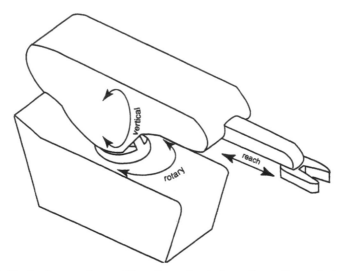

Figure 9-21. A pivot point enables the spherical configuration robot to move vertically. It also can rotate around a vertical axis.

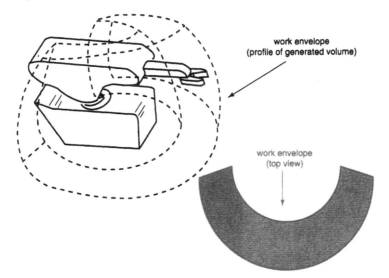

Figure 9-22. The work envelope of this robot takes the shape of a sphere.

- Injection molding.
- Forging.
- Machine tool loading.
- Heat treating.
- Glass handling.
- Parts cleaning.
- Dip coating.
- Press loading.
- Material transfer.
- Stacking/unstacking.

AUTOMATED ROBOTIC PROCESSES IN INDUSTRY

As robot technology advances, areas where a robot might be used are probably limited only by the user's imagination and creativity. A completely automated factory is still the ultimate goal of many scientists and engineers doing research today. Robots are limited by their ability to sense and make decisions about objects in their environment. As newer robots equipped with vision and other complex sensing systems are brought into the workplace, the potential uses for robots will increase dramatically.

Over the years, a number of industrial applications for robots have proven to be very suitable and economically sound. Many of the installed industrial robots are used in the automotive industry. As a general rule, industrial robots lift heavy loads, work with dangerous materials, function in dangerous or undesirable environments, or perform processes that is mundane and highly repetitive.

Some of the broad range of industrial process applications, from the classic uses to those that are more recent are included in the following sections. These processes do not, however, include all the uses of robots.

Pick-and-place Processes

The process of picking up parts at one location and moving them to another is one of the most common robot applications. Placing parts or removing them from a uniform series of positions palletizing or de-palletizing, for example is probably the most common form of pick-and-place. Some pick-and-place operations also are used for part orientation. Figure 9-23 shows the components used in a robot palletizing operation.

Less-sophisticated robots used for pick-and-place offer several advantages. They have been very successful in handling fragile parts made of glass or powder metal. Those used in lightweight pick-and-place applications offer excellent speed while maintaining good accuracy and repeatability. Robots are also useful for handling heavy and very hot or very cold items.

Machine Loading/Unloading

Possibly the second most popular use of robots in industry is machine loading and unloading. In production, robots load and unload parts associated with automated machining centers, such as CNC (computer numerically controlled) machines, Figure 9-24. A robot's machine loading and unloading skills can be applied to such operations as forging, injection molding, and stamping.

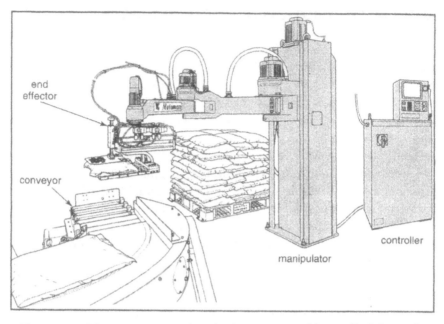

Figure 9-23. The components of a robotic system used in a palletizing task.

Industrial operations often expose workers to excessive heat, noise, dirt, and air pollution. Since these operations are considered unpleasant and are frequently dangerous, they are ideal choices for the use of robots. Replacing human operators with robots reduces the need for expensive safety equipment and eliminates costs involved in cleaning up the work environment. Also, unpleasant or dangerous jobs usually have high absenteeism rates. This contributes to low productivity and poor product quality. Substantial gains in productivity and product quality generally result when robots are properly used.

Die Casting

The first practical application of industrial robotics occurred many years ago when General Motors installed the Unimate robot in a die-casting operation. The use of robots in die casting has resulted in a tremendous increase in productivity. After robots proved useful in die casting, they were used in investment casting, forging, and welding operations.

Die casting involves the pumping of molten metals, such as zinc, aluminum, copper, brass, or lead, into closed dies to form specific shapes. After the metal solidifies, the die is opened and the casting is removed.

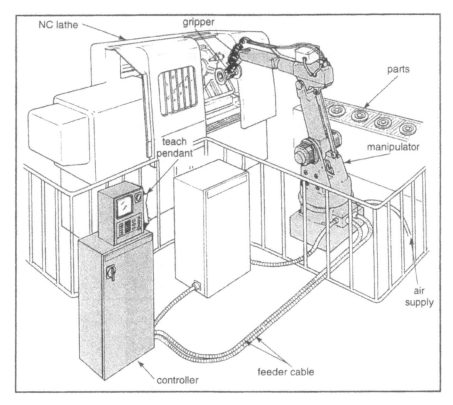

Figure 9-24. This robotic system is used for loading/unloading an NC lathe.

The function of the robot is to remove the hot casting from the die and dip it into a liquid, usually water. This operation is called quenching. Next, the robot transfers the cooled part to a trimming press where excess material, called flash is removed and the part is shaped. The robot then transfers the part to a pallet or conveyor belt, which takes it out of the work area.

Robots can work more consistently than humans with such hot castings. The increase in productivity is the result of decreasing the cycle time required to produce a finished casting. Another benefit is that the reduced cycle time results in a more uniform die temperature. Less flash material becomes attached to the casting, and trimming and scraping costs are reduced. The robots perform what is essentially a machine loading/unloading operation. It does not require critical or complex motions and is ideally suited for low-cost pick-and-place robots.

Welding Processes

The next largest group of robots in industry are those used in resistance welding and arc welding operations. In resistance welding, electric current is passed between two metals, causing them to heat and fuse together at that point. Spot and stud welding are classified as resistance welding. In arc welding, the weld is made along a joint rather than at one spot. The arc-welding process heats the metals until they melt at the joint and fuse into a single piece. Stick, MIG, and TIG are considered arc-welding applications.

Arc Welding

MIG (metal-inert gas) is the most common arc-welding process. It requires an arc-welding machine, a wire (electrode) feeder, a shielding gas, a gas flow meter, and a welding gun, Figure 9-25. The wire is a consumable metal electrode. Current passing through the electrode is controlled by the electrode's speed of consumption. An inert gas is used to shield the weld from the atmosphere.

Many of the problems associated with robotic MIG welding are

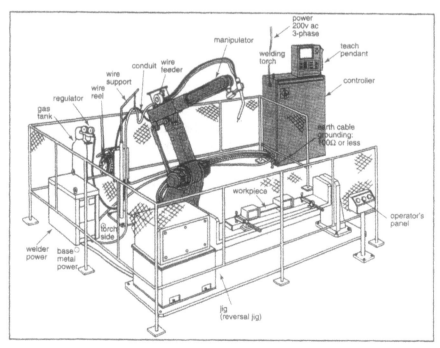

Figure 9-25. Components used in a typical robotic arc welding station.

mainly due to poor joint fit-up. Robotic arc welding requires more accurate fixtures and better-fitting joints than welding done by human operators. If joint location or the gap between parts varies, a human being can make adjustments, but most robots cannot.

Resistance Welding

Of all the welding processes, spot welding is the easiest to perform and the most common. It can be done on stationary parts or on parts moving down a production line. Robots used for spot welding must return repeatedly to an exact spot. Spot welding is used extensively in the automotive industry to weld body sections and other parts together. A large work envelope is generally needed for these large and complex workpieces.

The workpieces must be pressed together at the spot where the weld is applied. This is done by the welding electrodes, which exert a force of approximately 800 to 1000 psi (pounds per square inch). Next, a low-voltage direct current is passed through the parts, causing them to fuse. The electrodes themselves do not melt. Figure 9-26 shows the components found in a typical spot-welding work station.

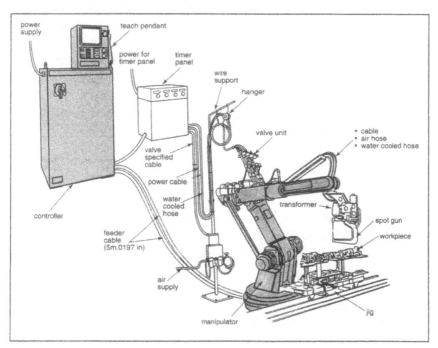

Figure 9-26. This robotic work station is set up for spot welding.

Spraying Processes

Painting or spray finishing involves the application of a variety of paints, polyurethanes, or other protective coatings to the surface of a part or product. Other spraying operations are sealing and gluing. The spray nozzle is mounted on the robot's wrist. Continuous-path programming must be done by an experienced operator because positioning is recorded on a time sample basis. Both intentional and unintentional moves will be recorded.

Several benefits can be realized by using robots in spraying applications. Spray vapors are often toxic and explosive, so expensive ventilation systems must be installed for human workers' health and safety. Since robots do not need fresh air to breathe, greater concentrations of solvent can be used. With the reduction of ventilation, energy costs are lowered. However, there is the danger of an explosion caused by a spark from the robot's electrical system. In the past, most painting robots have been hydraulically powered for this reason. Today, spark-free electrical robots are available for explosive environments.

The use of robots for sealing and gluing is becoming more widespread as the quality of sealing and gluing agents and application methods has improved. The components associated with a typical robotic sealing/gluing work station are shown in Figure 9-27. Path control is crucial in sealing and gluing work. Robot requirements are the same as for continuous-path arc welding applications. Sealing and gluing operations require specialized end-of-arm tooling for three-dimensional surfaces. Also, the robot's speed must not change between programmed points or glue will be dispensed unevenly.

Machining Processes

Routing, cutting, drilling, milling, grinding, polishing, deburring, riveting, and sanding are some of the machining processes performed by robots. These robots must have a high degree of repeatability. They must be combined with quick-change tooling, better fixtures, and improved sensory and adaptive control. Many older generation robots do not meet these requirements.

Cutting Processes

Cutting applications require flexible robots with tight servo control loops, directly linked drives, and dedicated cutting axes. Four common types of cutting are: gas, plasma, waterjet, and laser. The components used

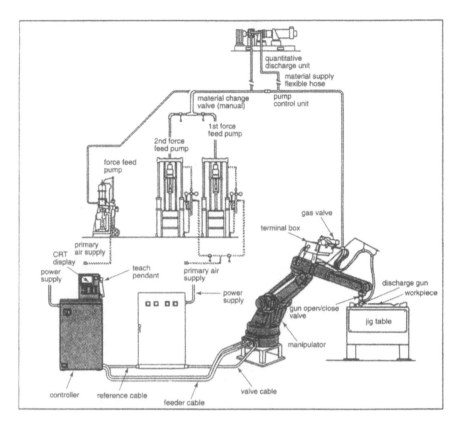

Figure 9-27. Components used in a typical sealing/gluing robotic work station.

in a typical gas cutting work station are illustrated in Figure 9-28. Many cutting operations require special safety precautions.

Deburring and Polishing Processes

Deburring and polishing require robots with smooth movements and rigid wrist designs. The robot control system must be capable of automatically adjusting for such things as tool wear, burr size variation, workpiece variation, and product positioning. See Figure 9-29. Force-sensing equipment can identify contact with a workpiece and adjust tools for constant pressure. See Figure 9-30.

Assembly Processes

In recent years, more attention has been focused on the use of robots for assembly. The potential is great since assembly operations may account

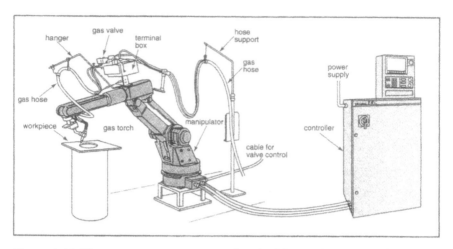

Figure 9-28. These components are associated with a typical gas-cutting robotic work station.

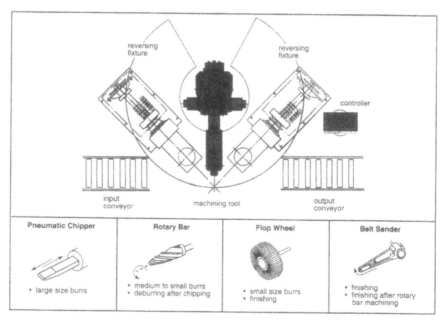

Figure 9-29. The components shown here are used for deburring and polishing.

for 50 percent of the labor cost involved in manufacturing a product. In 1980, an estimated 40 percent of the blue-color work force in the United States was engaged in assembly operations.

Workers who perform certain kinds of assembly tasks can develop

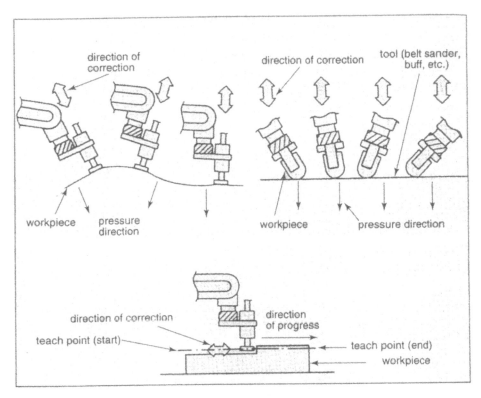

Figure 9-30. Force sensors allow for corrections in tool pressure against the workpiece.

work-related health problems. Problems might occur when workers use repetitive hand motions over long periods of time. Psychological problems due to job stress are also associated with highly repetitive work. Such health problems make the use of automated robotic assembly very desirable. Generally, the end result is an increase in productivity.

Robots used for assembly work are generally small and are designed to move small parts accurately at high speeds. Most operations involve the fitting and holding together of parts and assemblies. This is generally done by means of nuts, bolts, screws, fasteners, or snap-fit joints.

Newer robots are being equipped with sophisticated vision systems that can detect proper orientation of parts. Tactile sensing systems attached to the end effector may be used to detect misaligned, missing, or substandard parts. The robot in Figure 9-31 is equipped with a two-camera vision system. One camera, mounted on the manipulator, guides the robot around the work cell. The robot picks up components for air motors from various parts feeders and trays and assembles them. The second camera is used for precise placement of a component within 0.1 millimeter. The robot brings the components to a turntable where assembly takes place. Two sets of eight fixtures are located around the perimeter of the turntable. Each of the models of the air motor can be assembled using either set of fixtures

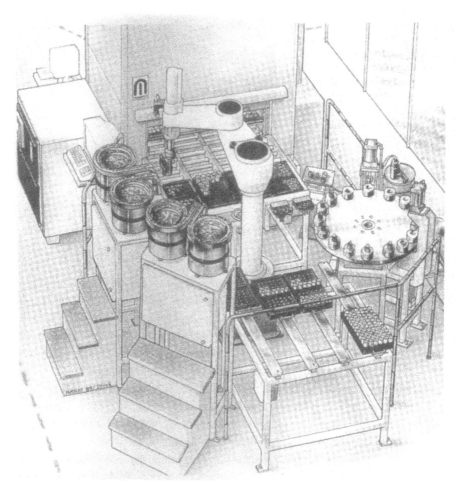

Figure 9-31. A two-camera vision system is used by this robot, which assembles air motors in a manufacturing plant.

Inspection

Quality assurance for producing the highest quality products remains an essential and costly part of robotic manufacturing. The demand for non-contact gauging and inspection systems have grown steadily in recent years. At the heart of these new systems is state-of-the-art electronic sensor technology. They check for tolerances, positioning, fixturing, and defects, among other things.

Robotic inspection systems consist of two subsystems. One scans and/or accumulates data. The other analyzes the data and presents it in a meaningful way. The electronic sensing units used in these systems are set up in one of two ways. They may be mounted on the robot's hand and the robot is programmed to move the sensor along the part. They may also be located within the work cell and inspect the parts as the parts move along the conveyor system.

Material Handling

Automated and robotic systems are often used for material handling. Automated control and tracking of inventory and handling parts and products increases efficiency and flexibility. An automated guided vehicle (AGV) is typically computer-controlled and battery-operated. AGVs may follow a guidepath in the floor. A guidepath generally consists of a wire laid in a sealed groove beneath the floor's surface. The path does not cause obstructions and floors remain clean and uncluttered. The vehicles receive instructions and report back to the command center through this path.

For safety, AGVs have bumpers and automatically stop if they contact any object in their path. AGVs are bi-directional and emit a beeping sound to warn humans out of the way. Some AGVs even signal their command center when their batteries are getting low. The computer then directs the vehicle to a battery charging station. AGVs are limited in mobility by the location of the guidepath.

SUMMARY

As you can see, automated process control, particularly with robotic systems plays a very important role in industry today. There are numerous other processes that can be automated using robotic systems. These applications are limited only by the imagination.

Industrial robots typically resemble an inverted human arm mounted on a base. They consist of subsystems that work together to perform some type of process. These subsystems are the controller, the manipulator, the power supply, the end effector and a means of programming.

Robots may be either servo or non-servo types. Servo robots are classified as either intelligent or highly intelligent while non-servo robots are the simplest types for performing pick and place and other simple automated operations.

There are several types of robot configurations. These include revolute, Cartesian, cylindrical and spherical. Automated robotic processes in industry are pick and place, machine loading/unloading, die casting, welding processes, spraying processes, machining processes, cutting processes, deburring/polishing processes, assembly processes, inspection and material handling processes.

Appendix A

U.S. TO METRIC

1 inch	= 25.4	millimeters
1 foot	= 0.3048	meter
1 yard	= 0.9144	meter
1 mile	= 1.609	kilometers
1 square inch	= 6.4516	square centimeters
1 square foot	= 0.0929	square meter
1 square yard	= 0.836	square meter
1 acre	= 0.4047	hectare
1 cubic inch	= 16.387	cubic centimeters
1 cubic foot	= 0.028	cubic meter
1 cubic yard	= 0.764	cubic meter
1 quart (liq)	= 0.946	liter
1 gallon	= 0.00378	cubic meter
1 ounce (avdp)	= 28.349	grams
1 pound (avdp)	= 0.4536	kilogram
1 horsepower	= 0.7457	kilowatt

METRIC TO U.S.

1 millimeter	= 0.03937	inch
1 meter	= 3.2808	feet
1 meter	= 1.0936	yards
1 kilometer	= 0.6214	mile
1 square centimeter	= 0.155	square inch
1 square meter	= 10.7639	square feet
1 square meter	= 1.196	square yards
1 hectare	= 2.471	acres
1 cubic centimeter	= 0.061	cubic inch
1 cubic meter	= 35.3147	cubic feet
1 cubic meter	= 1.308	cubic yards
1 liter	= 1.0567	quarts (liq)
1 cubic meter	= 264.172	gallons
1 gram	= 0.035	ounce (avdp)
1 kilogram	= 2.2046	pounds (avdp)
1 kilowatt	= 1.341	horsepower

Appendix B

ELECTRICAL / ELECTRONIC SYMBOLS

FIXED RESISTOR

TAPPED RESISTOR

VARIABLE RESISTOR

RHEOSTAT

THERMISTOR

FIXED CAPACITOR

VARIABLE CAPACITOR

POLARIZED CAPACITOR

TURNSTILE ANTENNA

DIPOLE ANTENNA

LOOP ANTENNA

BATTERY

GENERAL ALTERNATING-CURRENT SOURCE

PERMANENT MAGNET

PIEZOELECTRIC CRYSTAL

THERMOCOUPLE

THERMAL CUTOUT

WIRES CROSSING; NOT CONNECTED

WIRES CONNECTED

SHIELDED CONDUCTOR

SHIELDED 2-CONDUCTOR CABLE

GROUPING OF LEADS

CIRCUIT RETURN — GROUND — CHASSIS — COMMON

CONTACTS (NORMALLY CLOSED)

CONTACTS (NORMALLY OPEN)

CONTACTS (TRANSFER)

SWITCH (SINGLE-POLE, SINGLE-THROW)

SWITCH (SINGLE-POLE, DOUBLE-THROW)

SWITCH (DOUBLE-POLE, SINGLE-THROW)

SWITCH (DOUBLE-POLE, DOUBLE-THROW)

MULTIPOSITION SELECTOR SWITCH (ANY NUMBER OF POSITIONS MAY BE SHOWN.)

PUSH-BUTTON SWITCH (NORMALLY OPEN)

PUSH-BUTTON SWITCH (NORMALLY CLOSED)

PUSH-BUTTON SWITCH (2-CIRCUIT)

LIMIT SWITCH (SPRING RETURN, NORMALLY OPEN)

LIMIT SWITCH (SPRING RETURN, NORMALLY OPEN, HELD CLOSED)

LIMIT SWITCH (SPRING RETURN, NORMALLY CLOSED)

LIMIT SWITCH (SPRING RETURN, NORMALLY CLOSED, HELD OPEN)

OPEN SWITCH, TIME-DELAY CLOSING — TDC

CLOSED SWITCH, TIME-DELAY OPENING — TDO

OPEN SWITCH, TIME-DELAY OPENING — TDO

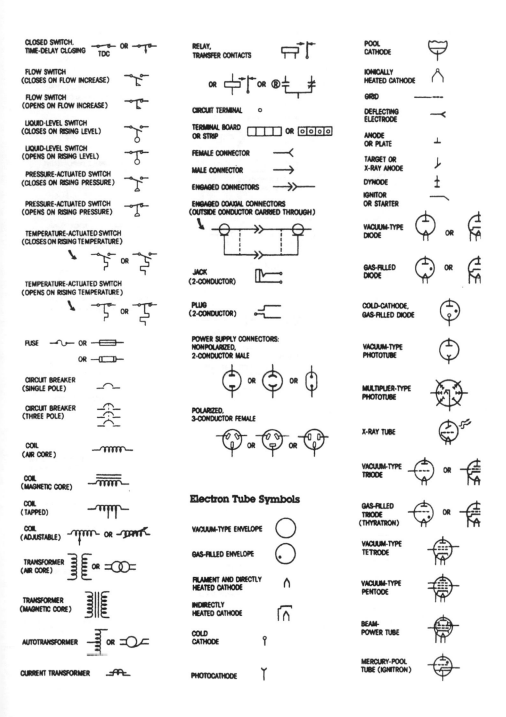

CLOSED SWITCH, TIME-DELAY CLOSING OR — TDC

FLOW SWITCH (CLOSES ON FLOW INCREASE)

FLOW SWITCH (OPENS ON FLOW INCREASE)

LIQUID-LEVEL SWITCH (CLOSES ON RISING LEVEL)

LIQUID-LEVEL SWITCH (OPENS ON RISING LEVEL)

PRESSURE-ACTUATED SWITCH (CLOSES ON RISING PRESSURE)

PRESSURE-ACTUATED SWITCH (OPENS ON RISING PRESSURE)

TEMPERATURE-ACTUATED SWITCH (CLOSES ON RISING TEMPERATURE) OR

TEMPERATURE-ACTUATED SWITCH (OPENS ON RISING TEMPERATURE) OR

FUSE OR OR

CIRCUIT BREAKER (SINGLE POLE)

CIRCUIT BREAKER (THREE POLE)

COIL (AIR CORE)

COIL (MAGNETIC CORE)

COIL (TAPPED)

COIL (ADJUSTABLE) OR

TRANSFORMER (AIR CORE) OR

TRANSFORMER (MAGNETIC CORE)

AUTOTRANSFORMER OR

CURRENT TRANSFORMER

RELAY, TRANSFER CONTACTS

OR OR ®

CIRCUIT TERMINAL

TERMINAL BOARD OR STRIP OR

FEMALE CONNECTOR

MALE CONNECTOR

ENGAGED CONNECTORS

ENGAGED COAXIAL CONNECTORS (OUTSIDE CONDUCTOR CARRIED THROUGH)

JACK (2-CONDUCTOR)

PLUG (2-CONDUCTOR)

POWER SUPPLY CONNECTORS: NONPOLARIZED, 2-CONDUCTOR MALE

OR OR

POLARIZED, 3-CONDUCTOR FEMALE

OR OR

Electron Tube Symbols

VACUUM-TYPE ENVELOPE

GAS-FILLED ENVELOPE

FILAMENT AND DIRECTLY HEATED CATHODE

INDIRECTLY HEATED CATHODE

COLD CATHODE

PHOTOCATHODE

POOL CATHODE

IONICALLY HEATED CATHODE

GRID

DEFLECTING ELECTRODE

ANODE OR PLATE

TARGET OR X-RAY ANODE

DYNODE

IGNITOR OR STARTER

VACUUM-TYPE DIODE OR

GAS-FILLED DIODE OR

COLD-CATHODE, GAS-FILLED DIODE

VACUUM-TYPE PHOTOTUBE

MULTIPLIER-TYPE PHOTOTUBE

X-RAY TUBE

VACUUM-TYPE TRIODE OR

GAS-FILLED TRIODE (THYRATRON) OR

VACUUM-TYPE TETRODE

VACUUM-TYPE PENTODE

BEAM-POWER TUBE

MERCURY-POOL TUBE (IGNITRON)

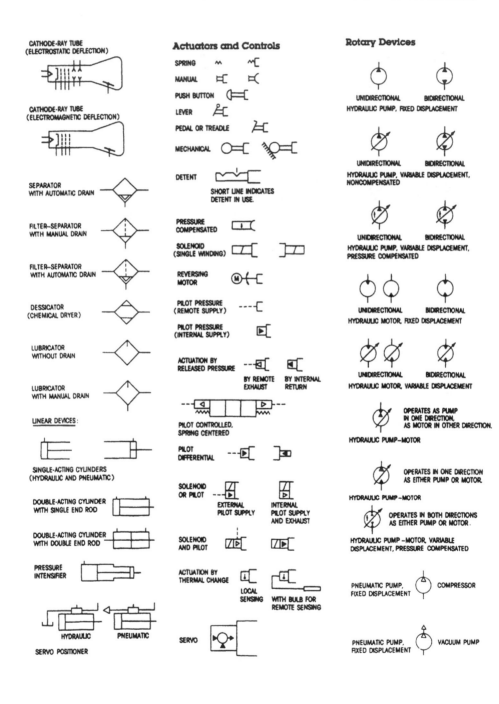

CATHODE-RAY TUBE
(ELECTROSTATIC DEFLECTION)

CATHODE-RAY TUBE
(ELECTROMAGNETIC DEFLECTION)

SEPARATOR
WITH AUTOMATIC DRAIN

FILTER-SEPARATOR
WITH MANUAL DRAIN

FILTER-SEPARATOR
WITH AUTOMATIC DRAIN

DESSICATOR
(CHEMICAL DRYER)

LUBRICATOR
WITHOUT DRAIN

LUBRICATOR
WITH MANUAL DRAIN

LINEAR DEVICES:

SINGLE-ACTING CYLINDERS
(HYDRAULIC AND PNEUMATIC)

DOUBLE-ACTING CYLINDER
WITH SINGLE END ROD

DOUBLE-ACTING CYLINDER
WITH DOUBLE END ROD

PRESSURE
INTENSIFIER

HYDRAULIC   PNEUMATIC

SERVO POSITIONER

Actuators and Controls

SPRING

MANUAL

PUSH BUTTON

LEVER

PEDAL OR TREADLE

MECHANICAL

DETENT

SHORT LINE INDICATES
DETENT IN USE.

PRESSURE
COMPENSATED

SOLENOID
(SINGLE WINDING)

REVERSING
MOTOR

PILOT PRESSURE
(REMOTE SUPPLY)

PILOT PRESSURE
(INTERNAL SUPPLY)

ACTUATION BY
RELEASED PRESSURE

BY REMOTE   BY INTERNAL
EXHAUST   RETURN

PILOT CONTROLLED,
SPRING CENTERED

PILOT
DIFFERENTIAL

SOLENOID
OR PILOT

EXTERNAL   INTERNAL
PILOT SUPPLY   PILOT SUPPLY
               AND EXHAUST

SOLENOID
AND PILOT

ACTUATION BY
THERMAL CHANGE

LOCAL
SENSING   WITH BULB FOR
           REMOTE SENSING

SERVO

Rotary Devices

UNIDIRECTIONAL   BIDIRECTIONAL
HYDRAULIC PUMP, FIXED DISPLACEMENT

UNIDIRECTIONAL   BIDIRECTIONAL
HYDRAULIC PUMP, VARIABLE DISPLACEMENT,
NONCOMPENSATED

UNIDIRECTIONAL   BIDIRECTIONAL
HYDRAULIC PUMP, VARIABLE DISPLACEMENT,
PRESSURE COMPENSATED

UNIDIRECTIONAL   BIDIRECTIONAL
HYDRAULIC MOTOR, FIXED DISPLACEMENT

UNIDIRECTIONAL   BIDIRECTIONAL
HYDRAULIC MOTOR, VARIABLE DISPLACEMENT

OPERATES AS PUMP
IN ONE DIRECTION,
AS MOTOR IN OTHER DIRECTION.
HYDRAULIC PUMP-MOTOR

OPERATES IN ONE DIRECTION
AS EITHER PUMP OR MOTOR.
HYDRAULIC PUMP-MOTOR

OPERATES IN BOTH DIRECTIONS
AS EITHER PUMP OR MOTOR.
HYDRAULIC PUMP-MOTOR, VARIABLE
DISPLACEMENT, PRESSURE COMPENSATED

PNEUMATIC PUMP,
FIXED DISPLACEMENT   COMPRESSOR

PNEUMATIC PUMP,
FIXED DISPLACEMENT   VACUUM PUMP

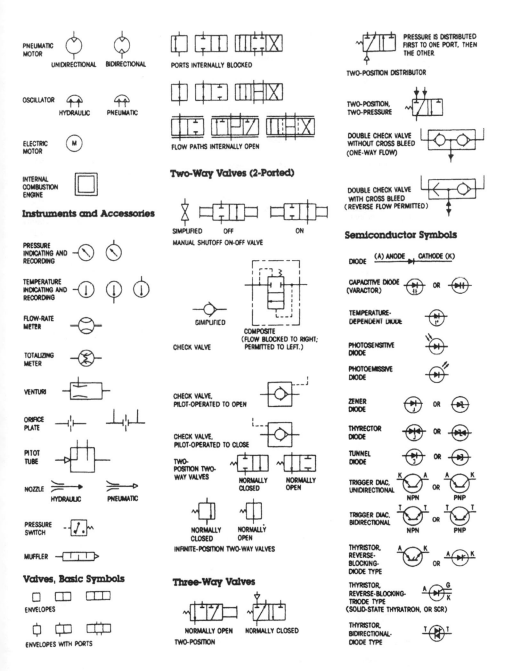

PNEUMATIC MOTOR
UNIDIRECTIONAL   BIDIRECTIONAL

PORTS INTERNALLY BLOCKED

PRESSURE IS DISTRIBUTED FIRST TO ONE PORT, THEN THE OTHER.

TWO-POSITION DISTRIBUTOR

OSCILLATOR
HYDRAULIC   PNEUMATIC

FLOW PATHS INTERNALLY OPEN

TWO-POSITION, TWO-PRESSURE

ELECTRIC MOTOR   M

Two-Way Valves (2-Ported)

DOUBLE CHECK VALVE WITHOUT CROSS BLEED (ONE-WAY FLOW)

INTERNAL COMBUSTION ENGINE

SIMPLIFIED   OFF   ON

MANUAL SHUTOFF ON-OFF VALVE

DOUBLE CHECK VALVE WITH CROSS BLEED (REVERSE FLOW PERMITTED)

Instruments and Accessories

Semiconductor Symbols

PRESSURE INDICATING AND RECORDING

DIODE   (A) ANODE ── CATHODE (K)

TEMPERATURE INDICATING AND RECORDING

SIMPLIFIED

COMPOSITE (FLOW BLOCKED TO RIGHT; PERMITTED TO LEFT.)

CAPACITIVE DIODE (VARACTOR)   OR

FLOW-RATE METER

CHECK VALVE

TEMPERATURE-DEPENDENT DIODE

TOTALIZING METER

PHOTOSENSITIVE DIODE

VENTURI

CHECK VALVE, PILOT-OPERATED TO OPEN

PHOTOEMISSIVE DIODE

ORIFICE PLATE

CHECK VALVE, PILOT-OPERATED TO CLOSE

ZENER DIODE   OR

PITOT TUBE

TWO-POSITION TWO-WAY VALVES
NORMALLY CLOSED   NORMALLY OPEN

THYRECTOR DIODE   OR

NOZZLE
HYDRAULIC   PNEUMATIC

TUNNEL DIODE   OR

PRESSURE SWITCH

NORMALLY CLOSED   NORMALLY OPEN

INFINITE-POSITION TWO-WAY VALVES

TRIGGER DIAC, UNIDIRECTIONAL
NPN   OR   PNP

MUFFLER

TRIGGER DIAC, BIDIRECTIONAL
NPN   OR   PNP

Valves, Basic Symbols

Three-Way Valves

THYRISTOR, REVERSE-BLOCKING-DIODE TYPE   OR

ENVELOPES

NORMALLY OPEN   NORMALLY CLOSED
TWO-POSITION

THYRISTOR, REVERSE-BLOCKING-TRIODE TYPE (SOLID-STATE THYRATRON, OR SCR)

ENVELOPES WITH PORTS

THYRISTOR, BIDIRECTIONAL-DIODE TYPE

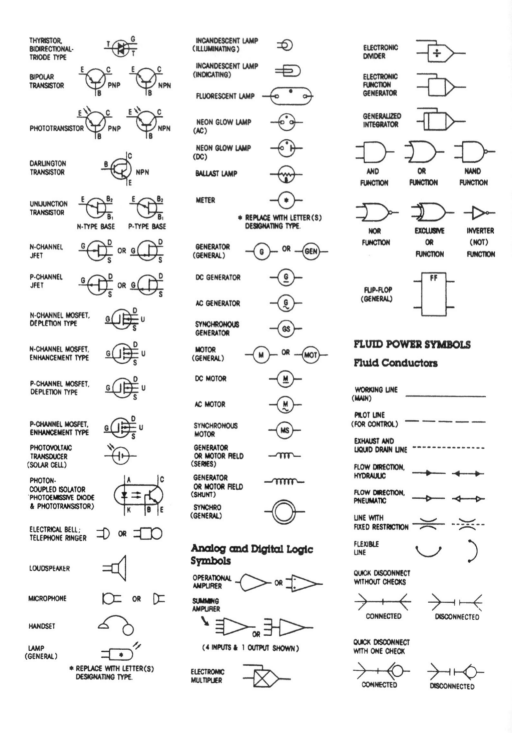

THYRISTOR,
BIDIRECTIONAL-
TRIODE TYPE

BIPOLAR
TRANSISTOR   PNP   NPN

PHOTOTRANSISTOR   PNP   NPN

DARLINGTON
TRANSISTOR   NPN

UNIJUNCTION
TRANSISTOR
                    N-TYPE BASE   P-TYPE BASE

N-CHANNEL
JFET   OR

P-CHANNEL
JFET   OR

N-CHANNEL MOSFET,
DEPLETION TYPE

N-CHANNEL MOSFET,
ENHANCEMENT TYPE

P-CHANNEL MOSFET,
DEPLETION TYPE

P-CHANNEL MOSFET,
ENHANCEMENT TYPE

PHOTOVOLTAIC
TRANSDUCER
(SOLAR CELL)

PHOTON-
COUPLED ISOLATOR
'PHOTOEMISSIVE DIODE
& PHOTOTRANSISTOR)

ELECTRICAL BELL;   OR
TELEPHONE RINGER

LOUDSPEAKER

MICROPHONE   OR

HANDSET

LAMP
(GENERAL)

* REPLACE WITH LETTER(S)
DESIGNATING TYPE.

INCANDESCENT LAMP
(ILLUMINATING)

INCANDESCENT LAMP
(INDICATING)

FLUORESCENT LAMP

NEON GLOW LAMP
(AC)

NEON GLOW LAMP
(DC)

BALLAST LAMP

METER

* REPLACE WITH LETTER(S)
DESIGNATING TYPE.

GENERATOR   OR
(GENERAL)   GEN

DC GENERATOR

AC GENERATOR

SYNCHRONOUS
GENERATOR

MOTOR   OR
(GENERAL)   MOT

DC MOTOR

AC MOTOR

SYNCHRONOUS
MOTOR

GENERATOR
OR MOTOR FIELD
(SERIES)

GENERATOR
OR MOTOR FIELD
(SHUNT)

SYNCHRO
(GENERAL)

Analog and Digital Logic Symbols

OPERATIONAL   OR
AMPLIFIER

SUMMING
AMPLIFIER
                          OR
(4 INPUTS & 1 OUTPUT SHOWN)

ELECTRONIC
MULTIPLIER

ELECTRONIC
DIVIDER

ELECTRONIC
FUNCTION
GENERATOR

GENERALIZED
INTEGRATOR

AND   OR   NAND
FUNCTION   FUNCTION   FUNCTION

NOR   EXCLUSIVE   INVERTER
FUNCTION   OR   (NOT)
               FUNCTION   FUNCTION

FLIP-FLOP
(GENERAL)

FLUID POWER SYMBOLS

Fluid Conductors

WORKING LINE
(MAIN)

PILOT LINE
(FOR CONTROL)

EXHAUST AND
LIQUID DRAIN LINE

FLOW DIRECTION,
HYDRAULIC

FLOW DIRECTION,
PNEUMATIC

LINE WITH
FIXED RESTRICTION

FLEXIBLE
LINE

QUICK DISCONNECT
WITHOUT CHECKS

CONNECTED   DISCONNECTED

QUICK DISCONNECT
WITH ONE CHECK

CONNECTED   DISCONNECTED

QUICK DISCONNECT
WITH TWO CHECKS

CONNECTED DISCONNECTED

Energy and Fluid Storage

VENTED
RESERVOIR

PRESSURIZED
RESERVOIR

RESERVOIR WITH
CONNECTING LINES

ABOVE FLUID LEVEL

BELOW FLUID LEVEL

SPRING-LOADED
ACCUMULATOR

GAS-CHARGED
ACCUMULATOR

WEIGHTED
ACCUMULATOR

RECEIVER FOR AIR
OR OTHER GASES

Fluid Conditioners

HEATER

INSIDE TRIANGLES INDICATE
THE INTRODUCTION OF HEAT.

HEATER

OUTSIDE TRIANGLES INDICATE
A LIQUID HEATING MEDIUM.

HEATER

OUTSIDE TRIANGLES INDICATE
A GASEOUS HEATING MEDIUM.

COOLER OR

INSIDE TRIANGLES INDICATE
HEAT DISSIPATION.

COOLER OR

OUTSIDE TRIANGLES INDICATE
A LIQUID OR GASEOUS COOLING MEDIUM.

TEMPERATURE
CONTROLLER OR

OUTSIDE TRIANGLES INDICATE A
LIQUID OR GASEOUS MEDIUM.

FILTER
OR STRAINER

SEPARATOR
WITH MANUAL DRAIN

Four-Way Valves

NORMAL

ACTUATED

TWO-POSITION

NORMAL

ACTUATED LEFT

ACTUATED RIGHT

THREE-POSITION

TYPICAL FLOW PATHS FOR CENTER CONDITION
OF THREE-POSITION, FOUR-WAY VALVES

TRANSITION
SYMBOL

TWO-POSITION, SNAP ACTION WITH TRANSITION

Infinite Positioning
(Between Open and Closed)

NORMALLY
CLOSED

NORMALLY
OPEN

Pressure Control Valves

PRESSURE
RELIEF

NORMAL ACTUATED
(RELIEVING)

SEQUENCE

PRESSURE
REDUCING

PRESSURE
REDUCING
AND RELIEVING

AIR LINE PRESSURE
REGULATOR (ADJUSTABLE, RELIEVING)

Infinite Positioning Valves

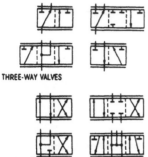

THREE-WAY VALVES

FOUR-WAY VALVES

Flow-Control Valves

ADJUSTABLE,
WITH BYPASS

FLOW CONTROLLED TO RIGHT,
FLOW TO LEFT BYPASSES CONTROL.

ADJUSTABLE AND PRESSURE
COMPENSATED, WITH BYPASS

ADJUSTABLE, NONCOMPENSATED
(FLOW CONTROL IN EACH DIRECTION)

ADJUSTABLE, TEMPERATURE
AND PRESSURE COMPENSATED

Air Line Accessories

SIMPLIFIED

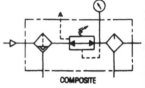

COMPOSITE

FILTER, REGULATOR, AND LUBRICATOR

Index